Foundations of Engineering Mechanics

Le xuan Anh

Springer
*Berlin
Heidelberg
New York
Hong Kong
London
Milan
Paris
Tokyo*

http://www.springer.de/engine-de/

Le xuan Anh

Dynamics of Mechanical Systems with Coulomb Friction

Translated by Alexander K. Belyaev

With 59 Figures

 Springer

Series Editors:

V. I. Babitsky
Department of Mechanical Engineering
Loughborough University
Loughborough, Leicestershire, LE11 3TU
Great Britain

J. Wittenburg
Institut für Technische Mechanik
Universität Karlsruhe (TH)
Kaiserstraße 12
76128 Karlsruhe / Germany

Author:
Le xuan Anh
Podolskaya ul. 45
app. 20
190013 St. Petersburg
Russian Federation

Translator:
Alexander K. Belyaev
Dept. of Mechanics and Control Processes
State Polytechnical University of St. Petersburg
Polytechnicheskaya ul. 29
195251 St. Petersburg
Russian Federation

ISBN 3-540-00654-0 Springer-Verlag Berlin Heidelberg New York

Cataloging-in-Publication data applied for
Bibliographic information published by Die Deutsche Bibliothek.
Die Deutsche Bibliothek lists this publication in the Deutsche Nationalbibliografie;
detailed bibliographic data is available in the Internet at <http://dnb.ddb.de>.

This work is subject to copyright. All rights are reserved, whether the whole or part of the material is concerned, specifically the rights of translation, reprinting, reuse of illustrations, recitation, broadcasting, reproduction on microfilm or in other ways, and storage in data banks. Duplication of this publication or parts thereof is permitted only under the provisions of the German Copyright Law of September 9, 1965, in its current version, and permission for use must always be obtained from Springer-Verlag. Violations are liable for prosecution under German Copyright Law.

Springer-Verlag Berlin Heidelberg New York
a member of BertelsmannSpringer Science+Business Media GmbH

http://www.springer.de

© Springer-Verlag Berlin Heidelberg 2003
Printed in Germany

The use of general descriptive names, registered names, trademarks, etc. in this publication does not imply, even in the absence of a specific statement, that such names are exempt from the relevant protective laws and regulations and therefore free for general use.

Typesetting: Camera-ready copy by translator
Cover-design: de'blik, Berlin
Printed on acid-free paper   62 / 3020 hu - 5 4 3 2 1 0

Dynamics of Mechanical Systems with Coulomb Friction

Le xuan Anh

Introduction

The first attempt to construct a general theory of motion for mechanical systems with Coulomb friction is attributed to the famous French mathematician P. Painlevé [117] at the end of the nineteenth century. Studying various real systems with friction he discovered situations when the solutions to the dynamical equations did not exist and were non-unique. These phenomena were later termed "Painlevé's paradoxes" and posed a major difficulty for scientists in the field of mechanics, who were developing theories of motion taking into account Coulomb friction.

Since the dynamical equations make no sense and are non-correct in these paradoxical cases, to construct a rigorous theory of motion it is vital that we can: i) derive the conditions under which the paradoxes occur; ii) establish reasons for the paradoxes; and iii) determine the true motions in the paradoxical cases.

Painlevé concluded that the law of Coulomb friction was logically inconsistent with the basic principles of classical mechanics. As a result, a number of critical discussions on Coulomb's law appeared in the literature involving French and German scientists L. Lecornue [95], De Sparre [29], R. Mises [110], G. Hamel [46], L. Prandtl [128] and Painlevé himself [118].

After Painlevé epoch, a voluminous literature on the theory of motion of systems with friction appeared during the twentieth century. The main attention of the authors were directed towards three problems: i) the derivation of the equations of motion; ii) the solution of "Painlevé's paradoxes"; and iii) construction of the theory of frictional self-excited oscillations. Nevertheless, a unified theory that provided a general solution to these specific problems was not forthcoming, until recently. These problems are the topic of the present book, which includes a derivation of the general differential equations of motion with removed constraint forces, calculation of the regions of paradoxes, determining the conditions for the state of rest and transition to motion, constructing a theory of self-braking, determining the feasibility of tangential impacts and dynamic seizure, and calculation of the efficiency.

The history of the development of the theory of motion of systems with friction is briefly outlined in the first chapter. Typical features of systems with friction are illustrated using simple examples. These features give rise to six specific problems on the theory of motion: 1) derivation of the equations of motion and the constraint forces; 2) non-correctness of these equations; 3) determination of the forces of friction acting on the system's particles; 4) determination of the condition for the system to remain at rest and its transition to motion; 5) determination of the property of self-braking of material systems; and 6) constructing the theory of frictional self-excited oscillations.

Solving these specific problems comprises the main content of Chapters 2-8 of the present book. In order to simplify the presentation, the model developed is introduced gradually. For example, Chapters 2-4 are concerned only with systems with a single degree of freedom and a single frictional constraint, whilst Chapter 5 deals with an arbitrary number of degrees of freedom and frictional constraints.

Chapters 6 and 7 are devoted to an experimental investigation of the law of friction, and the dependence of static friction on the rate of tangential loading is proved using test data.

The analysis of frictional self-excited oscillations of various types is carried out in Chapter 8 with the help of the experimental law of friction obtained in the previous two chapters.

The author is grateful to N.A.Zakharova, P.A.Zhilin and V.A.Palmov for their help in the preparation of this book. Sincere gratitude is expressed to Alexander Belyaev, who not only took the trouble of translating the manuscript into English, but who also made a number of important suggestions for its improvement. The help of Stewart McWilliam, from the University of Nottingham, UK who edited the translation is sincerely appreciated.

St. Petersburg, August 2002

Le xuan Anh

Contents

Introduction   3

1 Development of the theory of motion for systems with Coulomb friction   **11**
   1.1   Coulomb's law of friction 11
   1.2   Main peculiarities of systems with Coulomb friction and the specific problems of the theory of motion 12
       1.2.1   The principle peculiarity 13
       1.2.2   Non-closed system of equations for the dynamics of systems with friction and the problem of deriving these equations 14
       1.2.3   Non-correctness of the equations for systems with friction and the problem of solving Painlevé's paradoxes   15
       1.2.4   The problem of determining the forces of friction acting on particles 17
       1.2.5   Retaining the state of rest and transition to motion   18
       1.2.6   The problem of determining the property of self-braking   19
       1.2.7   Appearance of self-excited oscillations 19
   1.3   Various interpretations of Painlevé's paradoxes 20
   1.4   Principles of the general theory of systems with Coulomb friction 25
   1.5   Laws of Coulomb friction and the theory of frictional self-excited oscillations 32

2 Systems with a single degree of freedom and a single frictional pair   37
- 2.1 Lagrange's equations with a removed contact constraint .. 37
- 2.2 Kinematic expression for slip with rolling 44
  - 2.2.1 Velocity of slip and the velocities of change of the contact place due to the trace of the contact 44
  - 2.2.2 Angular velocity 45
- 2.3 Equation for the constraint force and Painlevé's paradoxes . 49
  - 2.3.1 Solution for the acceleration and the constraint force 50
  - 2.3.2 Criterion for the paradoxes 51
- 2.4 Immovable contact and transition to slipping 53
- 2.5 Self-braking and the angle of stagnation 57
  - 2.5.1 The case of no paradoxes 58
  - 2.5.2 The case of paradoxes ($\mu|L| > 1$) 64

3 Accounting for dry friction in mechanisms. Examples of single-degree-of-freedom systems with a single frictional pair   67
- 3.1 Two simple examples 67
  - 3.1.1 First example 67
  - 3.1.2 Second example 69
- 3.2 The Painlevé-Klein extended scheme 70
  - 3.2.1 Differential equations of motion, expression for the reaction force, condition for the paradoxes and the law of motion 72
  - 3.2.2 Immovable contact and transition to slip 74
  - 3.2.3 The stagnation angle and the property of self-braking in the case of no paradoxes 75
  - 3.2.4 Self-braking under the condition of paradoxes 77
- 3.3 Stacker 79
  - 3.3.1 Pure rolling of the rigid body model 79
  - 3.3.2 Slip of the driving wheel for the rigid body model . 82
  - 3.3.3 Speed-up of stacker 84
  - 3.3.4 Pure rolling in the case of tangential compliance .. 85
  - 3.3.5 Rolling with account of compliance 87
  - 3.3.6 Speed-up with account of compliance 88
  - 3.3.7 Numerical example 91
- 3.4 Epicyclic mechanism with cylindric teeth of the involute gearing 94
  - 3.4.1 Differential equation of motion, equations for the reaction force and the conditions for paradoxes. 95
  - 3.4.2 Relationships between the torques at rest and in the transition to motion 100
  - 3.4.3 Regime of uniform motion 103
- 3.5 Gear transmission with immovable rotation axes 103

		3.5.1	Differential equations of motion and the condition for absence of paradoxes	104

- 3.5.1 Differential equations of motion and the condition for absence of paradoxes 104
- 3.5.2 Regime of uniform motion 106
- 3.5.3 Transition from the state of rest to motion 109
- 3.6 Crank mechanism 110
  - 3.6.1 Equation of motion and reaction force 110
  - 3.6.2 Condition for complete absence of paradoxes 112
  - 3.6.3 The property of self-braking in the case of no paradoxes 114
- 3.7 Link mechanism of a planing machine 115
  - 3.7.1 Differential equations of motion and the expression for the reaction force 115
  - 3.7.2 Feasibility of Painlevé's paradoxes 119
  - 3.7.3 The property of self-braking 121
  - 3.7.4 Numerical example 123

4 Systems with many degrees of freedom and a single frictional pair. Solving Painlevé's paradoxes 125

- 4.1 Lagrange's equations with a removed constraint 125
- 4.2 Equation for the constraint force, differential equation of motion and the criterion of paradoxes 128
  - 4.2.1 Determination of the constraint force and acceleration 128
  - 4.2.2 Criterion of Painlevé's paradoxes 131
- 4.3 Determination of the true motion 132
  - 4.3.1 Limiting process 133
  - 4.3.2 True motions under the paradoxes 137
- 4.4 True motions in the Painlevé-Klein problem in paradoxical situations 141
  - 4.4.1 Equations for the reaction force 142
  - 4.4.2 True motions for the paradoxes 143
- 4.5 Elliptic pendulum 145
- 4.6 The Zhukovsky-Froude pendulum 148
  - 4.6.1 Equation for the reaction force and condition for the non-existence of the solution 150
  - 4.6.2 The equilibrium position and free oscillations 152
  - 4.6.3 Regime of joint rotation of the journal and the pin . 153
- 4.7 A condition of instability for the stationary regime of metal cutting 155
  - 4.7.1 Derivation of the equations of motion 155
  - 4.7.2 Solving the equations 157
  - 4.7.3 Relationship between instability of cutting and Painlevé's paradox 159
  - 4.7.4 Boring with an axial feed 161

5 Systems with several frictional pairs. Painlevé's law of friction. Equations for the perturbed motion taking account of contact compliance   163

- 5.1 Equations for systems with Coulomb friction 163
  - 5.1.1 System with removed constraints 163
  - 5.1.2 Solving the main system 166
  - 5.1.3 The case of $n = 1, m = 1$ 169
- 5.2 Mathematical description of the Painlevé law of friction . . 170
  - 5.2.1 Accelerations due to two systems of external forces . 170
  - 5.2.2 Improved Painlevé's equations 172
  - 5.2.3 Improved Painlevé's theorem 174
- 5.3 Forces of friction in the Painlevé-Klein problem 176
- 5.4 The contact compliance and equations of perturbed trajectories. 177
  - 5.4.1 Lagrange's equations for systems with elastic contact joints 177
  - 5.4.2 Equations for perturbed reaction forces 179
- 5.5 Painlevé's scheme with two frictional pairs 181
  - 5.5.1 Lagrange's equations, reaction forces and the equations of motion with eliminated reaction forces . . . 182
  - 5.5.2 Feasibility of Painlevé's paradoxes 184
  - 5.5.3 Expressions for the frictional force in terms of the friction coefficients 185
  - 5.5.4 Painlevé's scheme for compliant contacts 186
- 5.6 Sliders of metal-cutting machine tools 187
  - 5.6.1 Derivation of equations of motion and expressions for the reaction forces 187
  - 5.6.2 Signs of the reaction forces and feasibility of paradoxes 189
  - 5.6.3 Forces of friction 191
- 5.7 Concluding remarks about Painlevé's paradoxes 192
  - 5.7.1 On equations of systems with Coulomb friction . . . 192
  - 5.7.2 On conditions of the paradoxes 193
  - 5.7.3 On the reasons for the paradoxes 193
  - 5.7.4 On the laws of motion in the paradoxical situations 193
  - 5.7.5 On the initial motion of an immovable contact . . . 194
  - 5.7.6 On self-braking 194
  - 5.7.7 On the mathematical description of Painlevé's law . 195
  - 5.7.8 On examples 195

6 Experimental investigations into the force of friction under self-excited oscillations   197

- 6.1 Experimental setups 198
  - 6.1.1 The first setup 198
  - 6.1.2 The second setup 200
  - 6.1.3 The third setup 201

6.2	Determining the forces by means of an oscillogram	202
6.3	Change in the force of friction under break-down of the maximum friction in the case of a change in the velocity of motion	206
6.4	Dependence of the friction force on the rate of tangential loading	209
6.5	Plausibility of the dependence $F_+(f)$	213
	6.5.1 Control tests	213
	6.5.2 Estimating the numerical characteristics	213
	6.5.3 Statistical properties of the dependences	214
	6.5.4 Test data of other authors	215
6.6	Characteristic of the force of sliding friction	215

7 Force and small displacement in the contact — 217

- 7.1 Components of the small displacement — 217
  - 7.1.1 Definition of break-down and initial break-down — 217
  - 7.1.2 Reversible and irreversible components — 218
  - 7.1.3 Influence of the intermediate stop and reverse on the irreversible displacement — 220
  - 7.1.4 Dependence of the total small displacement on the rate of tangential loading — 222
  - 7.1.5 Small displacement of parts of the contact — 223
  - 7.1.6 Comparing the values of small displacement with existing data — 225
- 7.2 Remarks on friction between steel and polyamide — 226
  - 7.2.1 On critical values of the force of friction — 226
  - 7.2.2 Time lag of small displacement — 226
  - 7.2.3 Immovable and viscous components of the force of friction — 229
- 7.3 Conclusions — 230

8 Frictional self-excited oscillations — 231

- 8.1 Self-excited oscillations due to hard excitation — 231
  - 8.1.1 The case of no structural damping — 231
  - 8.1.2 Including damping — 237
- 8.2 Self-excited oscillations under both hard and soft excitations — 240
  - 8.2.1 Equations of motion — 240
  - 8.2.2 Critical velocities — 242
  - 8.2.3 Amplitude of auto-oscillation — 244
  - 8.2.4 Period of auto-oscillation — 246
  - 8.2.5 Self-excitation of systems — 247
- 8.3 Accuracy of the displacement — 249

References — 255

Index — 268

1
Development of the theory of motion for systems with Coulomb friction

1.1 Coulomb's law of friction

The phenomenon of sliding friction was first studied experimentally in the end of eighteenth century by the French physicist G. Amontons (1663-1705) who discovered the dependence of the force of friction on the surface characteristics of the contacting bodies. The laws of friction were formulated nearly a hundred years later by C.A. Coulomb (1736-1806), [99]. The essence of Coulomb's formulation is as follows.

Let a body be at rest on a horizontal plane under the action of the normal force $\mathbf{G}$ and tangential force $\mathbf{S}$, see Fig. 1.1. It is assumed that $\mathbf{G} \neq 0$. If $\mathbf{S} = 0$, the body is at rest and force $\mathbf{G}$ is reacted by the normal

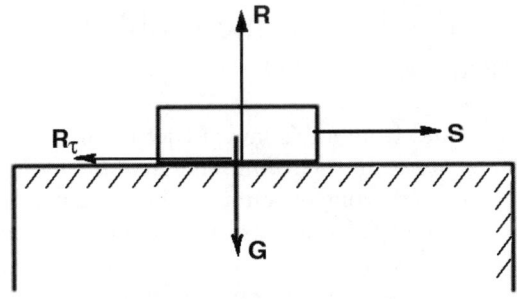

FIGURE 1.1.

reaction force **R** of the plane, that is, $\mathbf{R} = -\mathbf{G}$. If the shifting force is progressively increased, the state of rest is maintained due to the existing force of resistance to motion $\mathbf{R}_\tau$ from the plane and this force balances force **S**, i.e. $\mathbf{R}_\tau = -\mathbf{S}$. The state of rest is maintained until force **S** and $\mathbf{R}_\tau$ reach the critical value

$$R_\tau = R_{\tau\,\mathrm{max}} = S_{\mathrm{max}} = \mu_+ R. \tag{1.1}$$

The value $R_{\tau\,\mathrm{max}}$ is referred to as the maximum force of static friction or simply the static friction. As one can see from eq. (1.1) the force of static friction is proportional to the normal reaction force R. The proportionality coefficient μ_+ is called the coefficient of static friction.

When the value of the shifting force S reaches the critical value $\mu_+ R$ there comes a critical point of equilibrium. Provided that the value of the tangential force is equal to $\mu_+ R$ the equilibrium is not disturbed. However an infinitesimally small increment in force S is sufficient for the body to begin to move. When the body moves the coefficient of sliding friction can either be unchanged and equal to μ_+, or change depending upon the velocity of sliding and other factors.

The law of dry friction which assumes the coefficient of the sliding friction μ_+ to be constant throughout the sliding is extensively applied in classical mechanics. The force of sliding friction that remains constant and equal to the value of the friction at the beginning of motion is referred to as the force of dry friction and is represented by the formula

$$R_\tau = \begin{cases} < \mu_+ R, & v = 0, \\ = \mu_+ R, & v \neq 0, \end{cases} \tag{1.2}$$

where v denotes the sliding velocity.

While analysing the general questions and real systems we assume the force of dry friction to be given by eq. (1.2) unless stated otherwise.

1.2 Main peculiarities of systems with Coulomb friction and the specific problems of the theory of motion

The force of friction and Coulomb's law of friction are taught at the school and later at University while learning the fundamentals of physics and theoretical mechanics. Everyone encounters the cases when the influence of the frictional force on motion of mechanical systems is required for solving applied problems, for example, friction in the gear transmissions, friction in the plain bearings, friction in the Zhukovsky-Froude pendulum etc.

A convenient means of solving such problems is to consider the forces of friction as prescribed forces, then construct the equations of motion using

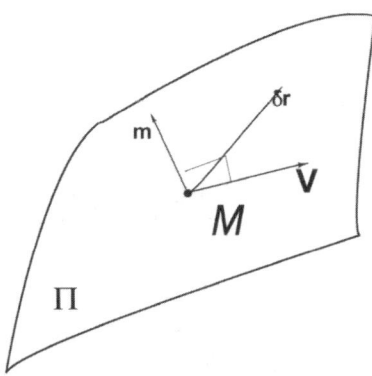

FIGURE 1.2.

standard methods (for instance Lagrange's equations), and then solve them. Seemingly, there is no need to carry out special investigations for developing the theory of motion taking account of the frictional forces.

However, experience in mechanical engineering indicates that actual systems with friction can exhibit peculiarities which give rise to specific problems and require a unified approach to their solution. It was Painlevé who realised the importance of this problem and stated, [116], "the problem of constructing the general theory for the systems with friction". We consider these peculiarities on simple examples with the aim of understanding the specific problems generated by the above peculiarities.

1.2.1 The principle peculiarity

The principle peculiarity for systems with friction is that the elementary work of the constraint forces due to arbitrary virtual displacements does not vanish in general. In order to illuminate this feature we consider sliding of a particle M on a rough surface Π with coefficient of friction μ, Fig. 1.2. The vector of the general reaction force $\mathbf{R}_\sigma$ is equal to the sum of the normal reaction force $\mathbf{m}R$ and the force of friction $-\mu\varepsilon_1\dfrac{\mathbf{v}}{|\mathbf{v}|}R$, that is

$$\mathbf{R}_\sigma = (\mathbf{m} - \mu\varepsilon_1\frac{\mathbf{v}}{|\mathbf{v}|})R, \quad \varepsilon_1 = \operatorname{sign} R. \qquad (1.3)$$

Here $\mathbf{m}$ denotes the unit vector of the normal to surface Π, $\mathbf{v}$ is the velocity of particle M, and R is the algebraic value of the normal reaction force which is taken to be positive if it is directed along $\mathbf{m}$, otherwise it is negative. Let us ascribe virtual displacement $\delta\mathbf{r}$ which must lie in the tangent plane ($\delta\mathbf{r} \perp \mathbf{m}$) due to the condition of the contact constraint. The elementary work of the reaction force $\mathbf{R}_\sigma$ done due to this displacement is as follows

$$\delta A = \mathbf{R}_\sigma \cdot \delta \mathbf{r} = -\mu\varepsilon_1 R \frac{\mathbf{v}}{|\mathbf{v}|} \cdot \delta \mathbf{r} = \begin{cases} = 0, & \text{for } \delta \mathbf{r} \perp \mathbf{v}, \\ \neq 0, & \text{for } \delta \mathbf{r} \not\perp \mathbf{v}. \end{cases} \quad (1.4)$$

As one can see, the elementary work of the constraint force is equal to zero only in the case when the virtual displacement is taken to be orthogonal to the vector of the velocity of sliding. Hence, in general

$$\delta A \neq 0. \quad (1.5)$$

Due to this feature for systems with Coulomb friction a number of well-known statements of mechanics (for example Lagrange's equations, Appell's equations, the variational principles of mechanics) derived under the assumption of zero work of the constraint forces need further development for systems with friction. This peculiarity of the systems under consideration leads to various specific problems of the theory of motion.

1.2.2 Non-closed system of equations for the dynamics of systems with friction and the problem of deriving these equations

By virtue of inequality (1.5) we can set the general elementary work of all forces acting on the system as the sum of two components: δA_1 which is the work of the prescribed forces applied to the particles and δA_2 which is the work of the forces of friction, hence

$$\delta A = \delta A_1 + \delta A_2 (R_1, \ldots, R_m), \quad (1.6)$$

where m designates the number of frictional pairs. Let us notice that δA_2 depends on normal forces of frictional constraints $R_1, \ldots, R_m$ as the frictional forces are proportional to these reaction forces.

Let us recall that the derivation of Lagrange's equations requires that the elementary work of the prescribed forces is as follows

$$\delta A_1 = Q_1 \delta q_1 + Q_2 \delta q_2 + \cdots + Q_n \delta q_n = \sum_{s=1}^{n} Q_s \delta q_s, \quad (1.7)$$

where δq_s denotes variation of the generalised coordinate q_s, n is the number of degrees of freedom and Q_s denotes the generalised force corresponding to q_s.

By analogy we can represent the work δA_2 of the forces of friction in the form

$$\delta A_2 = \sum_{s=1}^{n} P_s (R_1, \ldots, R_m) \delta q_s, \quad (1.8)$$

1.2 Main peculiarities of systems with Coulomb friction

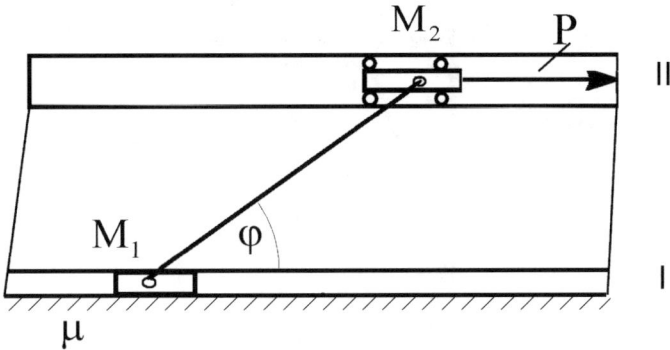

FIGURE 1.3.

where P_s denotes the generalised force of friction corresponding to q_s.

Referring to the derivation of Lagrange's equations we arrive at the following form for these equations

$$\frac{d}{dt}\frac{\partial T}{\partial \dot{q}_s} - \frac{\partial T}{\partial q_s} = Q_s + P_s(R_1, ..., R_m), \quad s = 1, 2, ..., n. \qquad (1.9)$$

As we can see, we obtained a non-closed system of n differential equations with $n+m$ unknown variables $q_1, ..., q_n, R_1, ..., R_m$. Hence, this gives rise to the problem of closure of the above system. A closed system of equations for the system with friction can be obtained if system (1.9) is combined with m equations for the reaction forces $R_1, ..., R_m$. The method of determining the constraint forces by means of the principle of removed constraints is suggested by Lurie, see [102] and [103].

1.2.3 Non-correctness of the equations for systems with friction and the problem of solving Painlevé's paradoxes

At the end of the nineteenth century Painlevé [116] discovered a problem with the dynamic equations for systems with friction. This problem manifested itself in the form of non-existence and non-uniqueness of solution of equations in a certain part of the phase space. In order to demonstrate this non-correctness we consider the Painlevé-Klein problem, Fig. 1.3, [116], [62]. Two particles M_1 and M_2 of unit masses ($m_1 = m_2 = 1$) are linked by an inextensible massless rod M_1M_2 moving along parallel guides. The angle between rod M_1M_2 and axis x is denoted by φ. Guide I is non-smooth and has a coefficient of friction μ whilst guide II is smooth. A horizontal force $P > 0$ acts on particle M_2. The motion of the system is governed by the following equation

$$2\ddot{x} = P - \mu |R| \operatorname{sign} \dot{x} \qquad (1.10)$$

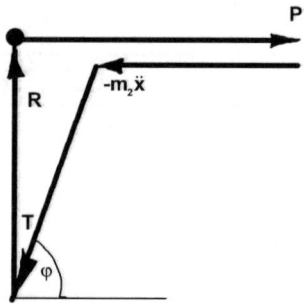

FIGURE 1.4.

which contains the normal reaction force R in addition to coordinate x.

In order to derive an additional equation we analyse the forces acting on particle M_2, see Fig. 1.4. The polygon of forces is closed and includes force P, the force of inertia $-m_2\ddot{x}$ and the force of tension (compression) T and the normal reaction force R. As we can see from Fig. 1.4

$$\ddot{x} = P - T\cos\varphi = P - \frac{R}{\tan\varphi}. \qquad (1.11)$$

Equations (1.10) and (1.11) form a closed system

$$\begin{cases} 2\ddot{x} = P - \mu|R|\operatorname{sign}\dot{x}, \\ \ddot{x} = P - \dfrac{R}{\tan\varphi}. \end{cases} \qquad (1.12)$$

In order to obtain the equation of motion and the equation for the reaction force R it is necessary to solve system (1.12) for R and $\ddot{x}$. To this aim, Painlevé considered the sign of the velocity, i.e. $\operatorname{sign}\dot{x}$, as an initial condition. Besides, Painlevé first assumed a sign of the reaction force ($\operatorname{sign} R$) and after resolving the system proved whether the sign obtained was coincident with the sign assumed. As a result, Painlevé proved the following condition for non-correctness of the equation of motion

$$\mu\tan\varphi > 2 \Rightarrow \begin{cases} \text{solution does not exist} & \text{for } \dot{x} > 0, \\ \text{solution is not unique} & \text{for } \dot{x} < 0. \end{cases} \qquad (1.13)$$

Indeed, for $\dot{x} > 0$ and $\mu\tan\varphi > 2$ we assume in system (2.10) that $\operatorname{sign} R = 1$, i.e. $|R| = R$, and obtain

$$R_+ = \frac{P\tan\varphi}{2 - \mu\tan\varphi} < 0. \qquad (1.14)$$

Assuming $\operatorname{sign} R = -1$, i.e. $|R| = -R$ yields

$$R_- = \frac{P\tan\varphi}{2 + \mu\tan\varphi} > 0. \qquad (1.15)$$

1.2 Main peculiarities of systems with Coulomb friction

Equations (1.14) and (1.15) show that in this case neither of the obtained signs of R coincides with the assumed signs of R. Hence, the solution of problem (1.12) does not exist.

In the case of $\dot{x} < 0$ and $\mu \tan \varphi > 2$, in accordance with two assumed signs of the reaction forces $\operatorname{sign} R = \pm 1$ we obtain the following values of R

$$R_+ = \frac{P \tan \varphi}{2 + \mu \tan \varphi} > 0, \quad R_- = \frac{P \tan \varphi}{2 - \mu \tan \varphi} < 0.$$

The signs are seen to coincide with the assumed ones, thus, for the case of $\dot{x} > 0$ and $\mu \tan \varphi > 2$ the non-uniqueness of the solution of problem (1.12) is observed.

Painlevé also showed that in the case of $\mu \tan \varphi < 2$, i.e. when the condition for paradoxes (1.13) does not hold true, the values of R and $\ddot{x}$ are uniquely determined from system (1.12)

$$R = \frac{P \tan \varphi}{2 - \mu \tan \varphi \operatorname{sign} \dot{x}} > 0, \quad \ddot{x} = \frac{P(1 - \mu \tan \varphi \operatorname{sign} \dot{x})}{2 - \mu \tan \varphi \operatorname{sign} \dot{x}}. \quad (1.16)$$

The situations of non-correctness of dynamics of systems with Coulomb friction discovered by Painlevé are named Painlevé's paradoxes and are nowadays the object of investigation of scientists of many generations, see [62], [94], [95], [100], [110], [116]- [118], [122], [125], [128], [29], [43], [44].

Thus, due to Painlevé's discovery, three problems related to the non-correctness of the dynamic equations were stated. These problems are: i) determination of the paradoxical regions in the space of dynamic parameters (in the expanded phase space), ii) explanation of the paradoxes and iii) establishing the true laws of motion in the paradoxical regions.

1.2.4 The problem of determining the forces of friction acting on particles

In Painlevé's view each particle of the system with friction is subjected to a single friction force. Hence, in a system of N particles there are N friction forces. Painlevé denoted these forces by vectors $\boldsymbol{\rho}_1, \ldots, \boldsymbol{\rho}_N$. In order to explain the existence of these forces let us consider the Painlevé-Klein problem, see Fig. 1.3. We restrict our consideration to the case when paradoxes do not occur (that is, unique solutions exist), i.e. the case of $\mu \tan \varphi < 2$, when the reaction force R and acceleration $\ddot{x}$ are found from eq. (1.16).

We notice that in the case of no friction, i.e. $\mu = 0$, the acceleration of each particle is

$$^0\ddot{x}_1 = {}^0\ddot{x}_2 = \frac{P}{2}. \quad (1.17)$$

Comparing eqs. (1.16) and (1.17) we see that the change in the acceleration due to friction is given by

$$\Delta \ddot{x}_i = \ddot{x} - {}^0\ddot{x}_i = \frac{P(1 - \mu \tan\varphi \operatorname{sign}\dot{x})}{2 - \mu \tan\varphi \operatorname{sign}\dot{x}} - \frac{P}{2} = -\frac{\mu P \tan\varphi \operatorname{sign}\dot{x}}{2(2 - \mu \tan\varphi \operatorname{sign}\dot{x})}$$
$$(i = 1, 2).$$

Due to the law of motion for the particles

$$m_i \ddot{\mathbf{x}}_i = \mathbf{F}_i + \boldsymbol{\rho}_i$$

we conclude that forces $\boldsymbol{\rho}_1$ and $\boldsymbol{\rho}_2$ exist and are equal such that

$$\boldsymbol{\rho}_1 = \boldsymbol{\rho}_2 = \frac{\mu P \tan\varphi \operatorname{sign}\dot{x}}{2(2 - \mu \tan\varphi \operatorname{sign}\dot{x})} \mathbf{i}_1, \qquad (1.18)$$

where $\mathbf{i}_1$ denotes the unit vector of axis Ox.

Next, using eq. (1.16) we notice that the Coulomb friction force between guide I and particle M_1 is given by

$$\mathbf{R}_\tau = \mu |R| \operatorname{sign}\dot{x} \mathbf{i}_1 = \frac{\mu P \tan\varphi \operatorname{sign}\dot{x}}{2 - \mu \tan\varphi \operatorname{sign}\dot{x}} \mathbf{i}_1. \qquad (1.19)$$

It follows from eqs. (1.18) and (1.19) that the sum of friction forces $\boldsymbol{\rho}_1$ and $\boldsymbol{\rho}_2$ acting on particles M_1 and M_2 is equal to the Coulomb friction force $\mathbf{R}_\tau$ between the guide I and particle M_1.

Thus, there appears a need to determine the forces of friction acting on each particle.

1.2.5 Retaining the state of rest and transition to motion

In contrast to the system without friction which can be shifted by any non-zero force exerted on the system, an external force can shift the system with friction only under certain conditions. To explain this peculiarity, we direct our attention again to the Painlevé-Klein system, see Fig. 1.3. As it was shown for the case when $\mu \tan\varphi < 2$, eq. (1.16) yields R and $\ddot{x}$.

Due to Coulomb's law for the beginning of motion it is necessary for the tangential force applied to particle M_1 to exceed the force of friction. This tangential force is equal to the projection of the force of tension $\mathbf{T}$ along axis Ox

$$X = T \cos\varphi = \frac{R}{\tan\varphi} = \frac{P}{2 - \mu \tan\varphi \operatorname{sign}\dot{x}}. \qquad (1.20)$$

By virtue of eqs. (1.19) and (1.20) we arrive at the formula for the shifting force

$$X = \frac{P}{2 - \mu \tan\varphi \operatorname{sign}\dot{x}} > R_\tau = \frac{-\mu P \tan\varphi \operatorname{sign}\dot{x}}{2 - \mu \tan\varphi \operatorname{sign}\dot{x}},$$

Thus

$$P(1 - \mu \tan\varphi \operatorname{sign} \dot{x}) > 0$$

which is equivalent to

$$\mu \tan\varphi \operatorname{sign} \dot{x} < 1 \qquad (1.21)$$

in the case of $P > 0$.

Provided that the initial velocity $\dot{x} = 0$, one should take $\operatorname{sign} \dot{x} = \operatorname{sign} \ddot{x} = 1$. Then for the considered system the condition of "getaway" is given by

$$\mu \tan\varphi < 1. \qquad (1.22)$$

It is easy to prove that if system of Fig. 1.3 is subjected to two forces P_1 and P_2 applied respectively to particles M_1 and M_2 then, instead of the latter condition we obtain the following condition for the "getaway"

$$\mu \tan\varphi < \frac{P_1 + P_2}{P_2}, \qquad (1.23)$$

which yields condition (1.22) in the particular case of $P_1 = 0$.

Thus, the problem of retaining the state of rest and transition to motion needs further investigation.

1.2.6 The problem of determining the property of self-braking

It is known, [148], that the frictional system manifests the property of self-braking. Equation (1.23) shows that if $P_1 = 0$ and $\mu \tan\varphi > 1$ no force P_2 is able to shift the mechanism, no matter how large this force is. Such a mechanism is referred to as self-braking, e.g. [64]. In what follows, particle M_2 is referred to as the point of self-braking. As follows from the condition for getaway, force P_1 shifts particle M_1 under the condition

$$P_1 > P_2 (\mu \tan\varphi - 1).$$

In particular, for zero force P_2 ($P_2 = 0$) the condition for getaway is fulfilled for any value of $P_1 > 0$. Such a point is conditionally termed the point of debraking.

Thus, the following problems should be studied: i) determination of the properties of self-braking of the frictional systems and ii) splitting the system in two subsets, namely the subsets of points of self-braking and debraking.

1.2.7 Appearance of self-excited oscillations

Experience suggests that the motion of elastic systems with friction is accompanied, as a rule, by self-excited oscillations, see [1], [12], [23], [151],

[153] etc. This gives rise to the necessity of developing the theory of self-excited oscillations.

We have already listed the properties of systems with friction which cause the six particular problems of the theory of motion: 1) derivation of equations; 2) non-correctness of the equations; 3) determination of the forces of friction acting on the particles of the system; 4) self-braking of the system with friction; 5) the condition for getaway; and 6) construction of the theory of frictional self-excited oscillations.

In the twentieth century, in the field of dynamics of systems with friction attention was paid to the following three problems from the above-listed: derivation of the equations, incorrectness of the equations and theory of frictional self-excited oscillations. For this reason, the last three section of the present chapter are devoted to the fundamentals of the various approaches to these problems.

1.3 Various interpretations of Painlevé's paradoxes

The condition for paradoxes for the simple example depicted in Fig. 1.3 is presented by the relationship (1.13) between the coefficient of Coulomb friction μ and the constant angle φ. Painlevé noted more complicated cases, for example elliptic pendulum, inhomogeneous disc etc., for which the condition for paradoxes contains not only the coefficient of friction and the sign of the velocity but also the coordinates of the system. The phase space is then split into two regions: the region of absence of paradoxes and the region of paradoxes. The region of paradoxes consists in turn of subregions of non-existence and non-uniqueness of the solution.

In the case of no paradoxes the motion of a system can be determined by means of the dynamic equations, however these equations lose their meaning and become incorrect in the paradoxical region. With this in view, constructing a general theory of motion of the system with Coulomb friction can not be separated from solving the following three problems: derivation of the criterion for determining the paradoxical region, understanding the reason for their appearance, and establishing the true motion in the paradoxical situations.

Because of the lack of general criterion for paradoxes one has to repeat similar derivations for determining the regions of non-existence and non-uniqueness for any particular problem and this requires considerable calculation and analysis in many cases. Apparently, the question of generalisation of the criterion of paradoxes is still open due to its complexity. P. Appell [6] was of the opinion that the mechanisms without paradoxes are widely spread and simple. However, the modern viewpoint, see [1], [36], [133], is that the appearance of such situations in the technology is rather a rule than an exception.

1.3 Various interpretations of Painlevé's paradoxes

P. Painlevé [117] illustrated the paradoxes by eight particular mechanisms. The examples considered later by other authors are either these mechanisms or modifications of them, [1], [6], [26], [7], [62], [94], [95], [100], [110], [125], [128], [29]. For frequently encountered mechanisms like gear transmission, crank mechanism, plain bearing, actuating systems of metal-cutting machine tools the conditions for the paradoxes are yet unknown and the very fact of their appearance is still unexplained.

The paradoxes provoked a lively discussion between the famous French and German scientists: L. Lecornue, De Sparre, F. Klein, R. Mises, G. Hamel, L. Prandtl, F. Pfeiffer and P. Painlevé himself, [46], [62], [94], [95], [110], [118], [125], [128], [29]. It is worth mentioning the papers by Bolotov [135], [20], Butenin [26], Skuridin [133], Abramov [1], Lötstedt [100], Ivanov [49] and many others.

Let us touch upon some principle aspects of these problems.

According to Painlevé, when the solution of the dynamic problem is not unique, then the system moves in such a way that the sign of the normal reaction force for the system with and without friction does not change. There still exists no rigorous proof of this principle. Painlevé considered the non-existence of the solution as a logical contradiction between the Coulomb law and the principles of mechanics of a rigid body. Seemingly, based upon this reasoning Painlevé suggested a new formulation for the frictional law for the general theory of motion. Section 1.4 addresses this formulation.

Lecornue [94], [95] studied motion of an inhomogeneous disc of unit mass on a horizontal rough plane under gravity and obtained the following expression for the reaction force

$$R = r_1^2 \frac{g + \omega^2(r - b)}{r_1^2 + a^2 - \mu ab}, \qquad (1.24)$$

where r denotes the radius of the disc, r_1 is the radius of gyration of the disc about the centre of gravity G, ω is the angular velocity and a and b denote respectively the distances from the centre of gravity G to the vertical central axis and the plane $(r > b)$.

Equation (1.24) is derived under the assumption that $R > 0$ which is natural in this case. However if

$$\mu > \frac{(r_1^2 + a_2)}{ab} \qquad (1.25)$$

this equation yields $R < 0$. In other words, under condition (1.25) the considered dynamic problem has no solution.

Lecornue suggested two explanations for the mentioned complication. In the framework of the first explanation, the dynamic equation was derived with account for the vertical compliance of the plane. It was proved that for $b = r$ under condition (1.25) and the initial condition $R = \dot{R} = 0$ the reaction force R increases drastically due to the exponential law

$$R = \frac{cg}{2s^2}(e^{st} + e^{-st} - 2),$$

where c is the rigidity of the plane and $s = c(\mu ab - k^2 - a^2)/r_1^2$. Hence the velocity of gliding vanishes nearly instantaneously. This process was referred to by Lecornue as the dynamic self-braking (nowadays this process is named dynamic seizure of the tangential impact). The second explanation was made possible by referring to the results of experiments on the preliminary shift which demonstrated that the coefficient of friction μ increases progressively from zero. Due to Lecornue, as μ increases the velocity of gliding becomes zero unless μ reaches a maximum (and correspondingly condition (1.25) holds). This explanation does not have a rigorous proof since the preliminary shift has not yet been studied in detail. Indeed, the elastic part of this shift occurs only in the phase of immovable contact and hence this explanation can not be generalised to the case of non-zero initial velocity of gliding.

The second explanation by Lecornue was used by De Sparre [29] for the study of the motion of the elliptic pendulum with two-sided contact constraint. The author pointed out that an impact caused by the force of friction (in the case of non-existence of the solution) which can stop motion of the slider instantaneously. Commenting on the situation of the non-uniqueness of solution, the author proposed that, under a zero initial velocity, the system moves in such a way that the sign of the reaction force at rest and under motion does not change. This conclusion is analogous to Painlevé's principle. Other initial conditions were not considered.

In the paper by Klein [62] the paradoxes were analysed to the example shown in Fig. 1.3. For the case of non-unique solutions ($\mu \tan \varphi > 2$; $\dot{x} > 0$) it was suggested that motion with a positive acceleration takes place. As follows from eq. (1.16)

$$\ddot{x} = P/(2 + \mu \tan \varphi), \quad \text{sign}\, R = 1.$$

Klein stated that the "second solution, if it existed, would be transformed immediately into the first solution under a small perturbation" since it is unstable. In the situation of non-existence of solution $\mu \tan \varphi > 2$ and $\dot{x} < 0$, an instantaneous stop occurs which was confirmed by experiments. Hence the coefficient of static friction takes the value $2/\tan \varphi$ which leads to an infinite reaction force and in turn leads to the dynamic seizure. "Coulomb's law is in conflict with neither the principles of mechanics, nor the real phenomena. The latter must be correctly interpreted".

The phenomenon of the dynamic seizure was first mentioned in the paper by Lecornue [94] on account of elastic contact deformations and is confirmed by numerous observations from practical mechanical engineering. Nonetheless, an explanation of the paradoxical situation suggested in

1.3 Various interpretations of Painlevé's paradoxes

[62] is not convincing if for no other reason than for non-zero initial velocity $\dot{x}(0) \neq 0$ the instantaneous stop should be considered as a result of the dynamic seizure rather than its reason. Hence, in the process of reducing the velocity to zero the coefficient of the friction of gliding must be greater than $2/\tan\varphi$.

R. Mises [110] took a somewhat different viewpoint while discussing the situation of non-existence of the solution for the Painlevé-Klein scheme, Fig. 1.3. The author wrote: "Painlevé is wrong when he calls Coulomb's laws logically unacceptable. He is however right when he states that these laws need improvement on the logical side". As such an improvement he adopted the second hypothesis by Lecornue, which is that the velocity of gliding is zero unless μ reaches its maximum value. Due to Mises's opinion in cases when the Lecornue hypothesis is unacceptable one can assume that the frictional coefficient μ becomes zero as the normal reaction force R tends to infinity. However, paper [110] proposed no dependence $\mu(R)$ which allowed the true motion to be calculated since the author held the view that this dependence should by obtained experimentally.

Hamel [46], on one hand, held the same viewpoint as Lecornue, that is, it is necessary to reject the hypothesis of a rigid body. On the other hand, he deemed the assumption by Klein of an instantaneous stop as being convenient from a methodological perspective since it enables one to retain the hypothesis of a rigid body.

The idea by Lecornue on elastic deformations (the first hypothesis) was applied in the papers by Prandtl [128] and Pfeiffer [125] for determining the true motion for the Painlevé-Klein scheme, Fig. 1.3. When the elastic modulus of the rod tends to infinity the dynamic seizure in the paradoxical situation of non-existence was obtained. In the case of non-existence, it was shown that among two possible motions the accelerated motion is stable whereas the decelerated motion is unstable. The results of the theoretical investigation were compared with the test results.

Butenin [26] used phase space to analyse this scheme. Considering various initial conditions, the author observed a discontinuous character of the motion in both paradoxical situations. It was established that one of the stationary values of the reaction forces corresponds to a stable centre whereas the second corresponds to an unstable saddle point. Due to a limiting passage to a rigid rod, the author showed that the hypothesis of a jump allows one to overcome the inherent contradiction of the scheme with a rigid rod.

Thus, if the analysis carried out by Lecornue allows us to explain the non-existence of the solution for the inhomogeneous disc, then the investigations by Prandtl, Pfeiffer and Butenin show that this idea can be utilised for interpretation of both non-existence and non-uniqueness of the solution of the Painlevé-Klein scheme.

The above papers are concerned only with those examples in which the phenomenon of tangential impact is permanently related to vanishing relative gliding. Bolotov [19], [20] proved that such a vanishing does not always take place. Generalisation of this approach to systems with many degrees of freedom and a single one-sided constraint is suggested in the paper by Ivanov [49]. The normal collisions due to roughness of the contact surface act as a perturbation factor. The true motion is chosen from the requirement of smoothness of the solution with respect to the value of collisions. It turns out that the generalised velocities experience jumps in both paradoxical situations, i.e. a tangential impact is observed. However, the result obtained differs from that determined by Painlevé's principle. The dependences of the reaction forces and the jumps in the velocities on the coefficient of friction in the case of the tangential impacts are not presented.

Not accounting for the elastic property of the contact Skuridin [133] suggests that maximum efficiency corresponds to the true motion in the case of non-uniqueness of the solution. The author gave no analytical proof for this statement but noticed that a proof can be carried out by means of the Gauss principle of least constraint. It is possible to accept Skuridin's suggestion if for no other reasons than the concept of efficiency is not generalised to the case of mechanical systems when the latter is subjected to many active and frictional forces rather than one or two forces.

In order to establish the true motion of planar mechanisms in the case of non-uniqueness Abramov studied their stability [1]. The equations for the perturbed motion are presented in the form

$$\frac{d^2 H_i}{dt^2} = \sum_{k=1}^{m} W_{ik} C_k H_k, \qquad (1.26)$$

where m is the number of frictional pairs, W_{ik} denotes the normal component of the acceleration gained by the $i-th$ contact point under the action of a unit force at the $k-th$ contact point, C_k denotes the effective rigidity and H_k is the elastic displacement of the $k-th$ contact point. General expressions for W_{ik} in terms of the frictional coefficients, generalised coordinates and velocities are not given, however equations in the form of (1.12) are constructed for a number of mechanisms. The results of solving these equations are in agreement with the suggestion by Skuridin.

In summarising the considerations of papers on explanations of the paradoxes it is necessary to mention a considerable evolutionary process since the time of Painlevé. As a result we can think that the hypothesis on accounting for the elastic deformation of the contacting bodies is acceptable for all studied cases. Applying this hypothesis to particular examples enables us to explain the seizure resulting in an instantaneous stop in the paradoxical regions. This result is in agreement with observations and confirms once again the necessity to prove the possibility of the paradoxical situations for any real mechanism and find a way to avoid such situations.

On the other hand, the existing studies which utilise this hypothesis are of a particular character. Because of the absence of rigorous proof in the general case the very question on plausibility of the hypothesis stands open. A general criterion on determining the feasibility of paradoxical situations in mechanisms has not yet been established.

It is also necessary to stress the difference in opinions about the true motions. For instance, according to the original Painlevé principle that motion is realised in the case of the non-uniqueness of the solution for which the sign of the reaction forces is the same in the case of both zero and non-zero friction. However, due to the results of analysis of the scheme of Fig. 1.3 performed by Pfeiffer and Butenin, the motion is dependent on the initial value of the reaction force. At the same time, Ivanov states that the impact occurs in the situations of both non-existence and non-uniqueness of the solution. What viewpoint is correct? Do the above viewpoints contradict or simply complement each other? The answer to these questions is possible only as a result of investigations of a more general character than the state-of-the-art.

1.4 Principles of the general theory of systems with Coulomb friction

Following Painlevé, under the general theory of mechanical systems with Coulomb friction we understand a theory "analogous to that which Lagrange's equations yield for systems without friction". As mentioned above, constructing such a theory is inseparable from elucidation of three problems of the paradoxes discussed above. Besides, it is necessary to derive equations for the reaction forces and the differential equations of motion and solve a series of problems typical for systems with friction. It is worth mentioning numerous investigations [148], [149], [151], [150], [36], [37], [38], [130], [131], [95], [116].

Let us consider first the elements of the theory of non-ideal systems outlined in papers by Painlevé [116], [117] and Rumyantsev [130], [131].

Let a system of N particles be subjected to $3N - n$ holonomic stationary constraints. If the constraints are ideal, then the sum of the elementary work of the reaction forces $^0\mathbf{K}_1, ..., ^0\mathbf{K}_N$ vanishes. Provided that there exists at least one force of friction (one frictional pair) the reaction forces $\mathbf{K}_1, ... \mathbf{K}_N$ differ from $^0\mathbf{K}_1, ... ^0\mathbf{K}_N$ and the sum of their work is no longer zero

$$\tau = \sum_{i=1}^{N} \mathbf{K}_i \cdot \delta \mathbf{r}_i = \sum_{i=1}^{N} \left(\mathbf{K}_i - {}^0\mathbf{K}_i \right) \cdot \delta \mathbf{r}_i \neq 0. \qquad (1.27)$$

Here $\mathbf{r}_i$ denotes the position vector of the $i - th$ particle.

It follows from equation $M_i \mathbf{W}_i = \mathbf{F}_i + \mathbf{K}_i$ that the change in the inertia force $-M_i \mathbf{W}_i$ due to friction is as follows

$$\boldsymbol{\rho}_i = M_i \mathbf{W}_i - M_i {}^0\mathbf{W}_i = \mathbf{K}_i - {}^0\mathbf{K}_i \qquad (1.28)$$

which Painlevé termed the force of friction. In eq. (1.28) M_i denotes the mass of the particle whilst ${}^0\mathbf{W}_i$ and $\mathbf{W}_i$ are its acceleration in the case of absent and present friction respectively. It is known that the reaction forces of the constraints can be expressed in terms of Lagrange multipliers in the form

$$ {}^0\mathbf{K}_i = \sum_{k=1}^{3N-n} \lambda_k \operatorname{grad}_i \Phi_k , \qquad (1.29)$$

where $\Phi_k(x_1, y_1, z_1, ..., x_N, y_N, z_N) = 0$ $(k = 1, ...3N)$ are the constraint equations. The forces of friction $\boldsymbol{\rho}_i$ were presented in [116] in terms of some coefficients ν_j

$$\boldsymbol{\rho}_i = \sum_{j=1}^{n} \nu_j \frac{\partial \mathbf{r}_i}{\partial q_j}. \qquad (1.30)$$

From these Painlevé arrived at Lagrange's equations of the first kind for systems with friction

$$M_i \mathbf{W}_i = \mathbf{F}_i + \sum_{j=1}^{n} \nu_j \frac{\partial \mathbf{r}_i}{\partial q_j} + \sum_{k=1}^{3N-n} \lambda_k \operatorname{grad}_i \Phi_k . \qquad (1.31)$$

In his opinion forces $\boldsymbol{\rho}_i$ are functions of ${}^0\mathbf{K}_i$ and coefficients ν_j are functions of λ_k and they are determined experimentally, the set of these functions representing the law of friction for the mechanical system.

Next, assuming the elementary work of reaction forces (1.27) in the virtual displacement Painlevé derived the following system of $3N$ equations

$$\sum_{i=1}^{N} \frac{\partial \mathbf{r}_i}{\partial q_k} \cdot \boldsymbol{\rho}_i = \sum_{i=1}^{N} \mathbf{K}_i \cdot \frac{\partial \mathbf{r}_i}{\partial q_k} \quad (k = 1, ..., n),$$

$$\sum_{i=1}^{N} \boldsymbol{\rho}_i \cdot \operatorname{grad}_i \Phi_j = 0 \quad (j = 1, ..., 3N - n) \qquad (1.32)$$

for $3N$ unknown variables $\rho_{ix}, \rho_{iy}, \rho_{iz}$. Due to these relationships, the following theorem was proved: among all systems of forces $\mathbf{K}'_i$ whose work is τ, the system $\boldsymbol{\rho}_i$ renders a minimum to the sum $\sum K'^2_i$. It is necessary to mention that while deriving eq. (1.32) values for $\boldsymbol{\rho}_i$ were taken as being proportional to a certain virtual displacement. However, as shown in Chapter 4, such an assumption is not valid in the general case.

1.4 Principles of the general theory of systems with Coulomb friction

Adopting Painlevé's definition of the forces of friction, Rumyantsev generalised the basic principles to the non-holonomic systems and established the Gauss principle in two forms, namely with explicit forces of friction and without explicit reactions, [130]. According to the first formulation, among all possible accelerations, the true acceleration renders a minimum to the function

$$A = \frac{1}{2}\sum_{i=1}^{N} M_i \left[\frac{\mathbf{F}_i}{M_i} + \frac{\boldsymbol{\rho}_i}{M_i} - \mathbf{W}_i\right]^2 \tag{1.33}$$

and vice versa, the condition of the minimum of this function with respect to the accelerations leads to the equations of motion.

The second formulation was established for the set of c-motions for which the work of the reaction forces is zero, i.e.

$$\sum_{i=1}^{N} \mathbf{K}_i \cdot \delta \mathbf{r}_i^c = \sum_{i=1}^{N} (\mathbf{F}_i - M_i \mathbf{W}_i) \cdot \delta \mathbf{r}_i^c = 0. \tag{1.34}$$

By using relationship (1.33) he proved the following theorem: deviation of the actual motion of the system with friction from an imaginary c-motion is less than the deviation of the latter from the motion of the system released from all constraints. In his next paper [131] by means of comparing the actual motion (d), an imaginary motion (δ) and the motion released from all or some of the constraints (∂) Rumyantsev derived the theorem: deviation of d-motion from any of δ-motion or ∂-motion plus the double work of the difference of the reaction forces $\mathbf{K}_i - \mathbf{K}_i^\partial$ in the virtual displacement $\delta \mathbf{r}_i = \frac{1}{2}(\delta \mathbf{v}_i - d\mathbf{v}_i)dt$ is less than deviation of δ-motion from ∂-motion.

Thus, the investigations by Painlevé and Rumyantsev created a theory of non-ideal systems which reduces the frictional forces $\boldsymbol{\rho}_1,...,\boldsymbol{\rho}_N$ to the particles. This theory is valid regardless of the experimental laws of friction, [130]. On the other hand, in order to apply this theory to systems with contact friction it is necessary to establish the relationship between Coulomb's law and forces $\boldsymbol{\rho}_i$. Such a relationship was suggested in [117] for some particular case, however this question was not studied for the general case.

Let us proceed to consider existing approaches to constructing equations of dynamics with Coulomb friction being taken into account. In many textbooks on the theory of mechanisms and machines, e.g. [156] and others, as well as in a number of studies [17], [54], [101], [139] etc. the normal reaction forces of the frictional contacts are taken to be equal to the reaction forces in the case of no paradoxes. Under such an assumption, the problems of the paradoxes disappear and the problem for systems with friction is essentially simplified. However, even the simple example of the Painlevé-Klein scheme indicates that this assumption may lead to a principle error in determining the reaction forces and accelerations and hence in estimation of the stresses and strains in the system and its law of motion.

In order to construct the general theory of motion of the systems with Coulomb friction it is primarily required to write down the exact equations of dynamics in the form which corresponds to and only to Coulomb's law. Such a position was suggested by Pozharnitsky [126]. Analysing a system with holonomic coordinates $q_1, ..., q_{n+k+l}$ subjected to the ideal holonomic constraints, l non-holonomic constraints and k removed constraints with dry friction, the author derived the following equation

$$\frac{\partial S}{\partial \ddot{q}_j} = Q_j + \sum_{i=1}^{m} \left(-\mu_i R_i \frac{v_{ix}\alpha_{ij}^1 + v_{iy}\alpha_{ij}^2}{\sqrt{v_{ix}^2 + v_{iy}^2}} + R_i \alpha_{ij}^3 \right) \quad (j = 1, ..., n+k). \tag{1.35}$$

Here S denotes the energy of accelerations, Q_j are the generalised active forces, m is the number of contacts, μ_i are the coefficients of Coulomb friction, x_i, y_i, z_i designates the $i-th$ moving coordinate system whose origin coincides with the $i-th$ contact point and axis z_i is directed along the normal to the contact surface. Moreover,

$$\delta x_i = \sum_{j=1}^{n+k} \alpha_{ij}^1 \delta q_j, \quad \delta y_i = \sum_{j=1}^{n+k} \alpha_{ij}^2 \delta q_j, \quad \delta z_i = \sum_{j=1}^{n+k} \alpha_{ij}^3 \delta q_j$$

are the virtual displacements of $i-th$ contact points, $\mathbf{v}_i$ are the relative velocities of gliding and R_i are the normal reaction forces of the contacts.

Provided that some velocities and acceleration are equal to zero $\mathbf{v}_1 = ... = \mathbf{v}_\tau = 0$ ($\tau \leq m$), $\dot{\mathbf{v}}_1 = ... = \dot{\mathbf{v}}_\nu = 0$ ($\nu < \tau$), at the initial instant considered, then, instead of eq. (1.35) we obtain the following system

$$\frac{\partial S}{\partial \ddot{q}_j} = Q_j + \sum_{i=\tau+1}^{m} \left(-\mu_i R_i \frac{v_{ix}\alpha_{ij}^1 + v_{iy}\alpha_{ij}^2}{\sqrt{v_{ix}^2 + v_{iy}^2}} + R_i \alpha_{ij}^3 \right) +$$

$$\sum_{i=\nu+1}^{\tau} \left(-\mu_i R_i \frac{\dot{v}_{ix}\alpha_{ij}^1 + \dot{v}_{ij}\alpha_{ij}^2}{\sqrt{v_{ix}^2 + v_{iy}^2}} + R_i \alpha_{ij}^3 \right) +$$

$$\sum_{i=1}^{\nu} (R_{1i}\alpha_{ij}^1 + R_{2i}\alpha_{ij}^2 + R_{3i}\alpha_{ij}^3), \quad (j = 1, ..., n+k). \tag{1.36}$$

Equations (1.35) and (1.36) explicitly contain the coefficients of friction μ_i and the normal reaction forces R_i and thus, are suitable for systems with Coulomb friction and only for these systems. They are constructed for the one-sided constraints and can not be used for analysis of mechanisms with two-sided constraints. Equations for two-sided constraints would be more powerful since they cover both cases. This can be observed through the examples of Chapters 3 and 4.

1.4 Principles of the general theory of systems with Coulomb friction

To derive eqs. (1.35) and (1.36) m moving coordinate systems were used, their origin coinciding with the contact points. In the general case, these points are not given and this makes eqs. (1.35) and (1.36) not very convenient in practical problems. For example, they do not allow one to determine the conditions for paradoxes, as was mentioned in [126].

In the second part of [126], i.e. in Chapters 6-8, the Gauss principle is considered for systems with Coulomb friction under the assumption of no paradoxes. The author found an isolated minimum of a certain function with respect to the accelerations for the true motion and established that there always exists at least one motion. Thus, according to the formulation of the principle suggested by Pozharnitsky, the question of the uniqueness of the motion remains open even in the case of no paradoxes.

In a number of papers, e.g. [49], [100], [122] Lagrange's equations for the systems with Coulomb friction are considered in the form

$$\ddot{q} = A + B\lambda, \tag{1.37}$$

where $A = A(q, \dot{q}, t) \in R^n$, $B = B(q,t) \in R^n \times R^n$, $q \in R^n$, $\lambda \in R^n$ denote the generalised constraint forces. In the particular case of a single one-sided frictional contact (among others) the contact condition and Coulomb's law are respectively given by

$$q_1 \geq 0, \quad \lambda \geq 0, \quad \lambda_j = f_j(q, \dot{q})\lambda_1 \quad (j = 2, ..., n). \tag{1.38}$$

Equations (1.37) take the form

$$\ddot{q} = A + B'\lambda_1, \quad B' = B'(q, \dot{q}, t) \in R^n. \tag{1.39}$$

It is easy to see that not only the force of Coulomb friction but also any non-dissipative force can be represented in the form of relationship (1.38). Indeed, applying arbitrary force $\mathbf{F} = F\mathbf{e}$ ($\mathbf{e}$ is the unit vector) to any point described by the position vector $\mathbf{r}(q)$ we obtain the following generalised forces

$$P_j = F\mathbf{e} \cdot \frac{\partial \mathbf{r}}{\partial q_j}.$$

Thus for $P_1 \neq 0$, i.e. for $\mathbf{e} \cdot \partial \mathbf{r}/\partial q_1 \neq 0$ we have

$$\frac{P_j}{P_1} = \frac{\mathbf{e} \cdot \dfrac{\partial \mathbf{r}}{\partial q_j}}{\mathbf{e} \cdot \dfrac{\partial \mathbf{r}}{\partial q_1}} = f_j(q).$$

Thus, according to the viewpoint of Pozharnitsky, expressions (1.37)-(1.39) are correct but are not specific for systems with Coulomb friction. Hence, these equations do not yield equations for the normal reaction forces, the differential equations of motion and the conditions for the paradoxes which explicitly contain the coefficients of friction, the coefficients of the kinetic energy and the generalised forces.

The papers by Smirnov [134]-[138] utilise Lagrange's equations derived in [102] and [103] for ideal systems with removed constraints as the general equations of dynamics of the systems with friction

$$\ddot{q}_\sigma = -\sum_{\rho=1}^{n}\sum_{\lambda=1}^{n}(\Gamma_{\rho\lambda}^{\sigma})^0 \dot{q}_\rho \dot{q}_\lambda + \sum_{\rho=1}^{n}(A_{\sigma\rho}^{-1})^0 Q_\rho^0, \quad (\sigma = 1,...,n),$$

$$\lambda_{n+r} = -\sum_{\sigma=1}^{n} A_{n+r,\sigma}^0 \sum_{\alpha=1}^{n}\sum_{\beta=1}^{n}(\Gamma_{\rho\lambda}^{\sigma})^0 \dot{q}_\rho \dot{q}_\lambda + \sum_{\sigma=1}^{n} A_{n+r}^0 \sum_{\rho=1}^{n}(A_{\sigma\rho}^{-1})^0 Q_\rho^0 +$$

$$\sum_{\sigma=1}^{n}\sum_{\rho=1}^{n}\Gamma_{\sigma\rho,n+r}^0 \dot{q}_\sigma \dot{q}_\rho - Q_{n+r}^0 \quad (r = 1,...,m),$$

where n denotes the number of degrees of freedom, m is the number of frictional pairs, $\Gamma_{mn,p}$ and Γ_{mn}^p are Christoffel's symbols of the first and second kind respectively, $A_{\sigma\rho}^{-1}$ are elements of the matrix inverse to the matrix of the quadratic form of the velocities, and superscript 0 corresponds to the zero values of the redundant coordinates. In the case of frictional constraints the generalised forces Q_ρ^0 and the constraint multipliers λ_{n+r} depend upon the normal reaction forces R_r and the coefficients of friction. The author does not propose dependences $\lambda_{n+r}(R_r)$ however he mentions that they can be non-linear with respect to the reaction forces. Since the equations for the reaction forces and the equations of motion are absent in [137], [138] Painlevé's paradoxes and other aspects are considered for particular examples rather than in the general form.

Papers [36], [37] by Dobroslavsky are devoted to the problem of stability with the contact compliance being taken into account. By analysing study [1] by Abramov and other examples the author arrived at the equation of perturbed motion in the form

$$A\ddot{q} + Bq = 0 \quad q \in R^n, \quad A, B \in R^n \times R^n.$$

It was stressed that matrix B is non-symmetric and contains "inertial parameters and properties of the regimes of motion" whereas matrix A represents the contact compliance. Stability of such systems was considered in [38] however the dependence of the general structure of the matrices on the coefficients of friction remained unstudied.

Let us touch upon the following two questions of the theory of systems with friction which are important for practical mechanical engineering, namely, the transition from the state of rest to motion and self-braking. A particle lying on the plane begins to move under a critical ratio of the tangential force to the normal pressure. Thus, the condition for transition from rest to motion is determined by Coulomb's law of friction. However,

1.4 Principles of the general theory of systems with Coulomb friction

more complex systems are not necessarily characterised in such a unique manner. For example, for the scheme shown in Fig. 1.3 Painlevé proved that under a zero initial velocity $\dot{x}(0) = 0$ and the condition for paradoxes $\mu \tan > 2$ the equation of motion admits simultaneously staying at rest and transition to motion. A general criterion of this non-uniqueness and the interrelation between this effect and the phenomenon of non-uniqueness and non-existence of motion (under zero initial velocity) have not yet been established. This problem can be solved apparently if one derives the relationship between the generalised forces for which the system begins to move, i.e. if Coulomb's law is expressed in terms of the generalised forces.

The expediency of this relationship is also due to the following reasoning. Let Lagrange's equations be derived for analysis of a mechanisms, that is the coefficients of the kinetic energy, the generalised active forces and the generalised forces of friction are given. This knowledge can be used for analysis of the state of rest and transition to motion only if the general condition for the transition is expressed in terms of these coefficients and forces. Nowadays, because of the absence of the general condition for the transition one has to calculate the normal reaction force and the force of static friction for each particular mechanism, and then study the critical case. In the majority of cases this procedure is very laborious.

It is likely that one of the most fruitful applications of the condition of staying at rest is the creation of the theory of self-braking mechanisms, [148]–[151], [64], [156]. These mechanisms are the subject of a rather developed branch of applied mechanics. The dynamics of such mechanisms has been studied in papers by Veits and his coworkers [148]–[151]. In particular, these papers deal with impacts in machines with self-braking mechanisms with and without dynamic seizure. Let us notice that applying the basic principles of the theory of self-braking mechanisms to systems of particles faces difficulties because of the absence of a rigorous definition for the concept of self-braking. It is known that a mechanism is termed self-braking if the beginning of motion is feasible when the force is transmitted from one chain to the other, but it is not feasible under the inverse transmitting. It is evident that this definition is not suitable for a mechanical system whose number of particles and thus the number of external forces exceeds two. On the other hand, it is not difficult to imagine that such systems may possess a property analogous to the property of self-braking under certain circumstances. Therefore, the question of self-braking for systems of particles needs further investigation for cases involving both the presence and absence of paradoxes.

Based upon the above we conclude that the theory of non-ideal systems was created with account for the Painlevé forces of friction applied to the particles. This theory can not be applied for the case of contact friction since the general interrelation between Coulomb's law and Painlevé's law has not yet been established.

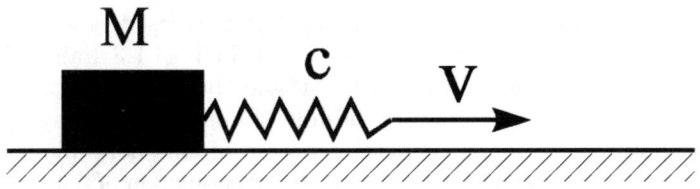

FIGURE 1.5.

As mentioned in Section 1.2 the peculiarities of Coulomb's law give rise to specific problems, whose solutions are required for constructing the general theory for mechanical systems with Coulomb friction. These are calculation of the normal reaction forces and eliminating them from the equations of motion, Painlevé's paradoxes, self-braking, transition from the state of rest to motion and so on. Although the literature on this subject is quite voluminous there is no complex investigation aimed at a unified approach to these problems.

In addition to the above mentioned methods of description of the dissipative systems, there exist a number of other approaches. It is worthwhile mentioning the papers by Agafonov [2], Appell [6], Bolotin [18], Bouligand [21], Do Shan [34], [35], Duvaut and Lions [41], Panovko [119], [120] Peres [122], Fufaev [43]-[45] and many others. However the problems considered in these works are beyond the scope of the present book.

1.5 Laws of Coulomb friction and the theory of frictional self-excited oscillations

The problem of frictional self-excited oscillations stands out from the special problems of the theory of mechanical systems with friction. This problem is frequently reduced to the scheme depicted in Fig. 1.5 with one degree of freedom.

In order to explain the frictional self-excited oscillations scientists put forward various hypotheses on the dependence of the force of friction at the beginning and throughout the motion upon the mechanical parameters. All these hypotheses are actually some refinements of Coulomb friction. For this reason, under the word "Coulomb friction" we understand the friction obeying both the classical and refined laws.

Let us consider the main hypotheses. A more widely held hypothesis is that of Kaidanovsky and Khaikin [50], [51], [60], [61]. According to this hypothesis, the origin of frictional self-excited oscillations is caused by a falling dependence of the force of sliding friction F_c on the velocity of the relative motion of the contacting bodies $\dot{x}$. The force required for the

1.5 Laws of Coulomb friction and frictional self-excited oscillations

beginning of motion F_+ is constant and equal to force F_c at zero velocity

$$F_c = F_c(\dot{x}), \quad F_+ = F_c(0).$$

Thus, this hypothesis is based upon the assumption by Rayleigh who suggested this law of friction between a violin string and a fiddle-stick. Many researches hold this viewpoint, e.g. [13], [10], [11], [14].

The hypothesis of Kaidanovsky and Khaikin is widespread in many works on non-linear mechanics [27], [155], [121], [141]. In [121] the dependence $F_c(\dot{x})$ is approximated by the third order polynomial

$$F_c = 3F_* \left(1 - \frac{\dot{x}}{\gamma} + \frac{\dot{x}^3}{3\gamma^3}\right),$$

where $F_* = \min F_c$, γ is the length of the falling part of the curve $F_c(\dot{x})$, i.e. the value of $\dot{x}$ for which $F_c = F_*$. The amplitude and the period of the quasi-harmonic self-excited oscillations in both a transient process and a stationary regime are determined for the above law of friction. It is found that the quasi-harmonic self-excited oscillations occur in a certain interval of the velocity v. A hard excitation of self-excited oscillations with periodic stops is considered in [119], [121] for the original hypothesis of Kaidanovsky and Khaikin. It is shown that the amplitude of the self-excited oscillation increases and the length of the part of the immovable contact decreases with the growth of velocity v.

It is worthwhile mentioning some cases of applications of this hypothesis to technological problems. Considering friction of the surfaces of the metal-cutting tool and the treated metal Kashirin [55] showed that vibrations appear at those values of the velocity of cutting for which the radial component of the force of cutting decreases with the growth of velocity. Constructing the differential equation of motion for cutting with account for the falling force-velocity dependence the author carried out a parametric study and compared the result with experimental data. Paper [55] is thus the first attempt to suggest a mathematical description of the oscillatory process for metal cutting.

In many papers by Murashkin and his coworkers [3], [111]-[114] self-excited oscillations are explained by means of a falling characteristic of the forces of friction and cutting. Paper [112] is concerned with investigating sliding in machine tools. The author used the Van-der-Pol equation for modelling the continuous sliding (without stops) and constructed the solution in the phase space. Under this formulation, the problem of the self-excited oscillation reduces to the problem derived, for example, in the book by Stoker [140] for an electronic generator.

Confirming the validity of the hypothesis of Kaidanovsky and Khaikin many authors mention that the curves $F_c(\dot{x})$ obtained in tests have falling parts of various forms, e.g. [76], [104], [112], [15].

Papers by Tolstoy and his coworkers [143]-[145] on microdisplacements of the contacting bodies in the normal direction suggest a reason for the decreasing force of friction with the growth of the velocity of gliding. The higher the velocity, the more intensive the normal components of the micropulses between the microbulges, the larger amplitude of oscillations in the normal direction and the higher the averaged level on which the slider glides. This decreases the actual area of the contact and finally results in a decrease in the force of friction.

Thus, according to the consensus of research the falling characteristic $F_c(\dot{x})$ exists. Nevertheless, taking account of the dependence of the force on the velocity of sliding does not account for the series of peculiarities observed in practice, for instance the property of hard excitation of self-excited oscillations, see [67]-[70], [74], [76], [78].

Let us consider now another explanation of self-excited oscillations. In 1939, Kragelsky [70] observed the dependence of the maximum of the force of friction F_+ on the duration T of the previous contact

$$F_+ = F_\infty - (F_\infty - F_0)e^{-\delta T}, \qquad (1.40)$$

where F_∞, F_0 and δ are constant parameters. Theoretical substantiations for formula (1.40) in the case of dry friction were given in [67], [68] and [52], whilst for the case of boundary friction it was suggested in [59].

Taking the law of friction (1.40) Ishlinsky and Kragelsky developed the theory of relaxation self-excited oscillations [47], [48]. The force of gliding friction F_c was taken as being constant and equal to F_0. In this case, the system is not self-excited but possesses the property of hard excitation of the self-excited oscillations with short-duration stops at velocity v which is less than the critical value. The amplitude of this oscillation decreases stepwise from movement to movement and tends to a stationary limiting value. For low velocities v this limiting amplitude value decreases with growth of v and the amplitude of the first jump increases.

The jump in the force under the transition from rest to the motion leads to the essentially serrated character of the self-excited oscillations which is confirmed by experiments. The fact that the amplitude of the first jump is larger than that of the other jumps is also reported in [39], [40], [67], [71].

Kragelsky and Kosterin [69] reported later that, in addition to the dependence $F_+(T)$, the falling characteristic $F_c(\dot{x})$ is also the reason for the self-excited oscillations. Tolstoy and Pan Bin-Yaao [146] noted that the force F_+ in the case of zero duration is equal to the force F_c for zero velocity, i.e. $F_+(0) = F_c(0)$. The papers, e.g. [32], [33], [71], [129], were devoted to generalising the problem of self-excitation accounting for the jump of the force of friction under the transition to gliding and the further smooth dependence on the velocity of gliding.

Discussing the plausibility of the hypotheses Kragelsky wrote:" The conclusions can not be considered as being final, they should be considered as preliminary ones...". In this regard, we point out the fact to which the authors have not paid attention. As the results of the experiments, the authors of [67], [24], [59], [56] detected that the value of F_+ decreases as velocity v increases. They concluded that the increase in F_+ is a direct consequence of the increasing dependence $F_+(T)$. In reality, an increase in F_+ with growth of v indicates only an existence of at least one of the dependences $F_+(T)$ and $F_+(cv)$. Judging from the decrease of the values of jumps after the first movement, dependence $F_+(T)$ exists. However, this does not imply the absence of the influence of the velocity of loading $f = cv$ on the value of F_+.

Along with the hypothesis of the time-dependence $F_+(T)$, it is worth mentioning the paper by Bouden and Leben [22] which explains break-down of the weld in bridges by jumps in the friction force in the contact regions. This explanation can be accepted only for the cases of molecular setting and welded surfaces. The practice however suggests that "relaxational self-excited oscillations are observed for material pairs like wood - steel, cast iron - asbestos etc. which exclude the very possibility of welding due to their nature", [71]. With this in mind, the above explanation can not be accepted as being general.

An interesting idea was advanced by Elyasberg [42] who analysed experimental results and proposed that the force of sliding friction F_c has two components F_1 and F_2 depending respectively on the velocity and acceleration. Not denying the existence of a jump in the force under the transition from the rest to motion Elyasberg however was of the opinion that "short intervals of rest" do not affect the value of the jump.

By using this idea Elyasberg [42] and Veits [147] suggested a method of determining the critical velocity ensuring transition from motion with stops to a continuous sliding. They also derived formulae for calculating the amplitude and the period of relaxation self-excited oscillations.

Papers [153], [154] by Vulfson adopted such a friction force characteristic which simultaneously reflects a jump in the force at the getaway and the phase shift of the force of sliding with respect to the velocity. In particular, it was shown that the frequency of quasi-harmonic oscillations is less than the eigenfrequency of the system.

Of considerable interest are the papers by Kudinov [73], [74] in which the self-excited oscillations in the tangential direction are considered together with displacement in the normal direction. The author is of the opinion that due to the hydrodynamic lubrication force the displacement along the normal increases as the velocity of gliding increases and thus, the force of friction decreases. However, under rapid changes in velocity the slider has no time to emerge, as it possesses a certain inertia. This is why the force of the sliding friction does not change as much as the experimental curve $F_c(\dot{x})$. Papers [73], [74] also discussed the influence of the "coordinate and inertial constraints" caused by details in the mass distribution and the contact compliance on the appearance of the frictional self-excited oscillations. A number of approaches to frictional self-excited oscillations are suggested in [9], [12], [24]-[27], [155], [65], [69], [97], [98], [123], [124], [142], [57]. The general tendency is to construct a theory by simultaneously taking account of several from the above hypotheses.

Finally, one should draw attention to the phenomenon of the so-called preliminary displacement. If under the word "displacement" we understand any relative motion of the contacting bodies with sliding over the whole contact surface, then the preliminary displacement is defined as the motion for which the sliding occurs only on a part of the contact surface, [72], [106].

Lecornue wrote in 1905 that "at the time instant when the bodies begin to contact each other the asperities of both surfaces catch each other. The outer layer of each body experiences an increasing displacement with respect to the deeper layers which is proportional to the actual tangential force. A relative displacement becomes possible only when the displacement reaches a certain value". Later on, fundamental investigations of the displacement were carried out by Verkhovsky [152], Demkin and Kragelsky [31], [71], Konyakhin [66], Maksak [106], Mikhin [108], [109], Courtney-Pratt and Eisner [28]. A complete historical overview of the problem is given in [106]. In the framework of the theory of frictional self-excited oscillations there should be a certain small displacement of the contacting bodies on the contact surface at each cycle of oscillation. This displacement has not yet been studied, and thus its dependence on the parameters of the oscillatory systems is still unknown.

Summarising, we considered the basic principles of the theory of frictional self-excited oscillations. As a result we can make the following conclusions.

Experience shows that the force of friction decreases at the initial movement as the velocity increases. This phenomenon can be explained by the dependence of the force on both the duration of the contact and the velocity of the tangential force. Since the first jump is always larger than the following ones, there is good reason to think that the first of these dependences does exist. At the same time, there is not enough experimental data to make a conclusion on the existence of the second dependence.

According to the various hypotheses on the law of friction, the dependences of the amplitude and the period of self-excited oscillations on the velocity turn out to be different. On the other hand, there is no experimental analysis in the literature which enables one to determine the character of the self-excited oscillations under the smooth change in velocity. For this reason, it is difficult to indicate a preferable hypothesis. When the frictional characteristic is given, the system should be self-excited, i.e. it should possesses the property of soft excitation. However this type of self-excitation has not yet been found in tests. Due to self-excitation along with the relaxation self-excited oscillations the stationary quasi-harmonic self-oscillations without stops are possible. The latter oscillations are still the subject of interest to the experimentalist. However not only their characteristics, but even their existence is still not known. An additional analysis is also required for the regime of motion with periodic stops which exposes local small displacements of the contacting bodies during the phase of the stop.

2
Systems with a single degree of freedom and a single frictional pair

For the above class of systems the following problems will be considered:

derivation of the equation for the normal reaction force in the frictional pair, as well as the differential equation of motion with the removed reaction force;

determination of the criterion of absence and non-uniqueness of solution for dynamical problem (criterion for Painlevé's paradoxes);

derivation of the condition for rest and the condition for transition to slip with and without paradox;

generalization of the concept of self-braking and the frictional cone to the case of systems of particles.

The theoretical results of this chapter will be illustrated by solving particular problems in the next chapter.

2.1 Lagrange's equations with a removed contact constraint

Let a system of N particles be subjected to $3N - 1$ stationary holonomic constraints. All of these constraints are assumed to be ideal except for a single pair for the Coulomb friction. The latter is a contact interaction of two bodies admitting their relative motion. The contact between these bodies is taken to be point-like or to be reduced to a point-like one. While adopting a model of a frictional pair it is necessary to choose between three types

38   2. Systems with a single degree of freedom and a single frictional pair

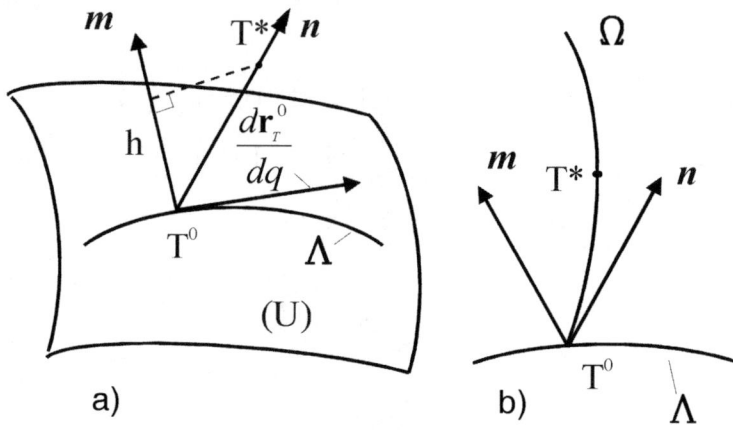

FIGURE 2.1.

of relative motion, namely pure slip, pure rolling and slip accompanied by rolling.

Relative motion under which a contact point on the surface of one body does not change is referred to as pure slip or slip without rolling. This body is considered as a point-like slider whereas the surface of the second body is viewed as the guiding surface, see Fig. 2.1. In this case, the velocity of the slip is equal to the rate of change of the contact point on the guiding surface (i.e. the velocity of the relative motion of the slider with respect to the guiding surface). Examples of pure slip are the Painlevé-Klein scheme, crank-slider mechanism, link gear and the systems studied in Chapters 3,4 and 5.

A motion, under which the contact place on the surfaces of both bodies are coincident, is referred to as the pure rolling or rolling without slip. In this case, the relative velocity of motion (the slip velocity) of the contact points is equal to zero. The motion of a railway wheel on a rail is an example of pure rolling.

The slip together with rolling occurs provided that the contact place of both bodies changes during the motion. However the relative velocity of the contact points is not zero. For example, such a mixed case is observed in a gear train.

For the case of pure rolling as well as slip together with rolling, the frictional pair is depicted by two curved surfaces, each surface having a tangential plane at each point, see Fig. 2.2.

In what follows, for mathematical modelling of the dynamics of systems with Coulomb friction we apply Lagrange's equations as suggested in [102], [103], [127] for the case of removed constraints. The generalised constraint forces with the generalised frictional forces included on the right hand sides of these equations are explicitly expressed in terms of the normal reaction

2.1 Lagrange's equations with a removed contact constraint   39

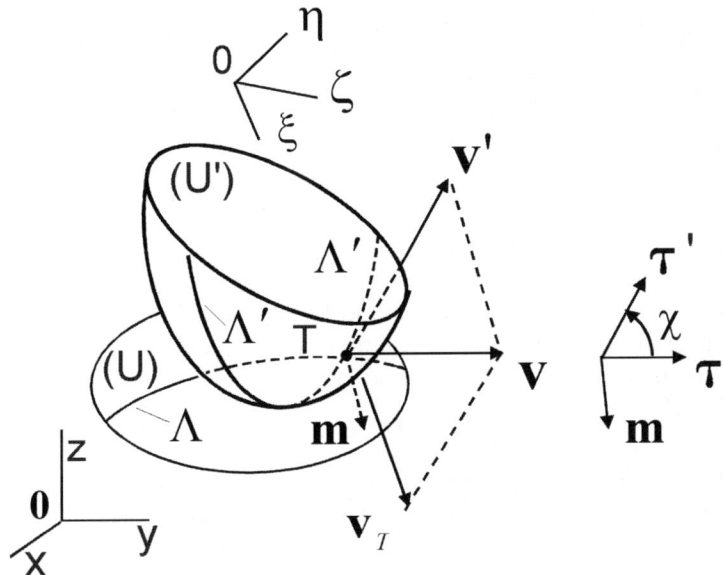

FIGURE 2.2.

force. Solving these equations for the reaction force and acceleration we obtain an equation for the reaction force and an equation of motion which does not contain the reaction force.

We begin our consideration with the case of slip without rolling. An additional treatment of slip with rolling is given in Sec. 2.2.

Let us write down the kinematic relationships which are needed for calculating the generalised reaction forces. Let the actual position of the slider be denoted by T^0 and the guiding surface with the Gaussian coordinates u^1, u^2 be denoted by U, see Fig. 2.1. For a single-degree-of-freedom system, a locus of point T^0 is some curve Λ on U. The position vectors of the slider $\mathbf{r}_T^0$ and the particles $\mathbf{r}_1^0, ..., \mathbf{r}_N^0$ are functions of the generalised coordinate q. Therefore,

$$\mathbf{r}_T^0 = \mathbf{r}_T^0(q) = \varphi[u^1(q), u^2(q)], \qquad \mathbf{r}_i^0 = \mathbf{r}_i^0(q), \quad (i = 1, ..., N)$$
$$\mathbf{v}_T^0 = \frac{d\mathbf{r}_T^0}{dq}\dot{q} = \left(\frac{\partial \varphi}{\partial u^1}\frac{du^1}{dq} + \frac{\partial \varphi}{\partial u^1}\frac{du^1}{dq}\right)\dot{q}, \quad \mathbf{v}_i^0 = \frac{d\mathbf{r}_i^0}{dq}\dot{q}, \qquad (2.1)$$

where $\mathbf{v}_T^0$ and $\mathbf{v}_i^0$ denote the velocity of the slider and the $i-th$ particle, respectively, a superscript 0 indicates that the quantities are determined before the contact constraint is removed and $\varphi(u^1, u^2)$ is the position vector of surface U.

Let us mentally remove the contact constraint. Let the slider gain an infinitesimally small additional displacement from T^0 to T^* along an a priori chosen direction $\mathbf{n} : |\mathbf{n}| = 1$; $\mathbf{n} \times \mathbf{v}_t^0 \neq 0$ admitted by constraints, see Fig. 2.1. In practice, an additional displacement can be prescribed by a certain curve Ω, crossing Λ at point T^0. Since the displacement is small it can be deemed as rectilinear, thus, vector $\mathbf{n}$ is the unit vector of the tangent to Ω at point T^0, Fig. 2.1b. The projection of $\overline{T^0 T^*}$ on the normal to U at T^0

$$h = (\mathbf{r}_T^* - \mathbf{r}_T^0) \cdot \mathbf{m} \tag{2.2}$$

is taken as the redundant coordinate. In the latter equation $\mathbf{r}_T^*$ denotes the position vector of point T^* and $\mathbf{m}$ is the unit base vector of the normal. It is evident that for a fixed q and a given $\mathbf{n}$ the value of h uniquely determines the position of the slider and all points of the system. It what follows, it is subjected to the constraint condition, see [102], [103],

$$F_h \equiv h = 0. \tag{2.3}$$

In the forthcoming analysis, all general statements are formulated in accordance with the redundant coordinate chosen by means of eq. (2.2).

It is clear that

$$\mathbf{r}_T^* = \mathbf{r}_T^*(q, h), \quad \mathbf{r}_T^*(q, 0) = \mathbf{r}_T^0(q), \tag{2.4}$$

$$\left(\frac{\partial \mathbf{r}_T^*}{\partial q}\right)_0 = \frac{d\mathbf{r}_T^0}{dq}. \tag{2.5}$$

By using the trihedron

$$\mathbf{m}, \quad \boldsymbol{\tau} = \frac{d\mathbf{r}_T^0}{dq} \left|\frac{d\mathbf{r}_T^0}{dq}\right|^{-1}, \quad \boldsymbol{\nu} = \mathbf{m} \times \boldsymbol{\tau}$$

we can represent vector $(\partial \mathbf{r}_T^*/\partial h)_0$ as follows

$$\left(\frac{\partial \mathbf{r}_T^*}{\partial q}\right)_0 = \left(\frac{\partial \mathbf{r}_T^*}{\partial h}\right)_0 \cdot \mathbf{mm} + \left(\frac{\partial \mathbf{r}_T^*}{\partial h}\right)_0 \cdot \boldsymbol{\tau\tau} + \left(\frac{\partial \mathbf{r}_T^*}{\partial h}\right)_0 \cdot \boldsymbol{\nu\nu}.$$

On the other hand, by virtue of eq. (2.2) we obtain

$$\mathbf{m} \cdot \left(\frac{\partial \mathbf{r}_T^*}{\partial h}\right)_0 = \lim_{T^* \to t^0} \frac{\mathbf{r}_T^* - \mathbf{r}_T^0}{(\mathbf{r}_T^* - \mathbf{r}_T^0) \cdot \mathbf{m}} = 1. \tag{2.6}$$

This yields the following expression for the derivative $(\partial \mathbf{r}_T^*/\partial h)_0$

$$\left(\frac{\partial \mathbf{r}_T^*}{\partial q}\right)_0 = \mathbf{m} + \lambda \frac{\partial \mathbf{r}_T^*}{\partial q} + \left(\frac{\partial \mathbf{r}_T^*}{\partial h}\right)_0 \cdot \boldsymbol{\nu\nu}, \tag{2.7}$$

where

$$\lambda = \left(\frac{\partial \mathbf{r}_T^*}{\partial h}\right)_0 \cdot \frac{d\mathbf{r}_T^0}{dq} \left/ \left|\frac{d\mathbf{r}_T^0}{dq}\right|^2\right.. \tag{2.8}$$

2.1 Lagrange's equations with a removed contact constraint

Thus, we obtained all geometric relationships needed for determining the generalised constraint forces.

Let us consider a two-sided contact constraint. The normal force of this constraint R, acting on the slider from the guide is considered to be positive (negative) if its direction is coincident with (opposite to) the direction of the unit vector $\mathbf{m}$, see Fig. 2.1. Then the vector of the normal reaction force $\mathbf{R}$ and the friction force $\mathbf{R}_\tau$ can be set in the form

$$\mathbf{R} = R\mathbf{m}, \quad \mathbf{R}_\tau = -\varepsilon_1 \mu \frac{\mathbf{v}_T^0}{|\mathbf{v}_T^0|} R, \qquad (2.9)$$

where $\varepsilon_1 = \operatorname{sign} R$ and μ denotes the friction factor. This leads to the expression for the general reaction force

$$\mathbf{R}_\sigma = \left(-\varepsilon_1 \mu \frac{\mathbf{v}_T^0}{|\mathbf{v}_T^0|} + \mathbf{m} \right) R. \qquad (2.10)$$

The scalar products

$$S_1 = \mathbf{R}_\sigma \cdot \frac{d\mathbf{r}_T^0}{dq}, \quad S_2^* = \mathbf{R}_\sigma \cdot \left(\frac{\partial \mathbf{r}_T^*}{\partial h} \right)_0$$

are the generalised reaction forces. Inserting the expressions for the vectors (2.1), (2.5), (2.7) and taking into account the following relationships

$$\mathbf{m} \cdot \frac{d\mathbf{r}_T^0}{dq} = 0, \quad \frac{\mathbf{v}_T^0}{|\mathbf{v}_T^0|} \cdot \frac{d\mathbf{r}_T^0}{dq} = \frac{\partial \mathbf{v}_T^0}{\partial \dot{q}} = \frac{\partial}{\partial \dot{q}} \left| \frac{d\mathbf{r}_T^0}{dq} \dot{q} \right| = \varepsilon_2 \left| \frac{d\mathbf{r}_T^0}{dq} \right|, \qquad (2.11)$$

where $\varepsilon_2 = \operatorname{sign} \dot{q}$, we arrive at the following equations for the generalised reaction forces

$$S_1 = -\varepsilon_1 \mu \beta \mathbf{v}_T^0 \dot{q} R = -\varepsilon_1 \varepsilon_2 \mu \left| \frac{d\mathbf{r}_T^0}{dq} \right| R,$$

$$S_2^* = \left[1 - \varepsilon_1 \mu \frac{\partial \mathbf{v}_T^0}{\partial \dot{q}} \right] R = \left(1 - \varepsilon_1 \varepsilon_2 \mu \lambda \left| \frac{d\mathbf{r}_T^0}{dq} \right| \right) R. \qquad (2.12)$$

Following Lurie [102], [103] and taking into account condition (2.3), we arrive at the following system of equations

$$A\ddot{q} + \frac{1}{2} \frac{dA}{dq} \dot{q}^2 = Q_1 - \varepsilon_1 \varepsilon_2 \mu \left| \frac{d\mathbf{r}_T^0}{dq} \right| R,$$

$$A_{12}^{*0} \ddot{q} + \left(\frac{\partial A_{12}^*}{\partial q} - \frac{1}{2} \frac{\partial A_{11}^*}{\partial q} \right) \dot{q}^2 = Q_2^* + \left(1 - \varepsilon_1 \varepsilon_2 \mu \lambda \left| \frac{d\mathbf{r}_T^0}{dq} \right| \right) R, \qquad (2.13)$$

where Q_1, Q_2 denote the generalised active forces, A_{ik}^* $(i, k = 1, 2)$ are the coefficients of the kinetic energy of the system with the removed constraint and A is the coefficient of the original system, that is the system which is not released from the contact force ($A = A_{11}^{*0}$).

Let us consider a particular case in which the trajectory Ω of the additional displacement of the slider is orthogonal to its trace Λ on U. The previous notation with the omitted superscript $*$ is used to this aim. Because in this case $(\partial \mathbf{r}_T/\partial h)_0 \perp (d\mathbf{r}_T^0/dq)$, it follows from eqs. (2.8) and (2.12) that

$$\left(\frac{\partial \mathbf{r}_T}{\partial h}\right)_0 = \mathbf{m} + \left(\frac{\partial \mathbf{r}_T}{\partial h}\right)_0 \cdot \boldsymbol{\nu}\boldsymbol{\nu}, \quad \lambda = 0, \quad S_2 = R. \qquad (2.14)$$

Instead of the system of equations (2.13) we have the following one

$$A\ddot{q} + \frac{1}{2}\dot{q}^2 = Q_1 - \varepsilon_1\varepsilon_2\mu \left|\frac{d\mathbf{r}_T^0}{dq}\right| R,$$

$$A_{12}^0 \ddot{q} + \left(\frac{\partial A_{12}}{\partial q} - \frac{1}{2}\frac{\partial A_{11}}{\partial q}\right)\dot{q}^2 = Q_2 + R. \qquad (2.15)$$

We have derived two equivalent systems of equations (2.13) and (2.15) according to two ways of describing the additional displacement. System (2.15) is used for investigation of the general problems whereas solving particular problem is carried out with the help of eqs. (2.13) and (2.15).

It is expedient to derive formulae for the coefficients in eqs. (2.13) and (2.15) and establish the relationship between the coefficients. To begin with, we express $(\partial \mathbf{r}_i^*/\partial h)_0$ in terms of $(\partial \mathbf{r}_i/\partial h)_0$ and $d\mathbf{r}_i^0/dq$. When the slider is subjected to small displacements in the tangential and normal directions $(d\mathbf{r}_T^0/dq)\delta_1$ and $(\partial \mathbf{r}_T^*/\partial h)_0 \delta_2$ (δ_1 and δ_2 are arbitrary small quantities) the position of the $i-th$ particle should change by $(d_T^0/dq)\delta_1$ and $(\partial \mathbf{r}_T/\partial h)_0 \delta_2$, respectively. Let the prescribed virtual displacement of the slider be

$$\delta \mathbf{r}_T^* = \frac{d\mathbf{r}_T^0}{dq}\delta q + \left(\frac{\partial \mathbf{r}_T^*}{\partial h}\right)_0 \delta h$$

which, due to eqs. (2.7) and (2.14), can be put in the following form

$$\delta \mathbf{r}_T^* = \frac{d\mathbf{r}_T^0}{dq}(\delta q + \lambda \delta h) + \left(\frac{\partial \mathbf{r}_T}{\partial h}\right)_0 \delta h.$$

According to the above mentioned correspondence, the particle under consideration experiences the displacement

$$\delta \mathbf{r}_i^* = \frac{d\mathbf{r}_i^0}{dq}(\delta q + \lambda \delta h) + \left(\frac{\partial \mathbf{r}_i}{\partial h}\right)_0 \delta h = \frac{d\mathbf{r}_i^0}{dq}\delta q + \left[\lambda \frac{d\mathbf{r}_i}{dq} + \left(\frac{\partial \mathbf{r}_i^*}{\partial h}\right)_0\right]\delta h. \qquad (2.16)$$

On the other hand

$$\delta \mathbf{r}_i^* = \frac{d\mathbf{r}_i^0}{dq}\delta q + \left(\frac{\partial \mathbf{r}_i^*}{\partial h}\right)_0 \delta h. \qquad (2.17)$$

2.1 Lagrange's equations with a removed contact constraint

It follows from eqs. (2.16) and (2.17) that

$$\left(\frac{\partial \mathbf{r}_i^*}{\partial h}\right)_0 = \left(\frac{\partial \mathbf{r}_i}{\partial h}\right)_0 + \lambda \frac{d\mathbf{r}_i^0}{dq}. \qquad (2.18)$$

Taking into account the equations for the generalised forces

$$Q_1 = \sum_{i=1}^{N} \mathbf{F}_i \cdot \frac{d\mathbf{r}_i^0}{dq}, \quad Q_2^* = \sum_{i=1}^{N} \mathbf{F}_i \cdot \left(\frac{\partial \mathbf{r}_i^*}{\partial h}\right)_0,$$

where $\mathbf{F}_i$ denotes an active force acting on the $i-th$ particle, we obtain

$$Q_2^* = \sum_{i=1}^{N} \mathbf{F}_i \cdot \left(\frac{\partial \mathbf{r}_i}{\partial h}\right)_0 + \lambda \sum_{i=1}^{N} \mathbf{F}_i \cdot \frac{d\mathbf{r}_i^0}{dq} = Q_2 + \lambda Q_1. \qquad (2.19)$$

By analogy, we arrive at the following expressions for the coefficients on the right hand sides of eqs. (2.13) and (2.15)

$$A = \sum_{i=1}^{N} \mathbf{M}_i \left|\frac{d\mathbf{r}_i^0}{dq}\right|^2, \quad \frac{1}{2}\frac{dA}{dq} = \sum_{i=1}^{N} \mathbf{M}_i \frac{d\mathbf{r}_i^0}{dq} \cdot \frac{d\mathbf{r}_i^0}{dq}, \qquad (2.20)$$

$$A_{12}^{*0} = \sum_{i=1}^{N} \mathbf{M}_i \frac{d\mathbf{r}_i^0}{dq} \cdot \left(\frac{\partial \mathbf{r}_i}{\partial h}\right)_0 + \lambda \sum_{i=1}^{N} \mathbf{M}_i \left|\frac{d\mathbf{r}_i^0}{dq}\right|^2 = A_{12}^0 + \lambda A, \qquad (2.21)$$

$$\left(\frac{\partial A_{12}}{\partial q} - \frac{1}{2}\frac{\partial A_{11}}{\partial q}\right)_0 = \sum_{i=1}^{N} \mathbf{M}_i \frac{d^2\mathbf{r}_i^0}{dq^2} \cdot \left(\frac{\partial \mathbf{r}_i}{\partial h}\right)_0 + \lambda \sum_{i=1}^{N} \mathbf{M}_i \frac{d\mathbf{r}_i^0}{dq} \cdot \frac{d^2\mathbf{r}_i^0}{dq^2}$$
$$= \left(\frac{\partial A_{12}^*}{\partial q} - \frac{1}{2}\frac{\partial A_{11}^*}{\partial q}\right)_0 + \frac{1}{2}\lambda\frac{dA}{dq}. \qquad (2.22)$$

Finally, by means of eqs. (2.12) and (2.14) we have

$$S_2^* = S_2 + \lambda S_1. \qquad (2.23)$$

Equalities (2.19)-(2.23) show that the second equation in system (2.13) is a linear combination of the equations in (2.15) with the coefficients λ and 1. Thus, these systems are equivalent.

Remark 1. In the case of a moving guide surface U expression (2.1) for the position vector $\mathbf{r}_T^0$ of the slider should be given in the moving coordinate system related to $\mathbf{U}$. Under quantities $d\mathbf{r}^0/dq$ and $\mathbf{v}_T^0$ we understand respectively the local derivative and the relative velocity of slip of the slider with respect to the guide.

2.2 Kinematic expression for slip with rolling

When rolling with slip occurs an infinite number of points of each body contact. Thus, function $\mathbf{r}_T^0(q)$ can not be prescribed in an explicit form and the quantities $\mathbf{v}_T^0$ and $d\mathbf{r}_T^0/dq$ appearing in eqs. (2.10), (2.12)-(2.15) can not be found by means of eq. (2.1). An additional study enabling determination of slip velocity $\mathbf{v}_T^0$, the derivative $d\mathbf{r}_T^0/dq$ and the angular velocity of rotation $\boldsymbol{\omega}$ is carried out below.

2.2.1 Velocity of slip and the velocities of change of the contact place due to the trace of the contact

Let the immovable guiding surface U and the moving surface U' be respectively described by the following equations

$$\mathbf{r} = \mathbf{r}\left(u^1, u^2\right), \quad \mathbf{r}' = \mathbf{r}'\left(p^1, p^2\right), \qquad (2.24)$$

where $\mathbf{r}$ is given with respect to the immovable coordinate system $Oxyz$, $\mathbf{r}'$ is given in the moving coordinate system $0\xi\eta\zeta$ related to U', u^1, u^2, p^1, p^2 are the Gaussian coordinates. It is assumed that the surfaces contact each other at a single point at any time instant and the contact traces on the surfaces are lines Λ and Λ', see Fig 2.2. At the contact point the following equality

$$\mathbf{r}\left(u^1, u^2\right) = \mathbf{r}_0 + \mathbf{r}'\left(p^1, p^2\right) \qquad (2.25)$$

holds, where $\mathbf{r}_0 = \overline{00}$. The velocity of rolling $\mathbf{v}_T$ (the superscript 0 is omitted) is the velocity of point T which moves on U' and which is the contact point at the considered time instant and is given by

$$\mathbf{v}_T = \frac{d\mathbf{r}_T^0}{dq}\dot{q} = \mathbf{v}_0 + \boldsymbol{\omega} \times \mathbf{r}'. \qquad (2.26)$$

Here $\mathbf{v}_0 = (d\mathbf{r}_0/dq)\dot{q}$ denotes the velocity of the origin 0 of the coordinate system and $\boldsymbol{\omega}$ designates the angular velocity of the slider U'.

Let us consider now the velocity of change of the contact point due to the traces Λ and Λ'. They are given by the following expressions

$$\mathbf{v} = \dot{\mathbf{r}} = \mathbf{r}_\alpha \dot{u}^\alpha = \dot{\sigma}\boldsymbol{\tau}, \qquad (2.27)$$

$$\mathbf{v}' = \dot{\mathbf{r}}' = \mathbf{r}'_\alpha \dot{p}^\alpha = \boldsymbol{\tau}'\dot{\sigma}', \qquad (2.28)$$

where

$$\mathbf{r}_\alpha = \frac{\partial \mathbf{r}}{\partial u^\alpha}, \quad \boldsymbol{\tau} = \frac{d\mathbf{r}}{d\sigma} = \mathbf{r}_\alpha \frac{dp^\alpha}{d\sigma}, \quad \boldsymbol{\tau}' = \frac{d\mathbf{r}'}{d\sigma'} = \mathbf{r}'_\alpha \frac{dp^\alpha}{d\sigma'}. \qquad (2.29)$$

Here τ and τ' denote the unit base vectors to the curves Λ and Λ', respectively, whilst σ and σ' denote the arc length of these curves, respectively. In the present subsection the summation sign with respect to repeated subscripts and superscripts is omitted. For example, quantity $\mathbf{r}_\alpha \dot{u}^\alpha$ in eq. (2.27) implies the sum $\sum_{\alpha=1}^{2} \mathbf{r}_\alpha \dot{u}^\alpha$.

We notice that the rate of change $\mathbf{v}$ of the contact place along the trace Λ is a sum of two velocities, namely the transition velocity coinciding with the slip velocity $\mathbf{v}_T$ and the relative velocity equal to the velocity of change of the contact $\mathbf{v}'$ along the trace Λ'. In other words,

$$\mathbf{v} = \mathbf{v}_T + \mathbf{v}'$$

which, due to eqs. (2.26), (2.27) and (2.28) yields

$$\mathbf{r}_\alpha \dot{u}^\alpha = \mathbf{v}_0 + \boldsymbol{\omega} \times \mathbf{r}' + \mathbf{r}'_\alpha \dot{p}^\alpha. \tag{2.30}$$

Thus, the slip velocity $\mathbf{v}_T$ can be represented in the form

$$\mathbf{v}_T = \mathbf{v}_0 + \boldsymbol{\omega} \times \mathbf{r}' = \tau\dot{\sigma}' - \tau'\sigma' = \mathbf{r}_\alpha \dot{u}^\alpha - \mathbf{r}'_\alpha \dot{p}^\alpha. \tag{2.31}$$

Since

$$\dot{u}^\alpha = \frac{du^\alpha}{dq}\dot{q}, \quad \dot{p}^\alpha = \frac{dp^\alpha}{dq},$$

equations (2.26)-(2.31) yield

$$\frac{d\mathbf{r}_T}{dq} = \mathbf{r}_\alpha \frac{du^\alpha}{dq} - \mathbf{r}'_\alpha \frac{dp^\alpha}{dq}. \tag{2.32}$$

Hence, we have obtained eqs. (2.31) and (2.32) for determining the slip velocity $\mathbf{v}_T$ and the derivative $d\mathbf{r}_T/dq$ in the case of slip with rolling. As follows from these equations, in the case of pure slip

$$\dot{p}^\alpha = 0; \quad \mathbf{v}_T = \mathbf{r}_\alpha \dot{u}^\alpha, \quad \frac{d\mathbf{r}_T}{dq} = \mathbf{r}_\alpha \frac{du^\alpha}{dq}, \tag{2.33}$$

whilst in the case of pure rolling

$$\frac{d\mathbf{r}_T}{dq} = \mathbf{v}_T = 0; \quad \mathbf{r}_\alpha \dot{u}^\alpha = \mathbf{r}'_\alpha \dot{p}^\alpha. \tag{2.34}$$

2.2.2 Angular velocity

The mutual position of the surfaces U and U' is determined when the Gaussian coordinates of the contact point u^α, p^α and the angle χ between the traces Λ and Λ' is found. Following this idea we express the angular velocity $\boldsymbol{\omega}$ of the moving surface U' in terms of the time derivatives of u^α, p^α, χ.

The unit normal vectors to the surfaces are given by the following expressions

$$\mathbf{m} = \frac{\mathbf{r}_1 \times \mathbf{r}_2}{\sqrt{|a|}}, \quad \mathbf{m}' = \frac{\mathbf{r}'_1 \times \mathbf{r}'_2}{\sqrt{|a'|}}, \qquad (2.35)$$

where $|a|$ and $|a'|$ denote the discriminants of the first quadratic form of the surfaces. Vectors $\mathbf{m}$ and $\mathbf{m}'$ are colinear at the contact point. Let us agree to number the Gaussian coordinates u^α, p^α such that the above vectors coincide, i.e.

$$\mathbf{m} = \mathbf{m}'. \qquad (2.36)$$

Taking the derivative of vectors $\mathbf{m}$ and $\mathbf{m}'$ in the immovable and moving coordinate systems, respectively, and taking into account equality (2.36), we can write

$$\dot{\mathbf{m}} = \frac{\delta \mathbf{m}'}{\delta t} + \boldsymbol{\omega} \times \mathbf{m}, \qquad (2.37)$$

where $\delta/\delta t$ denotes the local derivative with respect to time. Moreover, multiplying both sides of eq. (2.14) by $\mathbf{m}$ and taking into account the relationship

$$\mathbf{m} \times (\boldsymbol{\omega} \times \mathbf{m}) = \boldsymbol{\omega} - \Omega \mathbf{m},$$

where $\Omega = \boldsymbol{\omega} \cdot \mathbf{m}$ is the angular velocity of whirling, we obtain

$$\boldsymbol{\omega} = \mathbf{m} \times \dot{\mathbf{m}} - \mathbf{m} \times \frac{\delta \mathbf{m}'}{\delta t} + \Omega \mathbf{m}. \qquad (2.38)$$

It is known, see [103], that

$$\dot{\mathbf{m}} = \mathbf{m}_\alpha \dot{u}^\alpha = -b_{\alpha\beta} \mathbf{r}^\beta \dot{u}^\alpha. \qquad (2.39)$$

Here $b_{\alpha\beta} = -\mathbf{r}_\alpha \cdot \mathbf{m}_\beta$ are the coefficients of the second quadratic form of the surface U, and $\mathbf{r}^\beta (\beta = 1,2)$ are the covectors corresponding to the base vectors $\mathbf{r}_\alpha$.

Due to (2.35) and (2.39) we obtain

$$\mathbf{m} \times \dot{\mathbf{m}} = -\frac{1}{\sqrt{a}} b_{\alpha\beta} \dot{u}^\alpha (\mathbf{r}_1 \times \mathbf{r}_2) \times \mathbf{r}^\beta.$$

Carrying out elementary transformations we can set the latter equation in the form

$$\mathbf{m} \times \dot{\mathbf{m}} = -\frac{1}{\sqrt{a}} \dot{u}^\alpha (b_{\alpha 1} \mathbf{r}_2 - b_{\alpha 2} \mathbf{r}_1). \qquad (2.40)$$

2.2 Kinematic expression for slip with rolling

We can express $\mathbf{m} \times \delta\mathbf{m}'/\delta t$ in terms of the parameters of the surface U'

$$\mathbf{m} \times \frac{\delta \mathbf{m}'}{\delta t} = -\frac{1}{\sqrt{a}} \dot{p}^\alpha \left(b'_{\alpha 1} \mathbf{r}'_2 - b'_{\alpha 2} \mathbf{r}'_1 \right). \tag{2.41}$$

Substituting (2.40) and (2.41) into (2.38) yields

$$\boldsymbol{\omega} = \frac{\dot{u}^\alpha}{\sqrt{a}} (b_{\alpha 2} \mathbf{r}_1 - b_{\alpha 1} \mathbf{r}_2) - \frac{1}{\sqrt{a}} \dot{p}^\alpha (b'_{\alpha 2} \mathbf{r}'_1 - b'_{\alpha 1} \mathbf{r}'_2) + \Omega \mathbf{m}. \tag{2.42}$$

It remains only to express the angular velocity of whirling Ω in terms of $\dot{u}^\alpha$, $\dot{p}^\alpha$ and $\dot{\chi}$. To this end, we use the rule of differentiation in a moving coordinate system

$$\dot{\boldsymbol{\tau}}' = \overset{*'}{\boldsymbol{\tau}} + \boldsymbol{\omega} \times \boldsymbol{\tau}'. \tag{2.43}$$

Taking the local derivative of the both sides of eq. (2.29) for $\boldsymbol{\tau}'$ with respect to time and applying the formulae of surface theory [103], we obtain

$$\overset{*'}{\boldsymbol{\tau}} = \left(\mathbf{r}'_\alpha + b'_{\alpha\beta} \frac{dp^\alpha}{d\sigma'} \frac{dp^\beta}{d\sigma'} \mathbf{m} \right) \dot{\sigma}', \tag{2.44}$$

where the following quantities

$$k^{*\alpha} = \frac{d^2 p^\alpha}{d\sigma'^2} + (\Gamma^\alpha_{\beta\gamma})' \frac{dp^\beta}{d\sigma'} \frac{dp^\gamma}{d\sigma'} \tag{2.45}$$

are the contravariant components of the vector of geodesic curvature $\mathbf{r}'_\alpha$ of the surface U', $(\Gamma^\alpha_{\beta\gamma})'$ denotes its Christoffel's symbols of the second kind. As the vector of the geodesic curvature is orthogonal to $\mathbf{m}$ and $\boldsymbol{\tau}'$, the following representation is valid

$$k'^{*\alpha} \mathbf{r}'_\alpha \dot{\sigma}' = \mathbf{r}'_\alpha \cdot (\mathbf{m} \times \boldsymbol{\tau}') \mathbf{m} \times \boldsymbol{\tau}'.$$

Calculating the scalar product $\mathbf{r}'_\alpha \cdot (\mathbf{m} \times \boldsymbol{\tau}')$ by virtue of conditions (2.29), (2.35) and the condition $\dot{\sigma}' = a'_{\alpha\beta} \dot{p}^\alpha \dot{p}^\beta$, we obtain

$$k'^{*\alpha} \mathbf{r}'_\alpha \dot{\sigma}' = \sqrt{|a'|} (k'^{*2} \dot{p}^1 - k'^{*1} \dot{p}^2) \mathbf{m} \times \boldsymbol{\tau}'. \tag{2.46}$$

In order to find the quantity $\boldsymbol{\omega} \times \boldsymbol{\tau}'$ which is the second terms in eq. (2.20), it is necessary to multiply eq. (2.38) by $\boldsymbol{\tau}'$. Taking into account that

$$(\mathbf{m} \times \dot{\mathbf{m}}) \times \boldsymbol{\tau}' = -\boldsymbol{\tau}' \cdot \dot{\mathbf{m}} \mathbf{m},$$

$$(\mathbf{m} \times \overset{*}{\mathbf{m}}) \times \boldsymbol{\tau}' = -\boldsymbol{\tau}' \cdot \overset{*}{\mathbf{m}} \mathbf{m}' = -m b'_{\gamma\alpha} \frac{dp^\gamma}{d\sigma'} \dot{p}^\alpha,$$

we obtain

$$\boldsymbol{\omega} \times \boldsymbol{\tau}' = \Omega \mathbf{m} \times \boldsymbol{\tau}' - \left(\boldsymbol{\tau}' \cdot \mathbf{m} + b'_{\gamma\alpha} \frac{dp^\gamma}{d\sigma'} \dot{p}^\alpha \right) \mathbf{m}. \tag{2.47}$$

2. Systems with a single degree of freedom and a single frictional pair

Inserting eqs. (2.44), (2.46) and (2.47) into the right hand side of eq. (2.43) yields the expression for τ' in terms of the parameters of surface U and the velocity of Ω

$$\dot{\tau}' = \left[\sqrt{|a'|}(k'^{*2}\dot{p}^1 - k'^{*1}\dot{p}^2) + \Omega\right] \mathbf{m} \times \boldsymbol{\tau}' - \boldsymbol{\tau}' \cdot \dot{\mathbf{m}}\mathbf{m}. \qquad (2.48)$$

Let us express $\dot{\tau}'$ in terms of the parameters of surface U. We notice that the unit tangent vector $\boldsymbol{\tau}'$ is obtained by means of a turn of vector $\boldsymbol{\tau}$ through angle χ about the axis directed along $-\mathbf{m}$. Then, see [103],

$$\boldsymbol{\tau}' = \boldsymbol{\tau} - \frac{2\mathbf{m}\tan\chi/2}{1+\tan^2\chi/2} \times (\boldsymbol{\tau} - \mathbf{m}\tan\chi/2 \times \boldsymbol{\tau}).$$

Elementary transformations result in the following formula

$$\boldsymbol{\tau}' = \boldsymbol{\tau}\cos\chi - \mathbf{m}\times\boldsymbol{\tau}\sin\chi. \qquad (2.49)$$

Taking the time-derivative yields

$$\dot{\boldsymbol{\tau}}' = -(\boldsymbol{\tau}\sin\chi - \mathbf{m}\times\boldsymbol{\tau}\cos\chi)\dot{\chi} - \dot{\mathbf{m}}\times\boldsymbol{\tau}\sin\chi - \mathbf{m}\times\dot{\boldsymbol{\tau}}\sin\chi + \dot{\boldsymbol{\tau}}\cos\chi. \qquad (2.50)$$

Clearly, there exist equations for surface U which are analogous to eqs. (2.44) and (2.46), i.e.

$$\dot{\boldsymbol{\tau}}' = \left(k^{*\alpha}\mathbf{r}'_\alpha + b_{\alpha\beta}\frac{du^\alpha}{d\sigma}\frac{du^\beta}{d\sigma}\mathbf{m}\right)\dot{\sigma}, \qquad (2.51)$$

$$k^{*\alpha}\dot{\sigma}\mathbf{r}_\alpha = \sqrt{|a|}(k^{*2}\dot{u}^1 - k^{*1}\dot{u}^2)\mathbf{m}\times\boldsymbol{\tau}. \qquad (2.52)$$

Making use of eqs. (2.29),(2.35), (2.39) and (2.49) we obtain, after a series of rearrangements, the following expressions

$$\dot{\mathbf{m}}\times\boldsymbol{\tau} = \frac{\mathbf{m}}{\sqrt{|a|}}(a_{1\beta}b_{2\alpha} - a_{2\beta}b_{1\alpha})\dot{u}^\alpha\frac{du^\beta}{d\sigma},$$

$$\mathbf{m}\times\dot{\boldsymbol{\tau}} = -\sqrt{|a|}\boldsymbol{\tau}(k^{*2}\dot{u}^1 - k^{*1}\dot{u}^2), \quad \mathbf{m}\times\boldsymbol{\tau}' = \mathbf{m}\times\boldsymbol{\tau}\cos\chi + \boldsymbol{\tau}\sin\chi,$$

$$\boldsymbol{\tau}'\times\dot{\mathbf{m}} = \left[b_{\alpha\beta}\cos\chi + \frac{1}{\sqrt{|a|}}(a_{2\alpha}b_{1\beta} - a_{1\alpha}b_{2\beta})\sin\chi\right]\frac{du^\alpha}{d\sigma}\dot{u}^\beta. \qquad (2.53)$$

Substituting eqs. (2.51)–(2.53) into eq. (2.50) we arrive at the expression for vector $\dot{\boldsymbol{\tau}}'$ in terms of the parameters of surface U and angle χ

$$\dot{\boldsymbol{\tau}} = \left[\sqrt{|a|}(k^{*2}\dot{u}^1 - k^{*1}\dot{u}^2) - \dot{\chi}\right]\mathbf{m}\times\boldsymbol{\tau}' - \boldsymbol{\tau}'\cdot\dot{\mathbf{m}}\mathbf{m}. \qquad (2.54)$$

Finally, comparing eqs. (2.48) and (2.54) we obtain the formula for the angular velocity of whirling in the form

$$\Omega = \sqrt{|a|}(k^{*2}\dot{u}^1 - k^{*1}\dot{u}^2) - \sqrt{|a'|}(k'^{*2}\dot{p}^1 - k'^{*1}\dot{p}^2) - \dot{\chi}. \qquad (2.55)$$

Thus, the angular velocity vector of the slider $\boldsymbol{\omega}$ is expressed in terms of the time-derivative of the Gaussian coordinates u^α, p^α and angle χ by means of formulae (2.42) and (2.55). For a single-degree-of-freedom system the quantities u^α, p^α, χ are functions of the generalised coordinate q. Inserting the relationships

$$\dot{u}^\alpha = \frac{du^\alpha}{dq}\dot{q}, \quad \dot{p}^\alpha = \frac{dp^\alpha}{dq}\dot{q}, \quad \dot{\chi} = \frac{d\chi}{dq}\dot{q}$$

into eqs. (2.42) and (2.55) results in expressions for the angular velocity vector of the slider $\boldsymbol{\omega}$ and the angular velocity of whirling Ω in terms of the generalised coordinate and generalised velocity.

Particularly, when $\chi \equiv 0$ and the guiding surface U is a plane

$$b_{\alpha\beta} = 0; \quad \dot{\chi} = 0; \quad \sqrt{|a|}(k^{*2}\dot{u}^1 - k^{*1}\dot{u}^2) = \dot{\theta}, \qquad (2.56)$$

where θ denotes the angle between the tangent to the curve Λ and a fixed direction on the plane U. Then equations (2.42) and (2.55) take the form

$$\boldsymbol{\omega} = \frac{\dot{p}^\alpha}{\sqrt{|a|}}(b'_{\alpha 2}\mathbf{r}'_1 - b'_{\alpha 1}\mathbf{r}'_2) + \Omega \mathbf{m}, \qquad (2.57)$$

$$\Omega = \dot{\theta} - \sqrt{|a'|}(k'^{*2}\dot{p}^1 - k'^{*1}\dot{p}^2). \qquad (2.58)$$

Expressions (2.57) and (2.58) are derived in [103] for the case of rolling on a plane without slip.

2.3 Equation for the constraint force and Painlevé's paradoxes

Two equivalent systems of equations, (2.13) and (2.15), were derived in Section 2.1. Each system consists of two equations with two unknown variables which are the generalised coordinate q and the normal reaction force R. These equations contain $\varepsilon_1 = \text{sign } R$. In order to obtain the differential equations of motion and the equations for the reaction force, it is necessary to resolve these equations for $\ddot{q}$ and R, and also find a way of determining $\varepsilon_1 = \text{sign } R$. The present Section is concerned with this problem.

2.3.1 Solution for the acceleration and the constraint force

Resolving eqs. (2.13) and (2.15) results in the following equations

$$(1 + \varepsilon_1\varepsilon_2\mu L)\ddot{q} + H\dot{q}^2 = G, \quad (1 + \varepsilon_1\varepsilon_2\mu L)R = R_0, \quad (2.59)$$

where

$$L = \frac{A_{12}^0}{A}\left|\frac{d\mathbf{r}_T^0}{dq}\right| = \left(\frac{A_{12}^{*0}}{A} - \lambda\right)\left|\frac{d\mathbf{r}_T^0}{dq}\right|, \quad (2.60)$$

$$\begin{aligned}H &= \frac{1}{A}\left[\frac{1}{2}\frac{dA}{dq} + \varepsilon_1\varepsilon_2\mu\left|\frac{d\mathbf{r}_T^0}{dq}\right|\left(\frac{\partial A_{12}}{\partial q} - \frac{1}{2}\frac{\partial A_{11}}{\partial h}\right)_0\right] \\ &= \frac{1}{A}\left[\frac{1}{2}\frac{dA}{dq} + \varepsilon_1\varepsilon_2\mu\left|\frac{d\mathbf{r}_T^0}{dq}\right|\left(\frac{\partial A_{12}^*}{\partial q} - \frac{1}{2}\frac{\partial A_{11}^*}{\partial h} - \frac{\lambda}{2}\frac{dA}{dq}\right)_0\right], \quad (2.61)\end{aligned}$$

$$G = \frac{1}{A}\left(Q_1 + \varepsilon_1\varepsilon_2\mu\left|\frac{d\mathbf{r}_T^0}{dq}\right|Q_2\right) = \frac{1}{A}\left(Q_1 + \varepsilon_1\varepsilon_2\mu\left|\frac{d\mathbf{r}_T^0}{dq}\right|(Q_1^* - \lambda Q_2)\right), \quad (2.62)$$

$$\begin{aligned}R_0 &= \frac{A_{12}^0}{A}Q_1 - Q_2 + \left(\frac{\partial A_{12}}{\partial q} - \frac{1}{2}\frac{\partial A_{11}}{\partial h} - \frac{1}{2}\frac{A_{12}}{A}\frac{dA}{dq}\right)_0 \dot{q}^2 \\ &= \frac{A_{12}^{*0}}{A}Q_1 - Q_2^* + \left(\frac{\partial A_{12}^*}{\partial q} - \frac{1}{2}\frac{\partial A_{11}^*}{\partial h} - \frac{1}{2}\frac{A_{12}^*}{A}\frac{dA}{dq}\right)_0 \dot{q}^2. \quad (2.63)\end{aligned}$$

Equation (2.59) yields the following relationships between the signs

$$\varepsilon_3 \operatorname{sign}(1 + \varepsilon_1\varepsilon_2\mu L) = \operatorname{sign}(g - H\dot{q}^2), \quad \varepsilon_1 \operatorname{sign}(1 + \varepsilon_1\varepsilon_2\mu L) = \varepsilon_0, \quad (2.64)$$

where the new notation $\varepsilon_0 = \operatorname{sign} R_0, \varepsilon_3 = \operatorname{sign} \ddot{q}$ is adopted.

We notice that the value of $-R_0$ given by eq. (2.63) is the normal reaction force for the case of no friction, i.e. when $\mu = 0$. Generally speaking, the product μL characterises the influence of Coulomb friction on the normal reaction force. For this reason, L is referred to as the influence coefficient of the contact constraint. As one can see from eq. (2.60) this coefficient depends on the system configuration. On the other hand, the quantity $L_0 = \varepsilon_2 L = L \operatorname{sign} \dot{q}$ is invariant under the choice of the generalised coordinate q. In what follows L_0 is referred to as influence factor.

Let us assume that vector $\mathbf{r}_T^0$ is not constant within any finite interval of the generalised coordinate q, that is

$$\frac{d\mathbf{r}_T^0}{dq} \neq 0. \quad (2.65)$$

2.3 Equation for the constraint force and Painlevé's paradoxes

It follows from eq. (2.60) that

$$\operatorname{sign} L = \operatorname{sign} A^0_{12}, \quad L \equiv 0 \iff A^0_{12} \equiv 0. \tag{2.66}$$

Given a model, the second condition in eq. (2.66) allows us to make a very simple decision whether the dry friction affects the contact reaction or not. For example, for a single-degree-of-freedom system, the influence is absent if and only if

$$\frac{d\mathbf{r}^0}{dq} \cdot \left(\frac{\partial \mathbf{r}}{\partial h}\right)_0 = 0, \tag{2.67}$$

where $\mathbf{r}$ denotes the position vector of the particle. The latter equation means that the particle moves perpendicular to its original trajectory under the additional displacement of the slider along the normal to the guidance. The simplest example of such a system is a particle slipping on a surface. For a system with N degrees of freedom ($N > 1$) a sufficient (but not necessary) condition of a trivial L is given by

$$\frac{d\mathbf{r}^0}{dq} \cdot \left(\frac{\partial \mathbf{r}_i}{\partial h}\right)_0 = 0 \quad (i = 1, \ldots, N). \tag{2.68}$$

2.3.2 Criterion for the paradoxes

Equations (2.59) can be resolved for $\ddot{q}$ and R only when $\varepsilon_1 = \operatorname{sign} R$ is found. If ε_1 does not change, then the solution of the dynamical problem exists and is unique. If for given q and $\dot{q}$, eqs. (2.59)-(2.64) render no value of ε_1 or two different values of ε_1 simultaneously, then the paradoxical situations of non-existence or non-uniqueness of solution take place. Therefore, the phase space $(q, \dot{q})$ is separated into three regions: the region of existence and uniqueness of solution, the region of non-existence of solution and the region of non-uniqueness of solution. These regions are determined by the following theorem.

Theorem 1. For

$$\mu |L| < 1 \tag{2.69}$$

the signs of R and R_0 coincide ($\varepsilon_1 = \varepsilon_0$) and the normal reaction force R and the acceleration $\ddot{q}$ are uniquely determined by eqs. (2.59)-(2.63). For

$$\mu |L| > 1 \tag{2.70}$$

the solutions are not unique ($\varepsilon_1 = \pm 1$) if

$$\varepsilon_2 = \varepsilon_0 \operatorname{sign} L \tag{2.71}$$

and they do not exist ($\varepsilon_1 = \pm i = \pm\sqrt{-1}$) if

$$\varepsilon_2 = -\varepsilon_0 \operatorname{sign} L. \tag{2.72}$$

Proof. If $\mu|L| < 1$, then $1 + \varepsilon_1\varepsilon_2\mu L > 0$. It follows from eq. (2.64) that $\varepsilon_1 = \varepsilon_0$. Substituting the latter equality into eq. (2.59) yields a single differential equations of motion and a single expression for the reaction force. If $\mu|L| > 1$, then $\text{sign}(1 + \varepsilon_1\varepsilon_2\mu L) = \varepsilon_1\varepsilon_2 \,\text{sign}\, L$. Then, due to eq. (2.64) we have $\varepsilon_1^2 = \varepsilon_0\varepsilon_2 \,\text{sign}\, L$. Hence,

$$\varepsilon_1 = \begin{cases} \pm 1, & \text{for } \varepsilon_0 = \varepsilon_2 \,\text{sign}\, L, \\ \pm i, & \text{for } \varepsilon_0 = -\varepsilon_2 \,\text{sign}\, L, \end{cases}$$

which completes the proof.

Based upon the proved theorem, we can make the following remarks.

Remark 1. As eqs. (2.66) and (2.70) show, the paradoxes can appear in any mechanism for a sufficiently large friction factor except for the case of $L \equiv 0$, i.e. $A_{12}^0 \equiv 0$. The larger is the value of $|L|$, the smaller is the critical value of μ. When the friction factor is greater than the critical value, the paradoxes take place.

Remark 2. According to eqs. (2.60) and (2.70), for a given value of μ the condition for appearance of the paradoxes are determined by the system configuration but does not depend on the velocity of motion and the active forces. However the form of the paradoxes (non-uniqueness and non-existence) changes depending upon the sign of the velocity $\varepsilon_2 = \text{sign}\,\dot{q}$ and the sign of the reaction force of the ideal system $\varepsilon_0 = \text{sign}\, R_0$ and thus depends on the velocity of motion and external forces. With this in view, the border of the region of paradoxes in the phase space $(q, \dot{q})$ is determined by the equation $\mu|L(q)| - 1 = 0$ and the border between the subregion of non-uniqueness and non-existence is axis $\dot{q} = 0$.

Remark 3. By virtue of Theorem 1 the coefficients in front of $\ddot{q}$ and R in eq. (2.59) can be represented in the form

$$1 + \varepsilon_1\varepsilon_2\mu L = 1 \pm \mu|L| \quad \text{for} \quad \begin{cases} \mu|L| < 1, & \varepsilon_2 = \pm\varepsilon_0 \,\text{sign}\, L, \\ \mu|L| > 1, & \varepsilon_2 = \varepsilon_0 \,\text{sign}\, L. \end{cases} \quad (2.73)$$

The ratio R/R_0 is displayed in Fig. 2.3. Curve 1 corresponds to the case of a plus sign in front of $\mu|L|$ whilst the curves 2 and 3 describe the cases of a minus sign for $\mu|L| < 1$ and $\mu|L| > 1$, respectively. In the vicinity of the limiting point $\mu|L| = 1$ we have

$$\lim_{\mu|L| \to 1-0} \left(\frac{R}{R_0}\right) = \begin{cases} 0.5, & \text{for } \varepsilon_2 = \varepsilon_0 \,\text{sign}\, L, \\ \infty, & \text{for } \varepsilon_2 = -\varepsilon_0 \,\text{sign}\, L, \end{cases}$$

$$\lim_{\mu|L| \to 1+0} \left(\frac{R}{R_0}\right) = \begin{cases} 0.5 \text{ and } \infty, & \text{for } \varepsilon_2 = \varepsilon_0 \,\text{sign}\, L, \\ \text{does not exists}, & \text{for } \varepsilon_2 = -\varepsilon_0 \,\text{sign}\, L. \end{cases}$$

As one can see from Fig. 2.3, the right limit corresponds to the paradoxical situations. Because of the unbounded growth of the reaction force at $\varepsilon_2 = -\varepsilon_0 \,\text{sign}\, L$ the left limit may also be deemed to be a paradoxical case. Thus, we can take that the boundary points $\mu|L(q)| = 1$ also belong to the region of paradoxes.

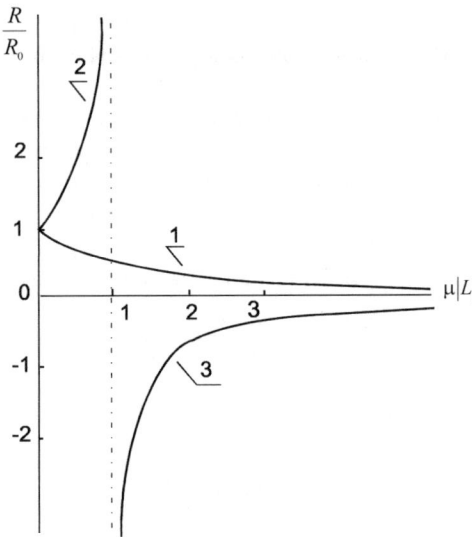

FIGURE 2.3.

2.4 Immovable contact and transition to slipping

According to Coulomb's law, the contacting bodies begin to slip relative to each other from an immovable state at the critical ratio of the tangential force to the normal load. Hence, to determine the condition for transition of a mechanism to slipping one often calculates the tangential force and normal forces at rest and then increases their ratio to the critical one. This approach leads to a cumbersome calculation. In order to decrease such a calculation and to establish the influence of the paradoxes on the immovable state let us express the general condition for transition to motion in terms of the generalised forces Q_1, Q_2 and friction factor μ.

A zero value of the slip velocity $\mathbf{v}_T^0 = (d\mathbf{r}_0/dq)\dot{q} = 0$ is realised in two cases: i) $(d\mathbf{r}_0/dq)\dot{q} = 0 \quad \dot{q} \neq 0$, and ii) $\dot{q} = 0$. In the first case the immovable contact (or rolling without slip) is of short duration (the case of $d\mathbf{r}_0/dq \equiv 0$ within a finite interval of q is not of interest) and the system becomes ideal momentarily since, according to eqs. (2.59)–(2.63), for $d\mathbf{r}_T^0/dq = 0$ Coulomb friction affects neither R nor $\ddot{q}$. Hence, in the case of $d\mathbf{r}_T^0/dq = 0$ and $\dot{q} \neq 0$ the problem of immobility and transition to motion is solved by means of the basic principles of mechanics of ideal systems.

The problem stated below is considered in accordance with the second case, i.e. the system is initially at rest due to a zero generalised velocity

$$\dot{q}(t)_0 = 0. \qquad (2.74)$$

It is clear that the sign of the acceleration $\varepsilon_3 = \text{sign}\,\ddot{q}$ is coincident with that of the velocity $\varepsilon_2 = \text{sign}\,\dot{q}$. For this reason, if it is follows from eqs.

(2.59)-(2.64) that $\varepsilon_3 = -\varepsilon_2$ for any given ε_2 at time instant t_0, then the tangential force in the contact zone is less than the limiting value and the system remains immovable.

Provided that there exists a value of ε_2 which when substituted into eqs. (2.59)–(2.64) ensures that $\varepsilon_3 = \varepsilon_2$, then the system begins to move in the direction which is determined by the obtained value of ε_2. With this reasoning in mind, let us prove the second and third theorems.

Theorem 2. In the case of no paradoxes, i.e. in the case of $\mu|L| < 1$, the system under the initial condition (2.74) remains at rest if

$$|Q_1| \leq \mu \left|\frac{d\mathbf{r}_T^0}{dq}\right| |Q_2| \qquad (2.75)$$

and begins to move in the direction of $\varepsilon_2 = \operatorname{sign} Q_2$, if

$$|Q_1| > \mu \left|\frac{d\mathbf{r}_T^0}{dq}\right| |Q_2| . \qquad (2.76)$$

Proof. It follows from eq. (2.60) and condition $\mu|L| < 1$ that

$$\mu \left|\frac{d\mathbf{r}_T^0}{dq}\right| < \left|\frac{A}{A_{12}^0}\right| .$$

Then, by virtue of eq. (2.75),

$$|Q_1| < \left|\frac{A}{A_{12}^0} Q_2\right| .$$

Taking into account this inequality and inserting condition (2.75) into eq. (2.63) yields the following expression for the sign of R_0

$$\varepsilon_0 = -\operatorname{sign} Q_2 . \qquad (2.77)$$

By using eqs. (2.61) and (2.62) and under conditions (2.74) and (2.75) we can write

$$\operatorname{sign}\left(G - H\dot{q}^2\right) = \operatorname{sign} G = \varepsilon_1 \varepsilon_2 \operatorname{sign} Q_2 . \qquad (2.78)$$

As $\operatorname{sign}(1 + \varepsilon_1\varepsilon_2\mu L) = 1$ for $\mu|L| < 1$, the sign relationship (2.64) for equalities (2.77) and (2.78) takes the form

$$\varepsilon_3 = \varepsilon_1\varepsilon_2 \operatorname{sign} Q_2, \quad \varepsilon_1 = -\operatorname{sign} Q_2,$$

thus $\varepsilon_3 = -\varepsilon_2$. According to the above reasoning the initial state of rest is retained which confirms the validity of the first part of the theorem.

It remains to show that eq. (2.76) is the condition for transition to motion. Indeed, it follows from eqs. (2.61) and (2.62) under conditions (2.74) and (2.76) that

$$\operatorname{sign}(G - H\dot{q}^2) = \operatorname{sign} G = \operatorname{sign} Q_1 . \qquad (2.79)$$

Then the first sign relationships in eq. (2.64) reduces to the form

$$\varepsilon_3 = \operatorname{sign} Q_1.$$

Thus, we have found the value of $\varepsilon_2 = \operatorname{sign} Q_1$, for which the criterion for transition to motion $\varepsilon_3 = \varepsilon_2$ is valid, and the system begins to move in the direction of $\varepsilon_2 = \operatorname{sign} Q_1$.

The theorem is proved.

Theorem 2 expresses Coulomb's law in terms of the generalised forces in the case of Painlevé's paradoxes. However in the general case, Q_1 and $(d\mathbf{r}_T^0/dq)Q_2$ in inequalities (2.75) and (2.76) are not the corresponding tangential and normal forces in the contact zone.

Theorem 3. Let the condition for paradoxes $\mu|L| > 1$ and the initial condition (2.74) be satisfied. Then i) the system is at rest if

$$|LQ_1| \leq \left|\frac{d\mathbf{r}_T^0}{dq}\right| |Q_2|, \qquad (2.80)$$

ii) provided that

$$\frac{1}{\mu}|Q_1| \leq \left|\frac{d\mathbf{r}_T^0}{dq}Q_2\right| < |LQ_1| \qquad (2.81)$$

the system begins to move in the direction of $\varepsilon_2 = \operatorname{sign} Q_1 = \varepsilon_0 \operatorname{sign} L$ under the condition

$$\operatorname{sign} Q_1 = \operatorname{sign}(LQ_2) \qquad (2.82)$$

and is at rest otherwise, and iii) if

$$|Q_1| > \mu \left|\frac{d\mathbf{r}_T^0}{dq}Q_2\right| \qquad (2.83)$$

the equation of dynamics admits simultaneously both maintaining immovable contact and transition to motion in the direction of $\varepsilon_2 = \operatorname{sign} Q_1 = \varepsilon_0 \operatorname{sign} L$. The sign $\varepsilon_1 = -\varepsilon_0$ corresponds to the immovable contact whereas the sign $\varepsilon_1 = \varepsilon_0$ corresponds to the motion.

Proof. Let us prove the first item of the theorem. Inserting L from eq. (2.60) into eq. (2.80) yields

$$\left|\frac{A_{12}^0}{A}Q_1\right| \leq |Q_2|.$$

Taking into account this inequality in eq. (2.63) we can find the sign of R_0

$$\varepsilon_0 = -\operatorname{sign} Q_2. \qquad (2.84)$$

On the other hand, it follows from eq. (2.80) and the condition for paradoxes $\mu|L| > 1$ that

$$|Q_1| \leq \frac{1}{|L|} \left|\frac{d\mathbf{r}_T^0}{dq} Q_2\right| \leq \mu \left|\frac{d\mathbf{r}_T^0}{dq} Q_2\right|.$$

Then by virtue of eqs. (2.61) and (2.62) under the initial condition (2.74) we obtain

$$\operatorname{sign}\left(G - H\dot{q}^2\right) = \varepsilon_1 \varepsilon_2 \operatorname{sign} Q_2. \tag{2.85}$$

Equalities (2.84) and (2.85) yield the values of the right hand side of the sign relationships (2.64). Under the condition

$$\operatorname{sign}(1 + \varepsilon_1 \varepsilon_2 \operatorname{sign} L) = \varepsilon_1 \varepsilon_2 \operatorname{sign} L \quad \text{for} \quad \mu|L| > 1 \tag{2.86}$$

these relationships reduce to the following ones

$$\varepsilon_3 \operatorname{sign} L = \operatorname{sign} Q_2, \quad \varepsilon_2 \operatorname{sign} L = -\operatorname{sign} Q_2.$$

Thus, $\varepsilon_3 = -\varepsilon_2$. In this case, as stated in item i) the system remains at rest.

Let us prove the second item of the theorem. By virtue of eqs. (2.61) and (2.62), the first inequality in eq. (2.81) and initial condition (2.74) we can write

$$\operatorname{sign}(G - H\dot{q}^2) = \varepsilon_1 \varepsilon_2 \operatorname{sign} Q_2. \tag{2.87}$$

At the same time, due to eq. (2.60) the second inequality in eq. (2.80) for the value of L can be set in the form

$$|Q_2| < \left|\frac{A_{12}^0}{A} Q_1\right|.$$

It follows from eqs. (2.63) and (2.66) that

$$\varepsilon_0 = \operatorname{sign}(A_{12}^0 Q_1) = \operatorname{sign}(LQ_1). \tag{2.88}$$

Under condition (2.82) we have

$$\varepsilon_0 = \operatorname{sign} Q_2. \tag{2.89}$$

Accounting for equalities (2.86), (2.87) and (2.89) in eq. (2.64) we obtain

$$\varepsilon_3 \operatorname{sign} L = \operatorname{sign} Q_2, \quad \varepsilon_2 \operatorname{sign} L = \operatorname{sign} Q_2. \tag{2.90}$$

Hence, $\varepsilon_3 = \varepsilon_2 \operatorname{sign}(LQ_2)$. By virtue of conditions (2.82) and (2.88) we have $\varepsilon_3 = \varepsilon_2 = \varepsilon_0 \operatorname{sign} L = \operatorname{sign} Q_1$. As one can see, the system begins to move in the direction of $\varepsilon_2 = \operatorname{sign} Q_1$.

If condition (2.82) is not fulfilled it follows from eq. (2.88) that $\varepsilon_0 = \operatorname{sign} Q_2$.

In this case, instead of eq. (2.90) we have

$$\varepsilon_3 \operatorname{sign} L = \operatorname{sign} Q_2, \quad \varepsilon_2 = \operatorname{sign} L = -\operatorname{sign} Q_2.$$

Hence, $\varepsilon_3 = -\varepsilon_2$ and the system remains at rest.

In order to prove the third item of the theorem we take into account conditions (2.74) and (2.83) in expressions (2.61) and (2.62). In this case

$$\operatorname{sign}(G - H\dot{q}^2) = \operatorname{sign} Q_1.$$

In addition to this, for $\mu|L| > 1$ we obtain from eqs. (2.60) and (2.83) that

$$Q_1 > \mu \left| \frac{d\mathbf{r}_T^0}{dq} Q_2 \right| > \left| \frac{A}{A_{12}^0} Q_2 \right|.$$

For this reason, the sign of R_0 is also determined by eq. (2.88). Then substituting eqs. (2.86), (2.88) and (2.90) into eq. (2.64) yields the following expressions

$$\varepsilon_1 \varepsilon_2 \varepsilon_3 \operatorname{sign} L = \operatorname{sign} Q_1,$$
$$\varepsilon_2 \operatorname{sign} L = \varepsilon_0 = \operatorname{sign}(LQ_1). \qquad (2.91)$$

Resolving the latter we can write

$$\varepsilon_3 = \varepsilon_0 \varepsilon_1 \varepsilon_2, \quad \varepsilon_2 = \varepsilon_0 \operatorname{sign} L = \operatorname{sign} Q_1. \qquad (2.92)$$

On the other hand, due to Theorem 1 in the case of $\mu|L| > 1$ and $\varepsilon_2 = \varepsilon_0 \operatorname{sign} L$ parameter ε_1 can simultaneously take the two values ± 1. Then the first equality in eq. (2.92) can be represented in the form

$$\varepsilon_3 = \pm \varepsilon_2 \quad \text{for} \quad \varepsilon_1 = \pm \varepsilon_0. \qquad (2.93)$$

Hence, there occur simultaneously two regimes: transition to motion in the direction of $\varepsilon_2 \operatorname{sign} Q_1$ for $\varepsilon_1 = \varepsilon_0$, and remaining at rest for $\varepsilon_1 = -\varepsilon_0$. The theorem is thus proved.

2.5 Self-braking and the angle of stagnation

In the theory of machines and mechanisms, a mechanism is referred to as being self-braking provided that its motion is feasible when the forces are transferred from one link to another and not feasible under the inverse transfer. Examples are worm gearing, the bolt-nut system etc. We generalise this concept to the general case of mechanical systems and establish the

58   2. Systems with a single degree of freedom and a single frictional pair

influence of Painlevé's paradoxes on the property of self-braking. Let us introduce four definitions:

1) The $i-th$ particle is called the point of self–braking if any force $\mathbf{F}_i$ exerted on this particle is not able to set the system into motion when $\mathbf{F}_k = 0$ for $k \neq i$.

2) The point of debraking is the point at which the brake releases.

3) Provided that no force $\mathbf{F}_i$ whose action line lies in a certain region is able to set the system into motion for $\mathbf{F}_k = 0$ $(k \neq i)$, then this region is referred to as the stagnation region for the $i-th$ particle.

4) That part of the space which does not contain a stagnation region for the $i-th$ particle is referred to as the region of motion for this particle. As follows from the definitions, the stagnation region for the point of self-braking occupies the whole space.

Let us prove the theorems which determine the regions of stagnation and motion, which allows one to decide whether the particle belongs to one of the introduced subsets.

2.5.1 The case of no paradoxes

Theorem 4. For $\mu|L| < 1$ the relationship

$$\frac{d\mathbf{r}_i^0}{dq} = \gamma\mu \left|\frac{d\mathbf{r}_T^0}{dq}\right| \left(\frac{\partial \mathbf{r}_i}{\partial h}\right)_0 \quad |\gamma| \leq 1 \tag{2.94}$$

is the necessary and sufficient condition for the $i-th$ particle to be a point of self-braking.

Proof. It is easy to see that condition (2.94) is equivalent to the following one

$$\frac{d\mathbf{r}_i^0}{dq} \parallel \left(\frac{\partial \mathbf{r}_i}{\partial h}\right)_0, \quad \left|\frac{d\mathbf{r}_i^0}{dq}\right| \leq \mu \left|\frac{d\mathbf{r}_T^0}{dq}\right| \left|\frac{\partial \mathbf{r}_i}{\partial h}\right|_0. \tag{2.95}$$

Let us first prove the necessary condition of (2.94) or (2.95). Guided by the adopted definition we take $\mathbf{F}_i \neq 0$, $\mathbf{F}_k = 0$ $(k \neq i)$. Then

$$Q_1 = \mathbf{F}_i \cdot \frac{d\mathbf{r}_i^0}{dq}, \quad Q_1 = \mathbf{F}_i \cdot \left(\frac{\partial \mathbf{r}_i}{\partial h}\right)_0. \tag{2.96}$$

These generalised forces must satisfy the immobility condition (2.75) which, in this particular case, takes the form

$$\left|\mathbf{F}_i \cdot \frac{d\mathbf{r}_i^0}{dq}\right| \leq \mu \left|\frac{d\mathbf{r}_T^0}{dq}\right| \left|\mathbf{F}_i \cdot \left(\frac{\partial \mathbf{r}_i}{\partial h}\right)_0\right|. \tag{2.97}$$

According to the definition, condition (2.97) is fulfilled for any direction of force $\mathbf{F}_i$ and, in particular, for that direction for which the vectors $\mathbf{F}_i$,

2.5 Self-braking and the angle of stagnation

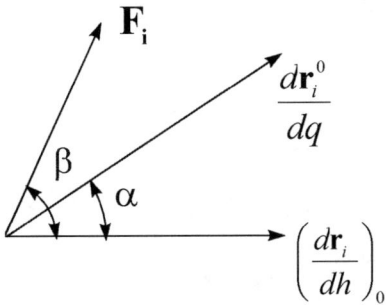

FIGURE 2.4.

dr_i^0/dq, $(\partial \mathbf{r}_i/\partial h)_0$ are coplanar and vectors $\mathbf{F}_i$ and $(\partial \mathbf{r}_i/\partial h)_0$ are orthogonal, i.e. $\beta = \left[\mathbf{F}_i, \widehat{(\partial \mathbf{r}_i/\partial h)_0}\right] = \pi/2$. It follows from eq. (2.97) that

$$\left| \frac{dr_i^0}{dq} \sin \alpha \right| \leq 0,$$

where

$$\alpha = \left[\widehat{\frac{dr_i^0}{dq}, \left(\frac{\partial \mathbf{r}_i}{\partial h}\right)_0}\right] = \beta - \left[\widehat{\mathbf{F}_i, \frac{dr_i^0}{dq}}\right],$$

see Fig. 2.4. However it is possible only for $\alpha = k\pi$ ($k = 1, 2$), i.e. under the first condition in eq. (2.95): $dr_i^0/dq \| (\partial \mathbf{r}_i/\partial h)_0$.

Furthermore, as $\alpha = k\pi$, we have

$$\left| \cos \left[\widehat{\mathbf{F}_i, \frac{dr_i^0}{dq}}\right] \right| = |\cos(\beta - \alpha)| = |\cos \beta| = \left| \cos \left[\widehat{\mathbf{F}_i, \left(\frac{\partial \mathbf{r}_i}{\partial h}\right)_0}\right] \right|.$$

For this relationship, the second condition in eq. (2.95) follows immediately from inequality (2.97). The necessary condition is thus proved.

In order to prove the sufficient condition we substitute the value of dr_i^0/dq from eq. (2.94) into eq. (2.96) for Q_1 and compare expressions for $Q_1 Q_2$. The result is

$$|Q_1| = \mu|\gamma| \left|\frac{dr_T^0}{dq}\right| \left|\mathbf{F}_i \cdot \frac{\partial \mathbf{r}_i}{\partial h}\right|_0 < \mu \left|\frac{dr_T^0}{dh} Q_2\right|,$$

hence, the immobility condition (2.75) is satisfied for any force $\mathbf{F}_i$. According to the definition, the $i-th$ particle is the point of self-braking, i.e. the theorem is proved.

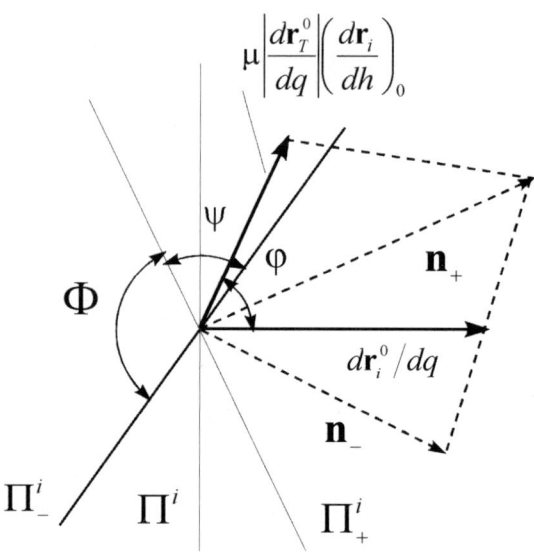

FIGURE 2.5.

Corollary. The $i-th$ particle is the point of debraking if and only if

$$\text{either} \quad \frac{d\mathbf{r}_i^0}{dq} \times \left(\frac{\partial \mathbf{r}_i}{\partial h}\right)_0 \neq 0 \quad \text{or} \quad \frac{d\mathbf{r}_i^0}{dq} = \gamma\mu \left|\frac{d\mathbf{r}_i^0}{dq}\right| \left(\frac{\partial \mathbf{r}_T^0}{\partial h}\right)_0. \quad (2.98)$$

Proof. Under condition (2.98) and only under this condition eq. (2.94) does not hold. Thus, the $i-th$ particle is not a point of self-braking. Thus, from the definition it is a point of debraking, which completes the proof.

Theorem 5. For $\mu|L| < 1$ the two planes

$$\Pi_\pm^i: \quad (\boldsymbol{\rho} - \mathbf{r}_i^0) \cdot \left[\frac{d\mathbf{r}_i^0}{dq} \pm \mu \left|\frac{d\mathbf{r}_i^0}{dq}\right| \left(\frac{\partial \mathbf{r}_T^0}{\partial h}\right)_0\right] = 1 \quad (2.99)$$

separates the space about the $i-th$ particle into the region of stagnation and the region of motion. The stagnation region consists of planes $\Pi_\pm^i$ and the angle between them which contains the plane $\Pi_\pm^i$ perpendicular to vector $d\mathbf{r}_i^0/dq$, see Fig. 2.5.

Proof. Inasmuch as $\mathbf{F}_k = 0$ for $k \neq i$, the generalised forces are determined by means of eq. (2.96). The immobility condition (2.75) and the condition of transition to motion (2.76) respectively take the form

$$\left|\mathbf{F}_i \cdot \frac{d\mathbf{r}_i^0}{dq}\right| - \mu \left|\frac{d\mathbf{r}_T^0}{dq}\right| \left|\mathbf{F}_i \cdot \left(\frac{\partial \mathbf{r}_i}{\partial h}\right)_0\right| \leq 0, \quad (2.100)$$

$$\left|\mathbf{F}_i \cdot \frac{d\mathbf{r}_i^0}{dq}\right| - \mu \left|\frac{d\mathbf{r}_T^0}{dq}\right| \left|\mathbf{F}_i \cdot \left(\frac{\partial \mathbf{r}_i}{\partial h}\right)_0\right| > 0. \quad (2.101)$$

2.5 Self-braking and the angle of stagnation

The first of these conditions must be fulfilled in the stagnation region, whilst the second one is satisfied in the motion region. On the border of these regions the following relationship

$$\left| \mathbf{F}_i \cdot \frac{d\mathbf{r}_i^0}{dq} \right| - \mu \left| \frac{d\mathbf{r}_T^0}{dq} \right| \left| \mathbf{F}_i \cdot \left(\frac{\partial \mathbf{r}_i}{\partial h} \right)_0 \right| = 0$$

is valid, which is equivalent to the following condition

$$\mathbf{F}_i \cdot \frac{d\mathbf{r}_i^0}{dq} \pm \mu \left| \frac{d\mathbf{r}_T^0}{dq} \right| \mathbf{F}_i \cdot \left(\frac{\partial \mathbf{r}_i}{\partial h} \right)_0 = 0. \qquad (2.102)$$

A plus sign is taken when the signs of the quantities $\mathbf{F}_i \cdot (d\mathbf{r}_i^0/dq)$ and $\mathbf{F}_i \cdot (\partial \mathbf{r}_i/\partial h)_0$ do not coincide, otherwise a minus sign is taken.

Let us make the origin of vector $\mathbf{F}_i$ and the particle coincident and denote the position vector of its end by $\boldsymbol{\rho}_i$, i.e. $\boldsymbol{\rho}_i = \mathbf{r}_i^0 + \mathbf{F}_i$. Then eq. (2.102) reduces to the form (2.99)

$$(\boldsymbol{\rho} - \mathbf{r}_i^0) \cdot \left[\frac{d\mathbf{r}_i^0}{dq} \pm \mu \left| \frac{d\mathbf{r}_i^0}{dq} \right| \left(\frac{\partial \mathbf{r}_T^0}{\partial h} \right)_0 \right] = 0.$$

To put it differently, any vector of force $\mathbf{F}_i$ lying in the planes Π_+^i and Π_-^i satisfies the limiting condition (2.102).

The left hand side of eqs. (2.100) and (2.101) is a continuous function of the direction cosines of force $\mathbf{F}_i$ and vanishes on planes $\Pi_\pm^i$. Hence, it does not change sign for any angle between these planes, i.e. only one of the conditions (2.100) and (2.101) is fulfilled. On the other hand, on the plane Π^i perpendicular to $d\mathbf{r}_i^0/dq$ we have

$$\mathbf{F}_i \cdot \frac{d\mathbf{r}_i^0}{dq} = 0$$

and condition (2.100) is met. Hence, it is not satisfied for any direction of force $\mathbf{F}_i$ within a solid angle Ψ, containing plane Π^i. Let us notice in passing that eq. (2.100) holds with equality sign on the boundary planes $\Pi_\pm^i$ and the stagnation region consists of angle Ψ and planes $\Pi_\pm^i$. Thus the theorem is proved.

The stagnation angle Ψ and the motion angle Φ are depicted in Fig. 2.5 where vectors $(\partial \mathbf{r}_i/\partial h)_0$ and $d\mathbf{r}_i^0/dq$ lie in the drawing plane whereas the planes Π_+^i, Π_-^i and Π^i are perpendicular to the drawing plane.

Corollary 1. Within and only within the motion angle any nontrivial force $\mathbf{F}_i$ at $\mathbf{F}_k = 0$ ($k \neq i$) can cause motion from the rest condition.

Proof. If the system were at rest under a certain force $\mathbf{F}_i \neq 0$ within Φ, then condition (2.100) would be satisfied. Then, according to the above said, this condition would be satisfied for any force $\mathbf{F}_i$ within Φ that contradicts the statement. In other words, any nontrivial force within the motion angle Φ can initiate the system into motion.

2. Systems with a single degree of freedom and a single frictional pair

Within the stagnation angle, condition (2.100) holds whereas condition (2.101) does not hold. Hence, the force can initiate the system from into motion only within the motion angle Φ. The corollary is proved.

Let us now fix the absolute values $|d\mathbf{r}_i^0/dq|$ and $|\partial \mathbf{r}_i/\partial h|_0$ of the corresponding vectors and study the dependence of the stagnation angle Ψ on the angle φ between these vectors

$$\varphi = \left[\widehat{\frac{d\mathbf{r}_i^0}{dq}, \left(\frac{\partial \mathbf{r}_i}{\partial h}\right)_0}\right].$$

Corollary 2. The stagnation angle changes within the limits

$$\Psi_{min} = 0, \qquad \text{for } \varphi = 0, \pi,$$

$$\Psi_{max} = \arccos \frac{\left|\dfrac{d\mathbf{r}_i^0}{dq}\right|^2 - \mu^2 \left|\dfrac{d\mathbf{r}_T^0}{dq}\right|^2 \left|\dfrac{\partial \mathbf{r}_i}{\partial h}\right|_0^2}{\left|\dfrac{d\mathbf{r}_i^0}{dq}\right|^2 + \mu^2 \left|\dfrac{d\mathbf{r}_i}{dq}\right|^2 \left|\dfrac{\partial \mathbf{r}_i}{\partial h}\right|_0^2}, \qquad \text{for } \varphi = \frac{\pi}{2}, \quad (2.103)$$

if

$$\left|\frac{d\mathbf{r}_i^0}{dq}\right| > \mu \left|\frac{d\mathbf{r}_T}{dq}\right| \left|\frac{\partial \mathbf{r}_i}{\partial h}\right|_0 \quad (2.104)$$

and within the following limits

$$\Psi_{min} = \pi - \arccos \frac{\mu^2 \left|\dfrac{d\mathbf{r}_T^0}{dq}\right|^2 \left|\dfrac{\partial \mathbf{r}_i}{\partial h}\right|_0^2 - \left|\dfrac{d\mathbf{r}_i^0}{dq}\right|^2}{\mu^2 \left|\dfrac{d\mathbf{r}_T^0}{dq}\right|^2 \left|\dfrac{\partial \mathbf{r}_i}{\partial h}\right|_0^2 + \left|\dfrac{d\mathbf{r}_i^0}{dq}\right|^2}, \qquad \text{for } \varphi = \frac{\pi}{2}, \quad (2.105)$$

$$\Psi_{max} = \pi, \qquad \text{for } \varphi = 0, \pi$$

if

$$\left|\frac{d\mathbf{r}_i^0}{dq}\right| \le \mu \left|\frac{d\mathbf{r}_T^0}{dq}\right| \left|\frac{\partial \mathbf{r}_i}{\partial h}\right|_0. \quad (2.106)$$

Proof. As one can see from Fig. 2.5, the normal vectors to the planes $\Pi_{\pm}^i$ are given by

$$\mathbf{n}_{\pm} = \frac{d\mathbf{r}_i^0}{dq} \pm \mu \left|\frac{d\mathbf{r}_T^0}{dq}\right| \left(\frac{\partial \mathbf{r}_i}{\partial h}\right)_0. \quad (2.107)$$

Hence,

$$\cos \Psi = \left[\left|\frac{d\mathbf{r}_i^0}{dq}\right|^2 - \mu^2 \left|\frac{d\mathbf{r}_T^0}{dq}\right|^2 \left|\frac{\partial \mathbf{r}_i}{\partial h}\right|_0^2\right] \times \quad (2.108)$$

$$\left[\left|\frac{d\mathbf{r}_i^0}{dq}\right|^4 + \mu^4 \left|\frac{d\mathbf{r}_T^0}{dq}\right|^4 \left|\frac{\partial \mathbf{r}_i}{\partial h}\right|_0^4 + 2\mu^2 \left|\frac{d\mathbf{r}_i^0}{dq}\right|^2 \left|\frac{d\mathbf{r}_T^0}{dq}\right|^2 \left|\frac{\partial \mathbf{r}_i}{\partial h}\right|_0^2 (1 - 2\cos^2 \varphi)\right]^{-1/2}.$$

2.5 Self-braking and the angle of stagnation

For $\varphi = 0, \pi$ we have

$$\cos\Psi = \frac{\left|\dfrac{d\mathbf{r}_i^0}{dq}\right|^2 - \mu^2 \left|\dfrac{d\mathbf{r}_T^0}{dq}\right|^2 \left|\dfrac{\partial \mathbf{r}_i}{\partial h}\right|_0^2}{\left|\dfrac{d\mathbf{r}_i^0}{dq}\right|^2 - \mu^2 \left|\dfrac{d\mathbf{r}_i}{dq}\right|^2 \left|\dfrac{\partial \mathbf{r}_i}{\partial h}\right|_0^2} = \begin{cases} 1 & \text{under condition (2.104)} \\ -1 & \text{under condition (2.106)} \end{cases}$$

For $\varphi = \pi/2$ we have

$$\cos\Psi = \frac{\left|\dfrac{d\mathbf{r}_i^0}{dq}\right|^2 - \mu^2 \left|\dfrac{d\mathbf{r}_T^0}{dq}\right|^2 \left|\dfrac{\partial \mathbf{r}_i}{\partial h}\right|_0^2}{\left|\dfrac{d\mathbf{r}_i^0}{dq}\right|^2 + \mu^2 \left|\dfrac{d\mathbf{r}_T}{dq}\right|^2 \left|\dfrac{\partial \mathbf{r}_i}{\partial h}\right|_0^2}.$$

Under conditions (2.104) and (2.106) one obtains values for Ψ_{max} (2.103) and Ψ_{min} (2.105).

Corollary 3. The point of self-braking can be treated as a particle for which the stagnation angle Ψ is equal to π, i.e. it occupies the whole space.

Proof. The value of Ψ is indeed equal to π only under condition (2.106) and only under $\varphi = 0, \pi$. But in this case condition (2.94) is satisfied, which means that the considered particle is a point of self-braking.

Corollary 4. In the case of an immovable guide surface and $(\partial \mathbf{r}_T/\partial h)_0 = \mathbf{m}$ the slider is the point of debraking for which the stagnation angle is equal to the opening of the friction cone

$$\Psi = 2\arctan\mu. \tag{2.109}$$

Proof. In this case

$$\frac{d\mathbf{r}_T^0}{dq} \perp \left(\frac{d\mathbf{r}_T}{dh}\right)_0, \quad \varphi = \frac{\pi}{2}, \quad \left|\frac{\partial \mathbf{r}_i}{\partial h}\right|_0 = 1.$$

By virtue of the corollary to Theorem 4, the slider is the point of debraking. To determine the stagnation angle Ψ it is necessary to replace $\mathbf{r}_i$ in eq. (2.108) by $\mathbf{r}_T$. The result is

$$\cos\Psi = \frac{1-\mu^2}{1+\mu^2}, \quad \sin\Psi = \frac{2\mu}{1+\mu^2}. \tag{2.110}$$

Indeed,

$$\tan\Psi = \frac{2\mu}{1-\mu^2} \quad \text{or} \quad \Psi = 2\arctan\mu, \tag{2.111}$$

which is required to complete the proof.

2.5.2 The case of paradoxes ($\mu|L| > 1$)

Theorem 6. In this case
  i) if

$$\frac{dr_i^0}{dq} = \gamma_0 \frac{A}{A_{12}^0} \left(\frac{\partial \mathbf{r}_i}{\partial h}\right)_0, \quad \gamma_0 \in [-1, 1] \tag{2.112}$$

and

$$\frac{1}{\gamma_1 \mu} \frac{dr_i^0}{dq} = -\left|\frac{d\mathbf{r}_T^0}{dq}\right| \left(\frac{\partial \mathbf{r}_i}{\partial h}\right)_0 \operatorname{sign} L = \gamma_2 |L| \frac{dr_i^0}{dq}, \quad \gamma_1, \gamma_2 \in (0, 1] \tag{2.113}$$

then the $i-th$ particle is a point of self-braking.
  ii) if

$$\frac{1}{\gamma_1 \mu} \frac{dr_i^0}{dq} = \left|\frac{d\mathbf{r}_T^0}{dq}\right| \left(\frac{\partial \mathbf{r}_i}{\partial h}\right)_0 \operatorname{sign} L = \gamma_2 |L| \frac{dr_i^0}{dq} \tag{2.114}$$

then the $i-th$ particle is a point of debraking.
  iii) if

$$\text{either} \quad \frac{dr_i^0}{dq} \times \left(\frac{\partial \mathbf{r}_i}{\partial h}\right)_0 \neq 0 \quad \text{or} \quad \frac{dr_i^0}{dq} = \gamma\mu \left|\frac{d\mathbf{r}_T^0}{dq}\right| \left(\frac{\partial \mathbf{r}_i^0}{\partial h}\right)_0, \gamma > 1 \tag{2.115}$$

then, due to the non-uniqueness of the solution of the problem of dynamics, this particle is simultaneously both the point of self-braking and the point of debraking.

Proof. Let us first prove the first item. Multiplying both sides of eq. (2.112) by $\left|d\mathbf{r}_T^0/dq\right| \mathbf{F}_i$ and assuming $\mathbf{F}_k = 0$ ($k \neq i$), we can write

$$\frac{A_{12}^0}{A} \left|\frac{d\mathbf{r}_T^0}{dq}\right| Q_1 = \gamma_0 \left|\frac{d\mathbf{r}_T^0}{dq}\right| Q_2. \tag{2.116}$$

According to (2.60), the coefficient in front of Q_1 in eq. (2.116) is the influence factor for L. Hence,

$$|LQ_1| \leq \left|\frac{d\mathbf{r}_T^0}{dq} Q_2\right|.$$

As one can see, the immobility condition (2.80) is satisfied for any force $\mathbf{F}_i$, thus the $i-th$ particle is the point of self-braking.

Let us consider now the case of double equality (2.113). Multiplying it by $\mathbf{F}_i$ and accounting for the condition $\mathbf{F}_k = 0$ ($k \neq i$), we obtain

$$\frac{1}{\gamma_1 \mu} Q_1 = -\left|\frac{d\mathbf{r}_T^0}{dq}\right| Q_2 \operatorname{sign} L = \gamma_2 |L| Q_1, \quad \gamma_1, \gamma_2 \in (0, 1].$$

2.5 Self-braking and the angle of stagnation

Hence

$$\frac{1}{\mu}Q_1 \le \left|\frac{d\mathbf{r}_T^0}{dq}Q_2\right| < |LQ_1|, \quad \text{sign}\, Q_1 = -\text{sign}(LQ_2).$$

In accordance with Theorem 3 the system remains at rest, and thus the $i-th$ particle is a point of self-braking.

Let us proceed to the proof of the second item of the theorem. Repeating the above transformations and taking into account eq. (2.114), we arrive at the following

$$\frac{1}{\mu}Q_1 \le \left|\frac{d\mathbf{r}_T^0}{dq}Q_2\right| < |LQ_1|, \quad \text{sign}\, Q_1 = +\text{sign}(LQ_2).$$

According to Theorem 3 the system begins to move from rest in the direction of $\varepsilon_2 = \text{sign}\, Q_1 = \varepsilon_0 \,\text{sign}\, L$ and the $i-th$ particle is the point of debraking.

Let us prove now the third item. As follows from the discussion following Theorem 4, eq. (2.115) is the necessary and sufficient condition for the inequality

$$\left|\mathbf{F_i} \cdot \frac{d\mathbf{r}_i^0}{dq}\right| > \mu \left|\frac{d\mathbf{r}_i^0}{dq}\right| \left|\mathbf{F}_i \cdot \left(\frac{\partial \mathbf{r}_i}{\partial h}\right)_0\right|.$$

In the case of $\mathbf{F}_k = 0$ $(k \ne i)$ the latter inequality implies that

$$|Q_1| > \mu \left|\frac{d\mathbf{r}_T^0}{dq}Q_2\right|.$$

According to Theorem 3 the system can be simultaneously both at rest and moving. Hence, the $i-th$ particle is both a point of self-braking and debraking.

The theorem is thus proved.

Finally, based upon eqs. (2.98) and (2.115) we notice that the condition for debraking ($\mu|L| < 1$) is transformed into the condition of non-uniqueness of self-braking ($\mu|L| > 1$).

Remark 1. While formulating Theorems 2-6 we used coefficients of the system of equations (2.15) derived for the case of $\mathbf{n} \cdot d\mathbf{r}_T^0/dq = 0$. This statement can be derived easily by using relationships (2.8), (2.18)-(2.23). For example, in this case conditions (2.75) and (2.94) take the form

$$|Q_1| \le \mu \left|\frac{d\mathbf{r}_T^0}{dq}\right| |Q_2^* - \lambda Q_1|, \quad \frac{d\mathbf{r}_i^0}{dq} = \gamma\mu \left|\frac{d\mathbf{r}_T^0}{dq}\right| \left(\frac{d\mathbf{r}_i^*}{dq} - \lambda \frac{d\mathbf{r}_i^0}{dq}\right)_0.$$

Remark 2. If another quantity $p \ne h$ is taken as the redundant variable, then

$$p = p(q,h), \quad \left(\frac{\partial \mathbf{r}_i}{\partial h}\right)_0 = \alpha_p \left(\frac{d\mathbf{r}_T^0}{dp}\right), \quad A_{12}^0 = \alpha_p \tilde{A}_{12}^0,$$

$$Q_2 = \alpha_p \tilde{Q}_2, \quad \lambda = \alpha_p \tilde{\lambda}; \quad \alpha_p \equiv \left(\frac{\partial p}{\partial h}\right)_0,$$

66   2. Systems with a single degree of freedom and a single frictional pair

where tilde implies that the coefficients in front of $\tilde{A}^0_{12}, \tilde{Q}_2$ and $\tilde{\lambda}$ are determined according to the chosen coordinate p. In this case the above mentioned conditions (2.75) and (2.94) reduce to the following ones

$$|Q_1| \leq \mu \left|\alpha_p \frac{d\mathbf{r}^0_T}{dq} \tilde{Q}_2\right| = \mu \left|\alpha_p \frac{d\mathbf{r}^0_T}{dq}\right| \left|\tilde{Q}^*_2 - \tilde{\lambda} Q_1\right|,$$

$$\frac{d\mathbf{r}_i}{dq} = \gamma\mu\alpha_p \left|\frac{d\mathbf{r}^0_T}{dq}\right| \left(\frac{\partial \mathbf{r}_i}{\partial p}\right)_0 = \gamma\mu\alpha_p \left|\frac{d\mathbf{r}^0_T}{dq}\right| \left(\frac{d\mathbf{r}^*_i}{dp} - \tilde{\lambda} \frac{d\mathbf{r}^0_i}{dq}\right)_0.$$

3
Accounting for dry friction in mechanisms. Examples of single-degree-of-freedom systems with a single frictional pair

The theoretical results explained in Section 2 for single-degree-of-freedom systems with a single frictional pair are applied in the present chapter. We consider first the construction of the equations, determining the feasibility of the paradoxes, establishing the conditions for the immovable contact and studying the property of self-braking. Integration of the equations of motion receives little attention. The two simple examples of Section 3.1 are of an auxiliary character. The further sections address six mechanisms, each of them is of practical interest.

To illustrate the approach developed in the present chapter, this approach is used for all mechanisms under consideration, despite the fact that a shorter way can be found in some cases.

3.1 Two simple examples

3.1.1 First example

A particle of mass M, see Fig. 3.1, moves along a straight line $0x$ of a rough horizontal plane with a friction factor μ under the actions of the force $\mathbf{F}$, whose direction comprises angle φ with the vertical.

As the generalised coordinate q we choose the abscissa of the considered point (its position on the axis $0x$ is denoted by T^0). When the contact is removed, the particle gains an additional displacement h and moves from

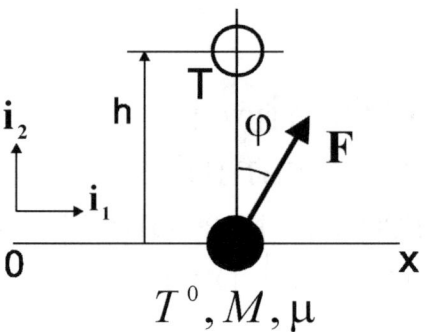

FIGURE 3.1.

point T^0 to T. Then

$$\mathbf{r}^0 = q\mathbf{i}_1, \quad \mathbf{r} = q\mathbf{i}_1 + h\mathbf{i}_2,$$
$$\frac{d\mathbf{r}^0}{dq} = \mathbf{i}_1, \quad \frac{\partial \mathbf{r}}{\partial h} = \mathbf{i}_2. \tag{3.1}$$

Here $\mathbf{r}^0$ and $\mathbf{r}$ denote the position vector before and after additional displacement, respectively. Applying standard formulae for the coefficients of the kinetic energy and the generalised forces, as well as formulae (2.59)-(2.62), we find that

$$A = A_{11} = M, \quad A_{12} = 0, \quad Q_1 = F\sin\varphi, \quad Q_2 = F\cos\varphi, \tag{3.2}$$
$$R_0 = -F\cos\varphi, \quad L = 0, \quad H = 0, \quad G = M^{-1}\left(F\sin\varphi - \varepsilon_2\mu|F\cos\varphi|\right),$$

where $\varepsilon_2 = \text{sign } \dot{q}$. Inserting these expressions into system (2.59) yields the differential equation of motion and the normal reaction force in the form

$$M\ddot{q} = F\sin\varphi - \varepsilon_2\mu|F\cos\varphi|, \quad R = -F\cos\varphi. \tag{3.3}$$

Expressions (3.1) and (3.3) are derived for the additional vertical displacement. It is easy to show that relationships (3.3) are obtained for example under displacement along a straight trajectory comprising angle θ with the vertical. Indeed, in this case instead of (3.1) and (3.2) we have

$$\mathbf{r}^0 = q\mathbf{i}_1, \quad \mathbf{r}^* = (q + h\tan\theta)\mathbf{i}_1 + h\mathbf{i}_2,$$
$$\frac{d\mathbf{r}^0}{dq} = \mathbf{i}_1, \quad \frac{\partial \mathbf{r}^*}{\partial h} = \mathbf{i}_1\tan\theta + \mathbf{i}_2, \quad A = A_{11}^* = M, \quad A_{12}^* = M\tan\theta,$$
$$\lambda = \tan\theta, \quad Q_1 = F\sin\varphi, \quad Q_2^* = F\sin\varphi\tan\theta + F\cos\varphi,$$
$$L = 0, \quad H = 0, \quad G = M^{-1}\left(F\sin\varphi - \varepsilon_2\mu|F\cos\varphi|\right), \quad R_0 = -F\cos\varphi.$$

As one can see, the values of L, H, G, R_0 coincide with those of the previous case, hence substituting them into system (3.1) also leads to relationships (3.3).

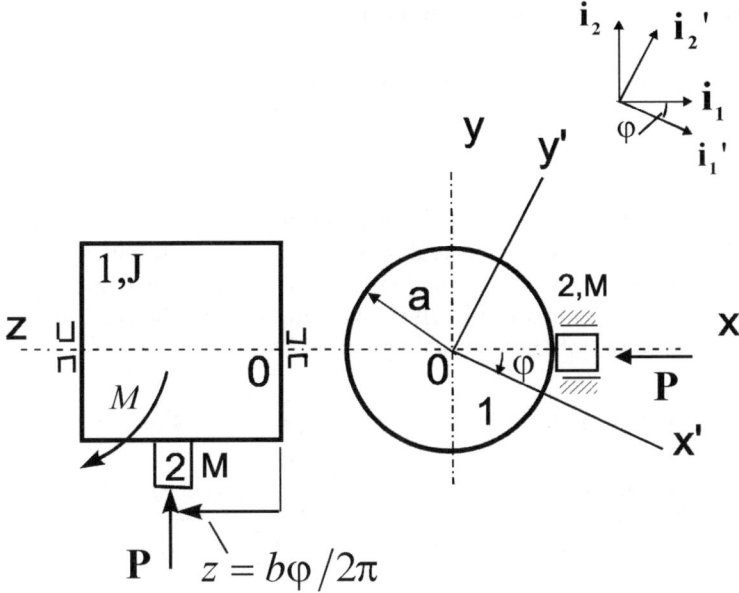

FIGURE 3.2.

Generally speaking, function $\mathbf{r}^*(q,h)$ is arbitrary, the only restriction is that it becomes $\mathbf{r}^0 = q\mathbf{i}_1$ at $h = 0$.

For example, if we take $\mathbf{r}^*$ in the form of $\mathbf{r}^*(q,h) = q(1+h)\mathbf{i}_1 + h\mathbf{i}_2$ then by comparing with Lagrange's equations we obtain relationships (3.3).

Next, by means of Theorems 1-5 and expressions (3.2) we obtain that,

firstly, Painlevé's paradoxes are absent as $L = 0$,

secondly, the condition for transition to motion is expressed in the form $\tan\varphi > \mu$ and

thirdly, the particle is the point of debraking with the stagnation angle $\Psi = 2\arctan\mu$.

It is needless to say that the results obtained are well-known.

3.1.2 Second example

A cylinder of diameter $2a$ rotates about axis $0z$, the moment of inertia about this axis being J, see Fig. 3.2. A slider 2 of mass M contacting with the cylinder translates in the $0z$ direction according to the law $z = b\varphi/2\pi$ where φ denotes the angle of rotation of the cylinder. The latter is subjected to moment M, whereas force $\mathbf{P} = -P\mathbf{i}_1$ ($P > 0$) acts on the cylinder.

Let $0xyz$ be a fixed coordinate system, whilst $0x'y'z'$ be a moving systems relative to the cylinder. The position vector of the slider in the moving coordinate system is given in the form

$$\mathbf{r}_T^0 = \mathbf{i'}_1 a \cos\varphi + \mathbf{i'}_2 a \sin\varphi + \mathbf{i'}_3 \frac{b}{2\pi}\varphi,$$

$$\mathbf{r}_T = \mathbf{i'}_1 (a+h) \cos\varphi + \mathbf{i'}_2 (a+h) \sin\varphi + \mathbf{i'}_3 \frac{b}{2\pi}\varphi, \quad (3.4)$$

where $\mathbf{r}_T^0$ denotes the position vector before the additional displacement whilst $\mathbf{r}_T$ denotes that after the displacement normal to the cylinder.

Under condition (3.4) the local derivative of $\mathbf{r}_T$ with respect to φ, the derivative of $\mathbf{r}_T$ with respect to h, the velocity of slip, the generalised forces and the coefficient of the contact influence are as follows

$$\frac{d\mathbf{r}_T^0}{d\varphi} = -\mathbf{i'}_1 a \sin\varphi + \mathbf{i'}_2 \cos\varphi + \mathbf{i'}_3 \frac{b}{2\pi}, \quad \frac{\partial \mathbf{r}_T}{\partial h} = \mathbf{i'}_1 = \mathbf{i'}_1 \cos\varphi + \mathbf{i'}_2 \sin\varphi,$$

$$\left|\frac{d\mathbf{r}_T^0}{d\varphi}\right| = (a^2 + b^2)^{1/2}, \quad \mathbf{v}_T^0 = \frac{d\mathbf{r}_T^0}{d\varphi}\dot\varphi, \quad A_{11} = J + M\frac{b^2}{4\pi^2},$$

$$Q_1 = M, \quad Q_2 = -P, \quad A_{12}^0 = L = 0.$$

Thus the system of equations (2.15) takes the form

$$\frac{4\pi^2 J + b^2 M}{2\pi}\ddot\varphi = 2\pi M - \varepsilon_2 \mu\sqrt{4a^2\pi^2 + b^2}\,P, \quad R = P, \quad (\varepsilon_2 = \operatorname{sign}\dot\varphi).$$

In addition to this, Painlevé's paradoxes and self-braking are absent for any μ whereas the transition to motion from rest occurs if

$$|M| > \mu \left(a^2 + \frac{b^2}{4\pi^2}\right)^{1/2} P.$$

3.2 The Painlevé-Klein extended scheme

The traditional Painlevé-Klein scheme has been studied by many authors, see e.g. [26], [64], [117], [62]. It consists of two particles M_1 and M_2 moving along two straight guides which are connected by a rigid massless rod of length l, see Fig. 1.3. The first guide is rough with friction coefficient μ and the second one is smooth. The angle between rod $M_1 M_2$ and the horizontal axis x is φ ($0 < \varphi < \pi/2$). Two tangential forces P_1 and P_2 are applied to the particles.

The Painlevé-Klein extended scheme differs from the traditional one in that a massless plate $ABCD$ is attached to the rod, see Fig. 3.3. Apart from the known results, we determine the points of self-braking and debraking, the stagnation angle for the plate points and demonstrate the influence of the paradoxical situations in the case of self-braking and at rest.

3.2 The Painlevé-Klein extended scheme 71

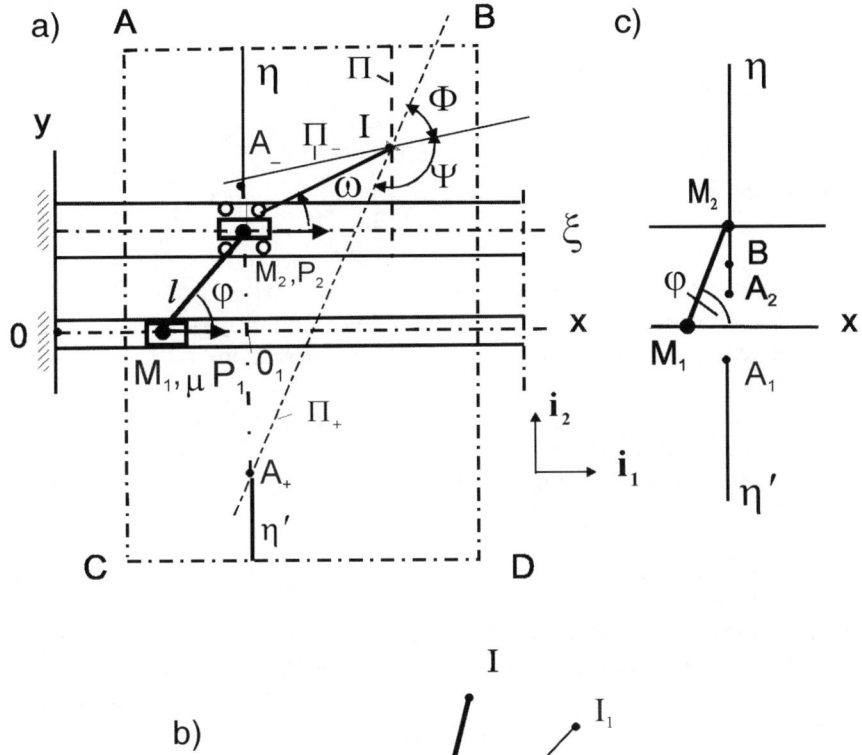

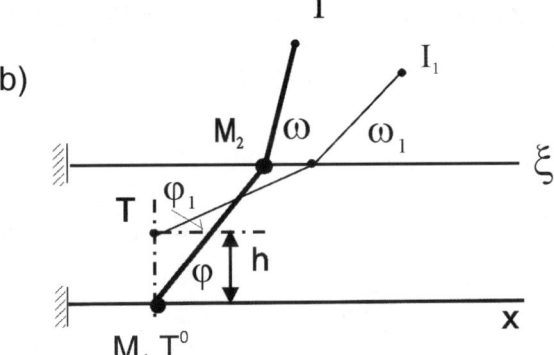

FIGURE 3.3.

3.2.1 Differential equations of motion, expression for the reaction force, condition for the paradoxes and the law of motion

In the system under consideration, the frictional slider is particle M_1. Taking the abscissa x_1 of particle M_1 ($q = x_1$) as the generalised coordinate, we have

$$\mathbf{r}_T^0 = \mathbf{r}_1^0 = q\mathbf{i}_1, \quad \mathbf{r}_2^0 = (q + l\cos\varphi)\mathbf{i}_1 + l\sin\varphi\,\mathbf{i}_2, \tag{3.5}$$

$$\mathbf{r}_T = \mathbf{r}_T = q\mathbf{i}_1 + h\mathbf{i}_2, \quad \mathbf{r}_2 = \left[q + \sqrt{l^2 - (l\sin\varphi - h)^2}\right]\mathbf{i}_1 + l\sin\varphi\,\mathbf{i}_2,$$

$$\frac{d\mathbf{r}_T^0}{dq} = \frac{d\mathbf{r}_k^0}{dq} = \mathbf{i}_1, \ (k = 1, 2), \quad \frac{\partial \mathbf{r}_T}{\partial h} = \frac{\partial \mathbf{r}_1}{\partial h} = \mathbf{i}_2, \quad \left(\frac{\partial \mathbf{r}_1}{\partial h}\right)_0 = \mathbf{i}_1 \tan\varphi,$$

where $\mathbf{r}_1^0, \mathbf{r}_2^0, \mathbf{r}_T^0$ denote the position vectors of particles M_1, M_2 and the slider, respectively, and $\mathbf{r}_1, \mathbf{r}_2, \mathbf{r}_T$ denote these quantities under an additional displacement of the slider along the vertical from T^0 to T, see Fig. 3.3b.

Let us consider an arbitrary point I of the plate which occupies a new position I_1 as a result of an orthogonal displacement of the slider. Its position vectors in the old and new positions are denoted by $\mathbf{r}^0$ and $\mathbf{r}$, respectively. Then

$$\mathbf{r}^0 = \mathbf{r}_2^0 + a\cos\omega\,\mathbf{i}_1 + a\sin\omega\,\mathbf{i}_2, \quad \mathbf{r} = \mathbf{r}_2 + a\cos\omega_1\,\mathbf{i}_1 + a\sin\omega_1\,\mathbf{i}_2, \tag{3.6}$$

where $a = |\overline{M_2 I}|$, ω designates the angle after the additional displacement, which is the angle between axis ξ and $M_2 I$, see Fig. 3.3b.

Under the shift, the rotation angle $(\omega_1 - \omega)$ of plate $ABCD$ is equal to the rotation angle $(\varphi_1 - \varphi)$ of rod $M_1 M_2$. Thus,

$$\omega_1(h) - \omega = \varphi_1(h) - \varphi, \quad \omega_1'(h) = \varphi_1'(h) \quad \left(' \equiv \frac{d}{dh}\right).$$

In addition to this, $l\sin\varphi_1 = l\sin\varphi - h$, $l\cos\varphi_1 \varphi_1'(h) = -1$. Therefore, for $h = 0$ we have

$$\omega_1'(0) = -(l\cos\varphi)^{-1}. \tag{3.7}$$

By using eqs. (3.6) and (3.7) we obtain the expressions for the derivatives such that

$$\frac{d\mathbf{r}^0}{dq} = \mathbf{i}_1, \quad \left(\frac{\partial \mathbf{r}}{\partial h}\right)_0 = \frac{l\sin\varphi + a\sin\omega}{l\cos\varphi}\mathbf{i}_1 + \frac{a\cos\omega}{l\cos\varphi}\mathbf{i}_2 = \frac{y\mathbf{i}_1 - \xi\mathbf{i}_2}{l\cos\varphi}, \tag{3.8}$$

where $\xi = a\cos\varphi$ denotes the abscissa of point I in the moving coordinate system $M_2\xi\eta$ and y is the ordinate of point I in system $0xy$.

3.2 The Painlevé-Klein extended scheme

Now one can apply the equations of Chapter 2. The coefficients of eqs. (2.15) and (2.59) are calculated by means of formulae (2.12), (2.14), (2.20)-(2.23) and (2.60)-(2.63). The result is as follows

$$A = M_1 + M_2; \quad A_{12}^0 = M_2 \tan \varphi, \quad \frac{dA_{12}^0}{dq} = \left(\frac{\partial A_{11}}{\partial h}\right)_0 = 0,$$

$$L = \frac{M_2}{M_1 + M_2} \tan \varphi; \quad Q_1 = P_1 + P_2; \quad Q_2 = P_2 \tan \varphi; \quad H = 0,$$

$$G = \frac{1}{M_1 + M_2}(P_1 + P_2 + \varepsilon_1\varepsilon_2\mu P_2 \tan \varphi), \quad R_0 = \frac{M_2 P_1 - M_1 P_2}{M_1 + M_2} \tan \varphi,$$

$$(\varepsilon_1 = \operatorname{sign} R, \quad \varepsilon_2 = \operatorname{sign} \dot{q}). \tag{3.9}$$

Substituting the obtained formulae for L, H, G and R_0 into system (2.59) leads to the differential equation of motion and the expression for the reaction force

$$\begin{aligned}(M_1 + M_2 + \varepsilon_1\varepsilon_2\mu M_2 \tan \varphi)\ddot{q} &= P_1 + P_2 + \varepsilon_1\varepsilon_2\mu P_2 \tan \varphi, \\ (M_1 + M_2 + \varepsilon_1\varepsilon_2\mu M_2 \tan \varphi) R &= (M_2 P_1 - M_1 P_2)\tan \varphi. \end{aligned} \tag{3.10}$$

The sign of the velocity $\varepsilon_2 = \operatorname{sign} \dot{q}$ is given as an initial condition. The sign of the normal reaction $\varepsilon_1 = \operatorname{sign} R$ is unknown and is found with the help of Theorem 1.

By virtue of Theorem 1, Painlevé's paradoxes do not occur for L from eq. (3.9) if

$$\mu \tan \varphi < \frac{M_1 + M_2}{M_2} \tag{3.11}$$

and occur if this inequality does not hold. When no paradox is present, the sign of the reaction force R is coincident with the sign of R_0. Then, inasmuch as $0 < \varphi < \pi/2$, it follows from eq. (3.9) for R_0 that

$$\varepsilon_1 = \varepsilon_0 = \operatorname{sign}(M_2 P_1 - M_1 P_2). \tag{3.12}$$

Equations (3.10) and (3.12) yield the reaction force

$$R = \frac{(M_2 P_1 - M_1 P_2)\tan \varphi}{(M_1 + M_2 + \varepsilon_1\varepsilon_2\mu M_2 \tan \varphi)} \tag{3.13}$$

and the law of motion $q(t)$ is determined by quadratures. For instance, for $P_1 = \text{const}, P_2 = \text{const}$ one obtains from eq. (3.10)

$$q(t) = \frac{1}{2}\frac{P_1 + P_2 + \varepsilon_0\varepsilon_2\mu \tan \varphi P_2}{M_1 + M_2 + \varepsilon_0\varepsilon_2\mu \tan \varphi M_2} t^2 + \dot{q}_0 t + q_0, \tag{3.14}$$

where q_0 and $\dot{q}_0$ are the initial values of q and $\dot{q}$, respectively. It is necessary to notice that expressions (3.13) and (3.14) are valid as long as

$$\varepsilon_2 = \operatorname{sign} \dot{q} = \operatorname{sign} \dot{q}_0,$$

i.e. unless the velocity of slip becomes zero. If $\dot{q}(t_0) = 0$ at a certain time instant t_0 then the law of motion and the reaction force $t > t_0$ are given by Theorems 2 and 3. This is discussed in what follows.

Let us consider the property of eq. (3.10) in the paradoxical situations when condition (3.11) is not satisfied, i.e. when

$$\arctan \frac{M_1 + M_2}{\mu M_2} < \varphi \leq \pi/2. \tag{3.15}$$

In this case, according to Theorem 1, the solution of eq. (3.10) is not unique ($\varepsilon_1 = \pm 1$), provided that the sign of the velocity $\dot{q}$ is coincident with the sign of R_0

$$\varepsilon_2 \varepsilon_0 = \text{sign}\,(M_2 P_1 - M_1 P_2).$$

When $\varepsilon_2 = -\varepsilon_0$ the solution does not exist. In the case of non-uniqueness, the normal reaction and the law of motion ($P_1, P_2 = \text{const}$) are determined by means of eq. (3.10) in the following way

$$R^{\pm} = \frac{(M_2 P_1 - M_1 P_2) \tan \varphi}{M_1 + M_2 \pm \mu M_2 \tan \varphi}, \tag{3.16}$$

$$q(t) = \frac{1}{2} \frac{P_1 + P_2 \pm \mu P_2 \tan \varphi}{M_1 + M_2 \pm \mu M_2 \tan \varphi} t^2 + \dot{q}_0 t + q_0. \tag{3.17}$$

Here a plus sign corresponds to $\varepsilon_1 = \varepsilon_0$ while a minus sign corresponds to the case $\varepsilon_1 = -\varepsilon_0$.

3.2.2 Immovable contact and transition to slip

Let at time instant t_0 the velocity of motion of the Painlevé-Klein scheme be zero, that is $\dot{q}(t_0) = 0$. Let us first consider the state of rest and the transition to motion in the case of no paradoxes, i.e. when the condition of absence of paradoxes (3.11) is fulfilled. According to Theorem 2, for values of $d\mathbf{r}_T^0/dq$, Q_1 and Q_2 from (3.5) and (3.9), the initial motionlessness is kept for $t > t_0$ if

$$\left| \frac{P_1 + P_2}{P_2} \right| \leq \mu \tan \varphi \tag{3.18}$$

and the motion $\varepsilon_2 = \text{sign}(P_1 + P_2)$ begins when inequality (3.18) holds.

In the case of paradoxes (3.15) the feasibility of retaining motionlessness and transition to motion can be determined by means of Theorem 3. Then for values $d\mathbf{r}_T^0/dq, L, A_{12}, Q_1$ and Q_2, from eqs. (3.5) and (3.9) it follows that the initial motionlessness is kept in the cases

$$\left| \frac{P_1 + P_2}{P_2} \right| \leq \frac{M_1 + M_2}{M_2} \tan \varphi \tag{3.19}$$

$$\frac{1}{\mu\tan\varphi}\left|\frac{P_1+P_2}{P_2}\right|\leq 1\leq \frac{M_2}{M_1+M_2}\left|\frac{P_1+P_2}{P_2}\right| \quad \text{for} \quad \text{sign}\,\frac{P_1+P_2}{P_2}=-1 \tag{3.20}$$

and the motion $\varepsilon_2 = \text{sign}\,P_2$ begins if

$$\frac{1}{\mu\tan\varphi}\left|\frac{P_1+P_2}{P_2}\right|\leq 1\leq \frac{M_2}{M_1+M_2}\left|\frac{P_1+P_2}{P_2}\right| \quad \text{for} \quad \text{sign}\,\frac{P_1+P_2}{P_2}=1. \tag{3.21}$$

In addition to this, in the case of

$$\left|\frac{P_1+P_2}{P_2}\right| > \mu\tan\varphi \tag{3.22}$$

the equation of dynamics (3.10) admits simultaneously both motionlessness and transition to motion.

3.2.3 The stagnation angle and the property of self-braking in the case of no paradoxes

Analysis of Section 2.5 shows that the particles in the system can be divided into two subsets, namely, a subset of points of self-braking and a subset of points of debraking. The points of self-braking have stagnation angle $\Psi = \pi$ whilst this angle for the points of debraking lies in the interval $0 \leq \Psi < \pi$.

For this reason, we start analysis of the properties of self-braking by determining the stagnation angle for the points of plate $ABCD$.

According to Theorem 5, the angles of stagnation Ψ and shift Φ for an arbitrary point I is determined by the straight lines (2.99), whose normal vectors are given by

$$\mathbf{n}_{\pm} = \frac{d\mathbf{r}^0}{dq} \pm \mu\left|\frac{d\mathbf{r}^0}{dq}\right|\left(\frac{\partial\mathbf{r}}{\partial h}\right)_0.$$

For these values of $d\mathbf{r}/dq$, $(\partial\mathbf{r}/\partial h)_0$ and $d\mathbf{r}_T^0/dq$ determined by eqs. (3.5) and (3.8), these vectors have the form

$$\mathbf{n}_{\pm} = \left(1 \pm \frac{\mu y}{l\cos\varphi}\right)\mathbf{i}_1 \mp \frac{\mu\xi}{l\cos\varphi}\mathbf{i}_2. \tag{3.23}$$

Inserting eqs. (3.6) and (3.23) into (2.99) yields the following equations for the straight lines $\Pi_{\pm}$

$$\Pi_+: \quad (\boldsymbol{\rho}-\mathbf{r}_2^0)\cdot\left[\left(1\pm\frac{\mu y}{l\cos\varphi}\right)\mathbf{i}_1 - \frac{\mu\xi}{l\cos\varphi}\mathbf{i}_2\right] - \xi(1+\mu\tan\varphi), \tag{3.24}$$

$$\Pi_-: \quad (\boldsymbol{\rho}-\mathbf{r}_2^0)\cdot\left[\left(1\pm\frac{\mu y}{l\cos\varphi}\right)\mathbf{i}_1 + \frac{\mu\xi}{l\cos\varphi}\mathbf{i}_2\right] - \xi(1-\mu\tan\varphi). \tag{3.25}$$

3. Accounting for dry friction in mechanisms

The value of Ψ is found using the following formula

$$\cos \Psi = \frac{l^2 \cos^2 \varphi - \mu^2 \left(y^2 + \xi^2 \right)}{\left\{ \left[l^2 \cos^2 \varphi - \mu^2 \left(y^2 + \xi^2 \right) \right]^2 + 4\mu^2 l^2 \xi^2 \cos^2 \varphi \right\}^{1/2}}. \quad (3.26)$$

Let us now establish a graphical method of constructing this angle. To this end, we find points A_+ and A_- of the straight lines $\Pi_\pm$ with the vertical axis $M_2\eta$. At points A_+ and A_-

$$\boldsymbol{\rho} - \mathbf{r}_2^0 = \eta \mathbf{i}_2, \quad (3.27)$$

where η is the ordinate of points A_+ and A_- in the moving system of axes $M_2\xi\eta$. Accounting for eq. (3.27) in eqs. (3.24) and (3.25), we obtain respectively

$$\eta_+ = -l\sin\varphi - \frac{1}{\mu}l\cos\varphi, \quad \eta_- = -l\sin\varphi + \frac{1}{\mu}l\cos\varphi. \quad (3.28)$$

Hence, the points of intersections A_+ and A_- in system $M_2\xi\eta$ are given by

$$A_+\left(0, -l\sin\varphi - \frac{1}{\mu}l\cos\varphi\right), \quad A_-\left(0, -l\sin\varphi = \frac{1}{\mu}l\cos\varphi\right). \quad (3.29)$$

As one can see, the intersection points A_+ and A_- of the straight line $Pi_\pm$ with axis $M_2\eta$ are the same for all points of the plate and are symmetric about axis $0x$, see Fig. 3.3. Thus, for an arbitrary point I of plate $ABCD$ the angles of stagnation Ψ and motion Φ are determined by the straight lines IA_+ and IA_-, and Ψ denotes the angle which contains the vertical itself. In other words, this angle contains the straight line Π which is perpendicular to vector $d\mathbf{r}^0/dq = \mathbf{i}_1$.

We proceed now to determine the points of self-braking. By Corollary 3 to Theorem 5, the points of self-braking are those particles for which $\Psi = \pi$, i.e. $\cos \Psi = -1$. In this case, it follows from eq. (3.26) that

$$l^1 \cos^2 \varphi - \mu^2(y^2 + \xi^2) = -\sqrt{[l^2 \cos^2 \varphi - \mu^2(y^2 + \xi^2)]^2 + 4\mu^2 l^2 \xi^2 \cos^2 \varphi} \quad (3.30)$$

which is possible only under the condition

$$\xi = 0. \quad (3.31)$$

Substituting eq. (3.31) into eq. (3.30), we can write

$$l^2 \cos^2 \varphi - \mu^2 y^2 = -|l^2 \cos^2 -\mu^2 y^2|,$$

hence

$$l^2 \cos^2 \varphi - \mu^2 y^2 \leq 0. \quad (3.32)$$

By using eqs. (3.31) and (3.32) we arrive at expressions for the coordinates of the points of self-braking

$$\xi = 0; \quad y \leq -\frac{1}{\mu} l \cos \varphi; \quad y \geq \frac{1}{\mu} l \cos \varphi. \tag{3.33}$$

Therefore, the set of the points of self-braking comprises two half-axes $A_+\eta'$ and $A_-\eta$.

It is pertinent to note that the upper half-space $A_-\eta$ contains point M_2 if $y(A_-) \leq y(A_2)$, i.e. if

$$\frac{1}{\mu} l \cos \varphi \leq l \sin \varphi \quad \text{or} \quad \tan \geq \frac{1}{\mu}. \tag{3.34}$$

This is the celebrated condition of self-braking for the traditional (non-extended) Painlevé-Klein scheme.

Expressions (3.33) for the coordinates of the points of self-braking are established by means of Corollary 3 to Theorem 5. Clearly, one can come to the same results by using Theorem 5. Indeed, for the considered scheme condition (2.94) of this theorem takes the form

$$\mathbf{i}_1 = \frac{\gamma \mu}{l \cos \varphi}(y \mathbf{i}_i - \xi \mathbf{i}_2) \quad |\gamma| \leq 1$$

which is equivalent to (3.33).

Let us next determine the points of debraking. By definition, we refer to the points which are not points of self-braking as points of debraking. Hence, the set of points of debraking for the extended Painlevé-Klein scheme is the plate $ABCD$ cut by two half-axes $A_+\eta'$ and $A_-\eta$.

For any point of debraking, the angle of shifting Φ differs from zero and the stagnation angle Ψ does not equal π. Moreover, as one can judge from the above graphical method of constructing these angles, $\Phi = \pi$ and $\Psi = 0$ on the part A_+A_- whereas $\Phi = \Psi = \pi/2$ on the circle with the centre at point 0_1 and passing through A_+ and A_-. While approaching the half-axes of self-braking $A_+\eta'$ and $A_-\eta$, as well as moving to infinity, the angle of shifting vanishes and the stagnation angle Ψ approaches π.

3.2.4 Self-braking under the condition of paradoxes

It follows from eq. (3.8) that or points lying off the vertical axis η the vectors $d\mathbf{r}^0/dq$ and $(\partial \mathbf{r}/\partial h)_0$ are not collinear, i.e. the first of the relationships (2.115) holds. Then by Theorem 6 these points are simultaneously the points of self-braking and the points of debraking.

It remains to consider the points on the axis η. The vectors of the derivatives $d\mathbf{r}^0/dq$ and $(\partial \mathbf{r}/\partial h)_0$ are collinear, thus the property of self-braking is determined according to conditions (2.112)-(2.114) and the second of

78   3. Accounting for dry friction in mechanisms

conditions (2.115). Hence, according to eq. (2.112), (3.5) and (3.9), the condition of self-braking takes the form

$$1 = \gamma_0 \frac{M_1 + M_2}{M_2 \, l \sin \varphi} y \qquad -1 \leq \gamma_0 \leq 1.$$

This relationship is possible if

$$y \leq -\frac{M_2 l \sin \varphi}{M_1 + M_2} \quad \text{or} \quad y \geq \frac{M_2 l \sin \varphi}{M_1 + M_2}. \tag{3.35}$$

Due to eqs. (2.113), (3.5) and (3.9) we have

$$\frac{1}{\gamma_1 \mu} = -\frac{y}{l \cos \varphi} = \gamma_2 \frac{M_2}{M_1 + M_2} \tan \varphi, \quad 0 < \gamma_1, \gamma_2 \leq 1.$$

The following values of y

$$-\frac{M_2 l \sin \varphi}{M_1 + M_2} \leq y \leq -\frac{l \cos \varphi}{\mu} \tag{3.36}$$

satisfies this condition. Unifying (3.35) and (3.36) we obtain that, on axis $\eta \eta'$, there are two intervals of self-braking

$$-\infty < y \leq -\frac{l \cos \varphi}{\mu}, \qquad \frac{M_2 l \sin \varphi}{M_1 + M_2} \leq y < \infty. \tag{3.37}$$

These intervals are respectively denoted by $A_1 \eta'$ and $B \eta$ in Fig. 3.3c. According to eqs. (2.114), (3.5) and (3.9)

$$\frac{1}{\gamma_1 \mu} = \frac{y}{l \cos \varphi} = \frac{\gamma_2 M_2}{M_1 + M_2} \tan \varphi, \quad 0 < \gamma_1, \gamma_2 < 1.$$

This yields the following expressions for the points of debraking on axis $\eta \eta'$

$$\frac{l \cos \varphi}{\mu} \leq y \leq \frac{M_2}{M_1 + M_2} l \sin \varphi. \tag{3.38}$$

In Fig. 3.3c these points form interval $A_2 B$. For the mechanism under consideration, the second of these conditions in eq. (2.115) takes the form

$$\mathbf{i}_1 = \gamma \mu \frac{1}{l \cos \varphi} (y \mathbf{i}_1 - \xi \mathbf{i}_2) \quad |\gamma| > 1,$$

which is equivalent to the following

$$\xi = 0, \quad -\frac{l \cos \varphi}{\mu} < y < \frac{l \cos \varphi}{\mu}. \tag{3.39}$$

For this reason, any point in interval A_1A_2 is simultaneously both a point of self-braking and a point of debraking.

Thus, when the conditions for paradoxes (3.15) are satisfied the set of points of self-braking consists of two half-axes $A_1\eta'$ and $B\eta$, whilst the set of points of debraking comprises interval A_2B. On the whole plate $ABCD$ cut by half axes $A_1\eta'$ and $A_2\eta$, the property of self-braking is not defined uniquely due to the non-uniqueness of the solution of the dynamical problem.

3.3 Stacker

Currently, stackers which are carts with engines are widely used in industry. Experience suggests that their motion is of a non-smooth character in many cases. By means of Painlevé's paradoxes we will demonstrate in Chapter 4 that the motion of mechanical systems with friction is not always smooth.

With this in mind, the analysis of possible paradoxical situations of the non-existence and the non-uniqueness of solutions for the stackers is of practical interest. The problem stated below is solved by using a rigid body model shown in Fig. 3.4 and a model accounting for elastic tangential deformation, Fig. 3.5.

3.3.1 Pure rolling of the rigid body model

The rigid body model used is depicted in Fig. 3.4. A driving wheel 1 of radius r is connected with the driven wheel 2 of the same radius by means of the rod $0_1 0_2$ of length $2l$. The effective mass m is placed at point A on the vertical central axis $0y$ at height h. The driving torque M acts on driving wheel 1, a clockwise torque being considered positive. The coefficient of Coulomb friction between the wheel and the rail is equal to μ. Friction in the joints and rolling friction are neglected. It is assumed that there is a gap between the upper rod of the frame and the upper guide which prevents the contact.

In the regime of pure rolling, both wheels rotate with angular velocity ω, which is equal to the velocity $\dot{x}$ of the stacker divided by r. Thus,

$$\dot{x} = \omega r. \tag{3.40}$$

This regime is realised under two conditions. Firstly, the normal reactions R_1 and R_2, acting on the wheel from the rail are positive, i.e. they are directed upwards

$$R_1 > 0, \quad R_2 > 0. \tag{3.41}$$

Secondly, the absolute value of the shifting force, which is the tangential reaction force of the driving wheel, does not exceed the limiting value of

3. Accounting for dry friction in mechanisms

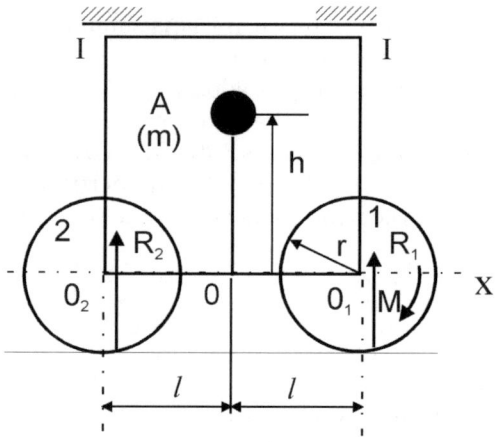

FIGURE 3.4.

the static force

$$|M| < \mu r R_1$$

which is equivalent to the relationship

$$-\mu r R_1 < M < \mu r R_1. \tag{3.42}$$

If at least one of the reaction forces R_1 and R_2 vanishes, the contact between the corresponding wheel and rail weakens.

Provided that the shifting torque is equal to one of the limiting values (3.42) the rolling of the driving wheel is accompanied by slip ($\omega r \neq \dot{x}$). Moreover for $M = \mu r R_1$ we have $\omega r > \dot{x}$, whereas for $M = -\mu r R_1$ we have $\omega r < \dot{x}$.

Let us transform eqs. (3.41) and (3.42) to the conditions of pure rolling expressed in terms of parameters h, l, m, μ and M. Let us start with the equations of kinetostatics

$$\begin{aligned} hm\ddot{x} - mgl + 2lR_1 + M &= 0, \\ m\ddot{x} - M/r &= 0, \\ R_1 + R_2 - mg &= 0. \end{aligned} \tag{3.43}$$

The first of these equations expresses the sum of the moments of the inertia force and the reaction force acting on the wheel from the rails about centre 0_2. The second and the third equations in eq. (3.43) are obtained by equating the projections of these forces to zero.

Resolving eq. (3.43) for R_1 and R_2 yields

$$R_1 = \frac{rlmg - (h+r)M}{2lr}, \quad R_2 = \frac{rlmg + (h+r)M}{2lr}. \tag{3.44}$$

Inserting these expressions for the reaction forces into eq. (3.41) and rearranging, we obtain the corresponding conditions

$$-M_1 < M < M_1, \quad \text{where} \quad M_1 = \frac{lrmg}{h+r}. \tag{3.45}$$

The conditions for absence of slip, eq. (3.42), reduce to the form

$$M_2^- < M < M_1^+, \quad \text{for} \quad h < h_*, \tag{3.46}$$

$$M < M_2^- \quad \text{and} \quad M < M_2^+ \quad \text{for} \quad h > h_*, \tag{3.47}$$

where

$$h_* = 2l\mu^{-1} - r, \quad M_2^\pm = \frac{\pm \mu lrmg}{2l \pm \mu(h+r)}.$$

Value of M_2^+ is positive and bounded for all negative h. Value of M_2^- has a pole $h = h_*$ and

$$M_2^- = \begin{cases} < 0, & h < h_*, \\ \pm\infty, & h = h_* \pm 0, \\ > M_2^+ > 0, & h > h_*. \end{cases} \tag{3.48}$$

As follows from eqs. (3.46) and (3.47), the transition to slip for $\text{sign}(\omega r - \dot{x}) = -1$ is expressed by the inequality $M = M_2^- < 0$ for $h < h_*$ and is expressed by inequality $M = M_2^- > 0$ for $h > h_*$. The inequality appears to be physically inexplicable. If it were the case then the stacker would begin to move with velocity $\dot{x}$, which is higher than the circumferential velocity of the wheel ωr due to the positive moment $M = M_2^-$. On the other hand, by means of the third relationship in eq. (3.48) we can note that the condition of absence of slip $M < M_2^-$ for $h > h_*$ is always satisfied since the shifting torque M can not exceed the value of M_2^+ ($M_2^+ < M_2^-$) causing slip with $\text{sign}(\omega r - \dot{x}) = 1$. As shown in the next subsection Painlevé's paradoxes are absent for slip if $h < h_*$ and present if $h > h_*$.

Let us consider the case $h < h_*$ and find the value of M that ensures pure rolling. This regime takes place when relationships (3.41) and (3.42) are fulfilled. There relationships are reduced to conditions (3.45) and (3.46), thus the value of M must satisfy both of these conditions, i.e.

$$\max(-M_1, M_2^-) < M < \min(M_1, M_2^+). \tag{3.49}$$

The values of $\max(-M_1, M_2^-)$ and $\min(M_1, M_2^+)$ can be found by means of eqs. (3.45) and (3.46) for M_1 and M_2^+. Then we have

$$M_1 = \left[1 + \frac{2l}{\mu(h+r)}\right] M_2^+ > M_2^+ > 0,$$

$$M_2^- + M_1 = \frac{2[l - \mu(h+r)]lrmg}{[2l - \mu(h+r)](h+r)} \begin{cases} \geq 0, & h \leq h_1 = l/\mu - r \\ \leq 0, & h_1 \leq h < h_* \end{cases}$$

and therefore

$$\min(M_1, M_2^+) = M_2^+,$$
$$\max(-M_1, M_2^-) = \begin{cases} M_2^-, & h \leq h_1 \\ -M_1, & h_1 \leq h < h_* \end{cases}. \quad (3.50)$$

Inserting eq. (3.50) into eq. (3.49) results in the condition for pure rolling in the form

$$\begin{array}{ll} M_2^- < M < M_2^+, & h \leq h_1 \\ -M_1 < M < M_2^+, & h_1 \leq h < h_* \end{array} \quad (3.51)$$

or

$$\begin{array}{ll} \dfrac{-\mu lrmg}{2l - \mu(h+r)} < M < \dfrac{\mu lrmg}{2l + \mu(h+r)} & \text{for} \quad h \leq \dfrac{l}{\mu} - r, \\ \dfrac{-lrmg}{h+r} < M < \dfrac{\mu lrmg}{2l + \mu(h+r)} & \text{for} \quad \dfrac{l}{\mu} - r \leq h < \dfrac{2l}{\mu} - r. \end{array} \quad (3.52)$$

Given a value of M within these limits, the law of motion is obtained by integrating the second equation in (3.43) whereas the reaction forces R_1 and R_2 are calculated by formulae (3.44). Let us recall that in the limiting cases $M = M_2^\pm$ inequalities (3.42) are not satisfied and rolling of the driving wheel is accompanied by slip, the slip velocity having respectively the sign $\varepsilon_2 = \text{sign}(\dot{x} - \omega r) = \mp 1$. In the limiting case $M = -M_1$ the reaction force is zero. Under condition (3.51) the reaction force R_1 is always positive because the right limit M_2^+ of condition (3.51) is less than the right limit M_1 of condition (3.45).

3.3.2 Slip of the driving wheel for the rigid body model

In this regime of motion the tangential force acting on driving wheel 1 from the rail is equal to

$$F_\tau = -\varepsilon_1 \varepsilon_2 \mu R_1, \quad (3.53)$$

where

$$\varepsilon_1 = \text{sign}\, R_1, \quad \varepsilon_2 = \text{sign}(\dot{x} - \omega r) = \mp 1 \quad \text{for} \quad M = M_2^\pm. \quad (3.54)$$

Taking this into account we can construct the following equations of kinetostatics

$$\left. \begin{array}{l} m\ddot{x} + \varepsilon_1 \varepsilon_2 \mu R_1 = 0, \\ hm\ddot{x} - mgl + (2l - \varepsilon_1 \varepsilon_2 \mu r) R_1 = 0, \\ M + \varepsilon_1 \varepsilon_2 \mu r R_1 = 0, \\ R_1 + R_2 - mg = 0. \end{array} \right\} \quad (3.55)$$

As contact of wheel 1 with the rail is a one-sided constraint, the case $\varepsilon_1 = -1$ is not feasible. Nevertheless, introducing the symbol $\varepsilon_1 = \operatorname{sign} R_1$ makes the analysis more comfortable.

The differential equations of motion and expressions for the normal reaction force are obtained by solving eq. (3.55) for $\ddot{x}$ and R_1

$$\left.\begin{aligned} [2l - \varepsilon_1\varepsilon_2\mu(r+h)]\ddot{x} &= -\varepsilon_1\varepsilon_2\mu g l, \\ [2l - \varepsilon_1\varepsilon_2\mu(r+h)]R_1 &= mgl. \end{aligned}\right\} \quad (3.56)$$

This system has the form of eq. (2.59) with the following expressions for factors L, H, G, R_0

$$L = -\frac{r+h}{2l}, \quad H = 0, \quad G = -\varepsilon_1\varepsilon_2\mu g, \quad R_0 = \frac{1}{2}mg. \quad (3.57)$$

By using Theorem 1 we can rewrite the condition for absence of paradoxes in the form

$$\mu\frac{r+h}{2l} < 1 \quad \text{or} \quad h < \frac{2l}{\mu} - r = h_*. \quad (3.58)$$

Let us notice that the same requirement ($h < h_*$) is obtained while analysing the regime of pure rolling with $\varepsilon_2 = 1$.

When the condition of absence of paradoxes (3.58) is fulfilled, the solution of system (3.56) can be obtained by substituting $\varepsilon_1 = \varepsilon_0 = 1$. Hence,

$$\dot{x}(t) = \frac{-\varepsilon_2\mu l g}{2l - \varepsilon_2\mu(r+h)} t + \dot{x}(0),$$

$$R_1 = \frac{mgl}{2l - \varepsilon_2\mu(r+h)}, \quad R_2 = \frac{l - \varepsilon_2\mu(r+h)}{2l - \varepsilon_2\mu(r+h)} mg. \quad (3.59)$$

When eq. (3.58) is not fulfilled, i.e. for

$$h \geq \frac{2l}{\mu} - r = h_*, \quad (3.60)$$

the solution of problem (3.56) is not unique for $\varepsilon_2 = -1$ (or $M = M_2^+$) and does not exist for $\varepsilon_2 = 1$ ($M = M_2^-$). In the case of non-uniqueness, it follows from eq. (3.56) that

$$\ddot{x} = \frac{\pm\mu l g}{2l \pm \mu(r+h)}, \quad R_1 = \frac{mgl}{2l \pm \mu(r+h)}. \quad (3.61)$$

However in practical applications, the minus sign in eq. (3.61) does not appear as the contact between the wheels and the rail is a one-sided constraint.

As will be shown in Chapter 4, the true motion is of a non-smooth character in the case of "non-existence of solution".

3.3.3 Speed-up of stacker

The previous subsections address the laws of motion and the normal reaction forces in the cases of both pure rolling and rolling with slip of the driving wheel. In the process of speed-up and deceleration the stacker can switch from one regime of motion to another. Matching solutions is needed for the complete law of motion. As an example let us consider speed-up in the positive direction caused by a motor with a flat characteristic ($\omega = $ const). The paradoxes are assumed to be absent, i.e. $h < h_*$. In this case, the normal condition can be expressed in the form

$$\omega(t) = \text{const} > 0, \quad \dot{x}(0) = 0, \quad \varepsilon_2(0) = -1. \quad (3.62)$$

Under this condition, the slip of driving wheel 1 is unavoidable. Indeed, in order to guarantee rolling without slip ($\dot{x} = \omega r$) at time instant $t = 0 + 0$ an infinite acceleration implying an infinite torque M, is needed. This is, however, is not feasible by virtue of the condition of rolling without slip, eqs. (3.42).

The law of speed-up is determined in the form of eq. (3.59) for the velocity sign $\varepsilon_2 = -1$. This regime continues until velocity $\dot{x}$ is equal to ωr. For $\dot{x} = \omega r$ the stacker rolls with a constant velocity without slip. We have the following expressions for the velocity and the normal reactions forces

$$\dot{x} = \begin{cases} \dfrac{\mu l g}{2l + \mu(r+h)} t, & 0 \leq t < t_1 \\ \omega r & t \geq t_1 \end{cases}, \quad t_1 = \dfrac{2l + \mu(r+h)}{\mu l g} \omega r \quad (3.63)$$

$$R_1 = \begin{cases} \dfrac{mgl}{2l + \mu(r+h)}, & t < t_1 \\ \dfrac{mg}{2}, & t \geq t_1 \end{cases}, \quad R_2 = \begin{cases} \dfrac{l + \mu(r+h)}{2l + \mu(r+h)} mg, & t < t_1 \\ \dfrac{mg}{2}, & t \geq t_1 \end{cases}$$

These equations suggest that the value t_1 is the duration of the slip and simultaneously the duration of the speed-up after which the stacker moves uniformly and without slip. Secondly, the normal reaction forces R_1 and R_2 are piecewise-constant functions for $0 < t < t_1$ and $t > t_1$ and experience jumps at instant t_1. Besides, according to the second of the equations in (3.43) for $t \geq t_1$ the shifting torque $M = 0$ since $\ddot{x} = 0$ (in practice, in order to take into account the rolling friction we have $M \neq 0$ for $\ddot{x} = 0$). Here the problem of speed-up is considered under the assumption that the angular velocity of the driving wheel is constant, that is, at the initial time instant, slip is unavoidable. Provided that the motor has a drooping characteristic then $x(t)$, R_1 and R_2 are determined with the help of the second equation in (3.43) and formulae (3.44) in the case $0 < M < M_2^+$, as well as (3.59) in the case of $M = M_2^+$.

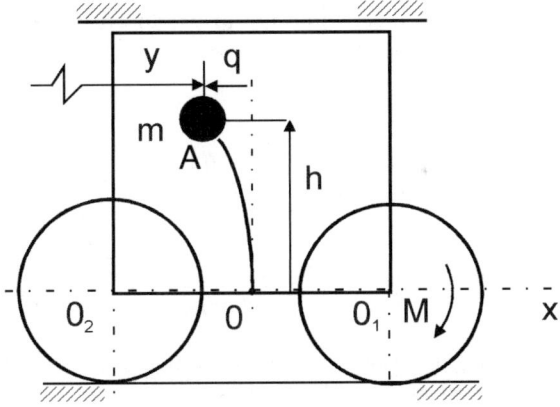

FIGURE 3.5.

3.3.4 Pure rolling in the case of tangential compliance

The model displayed in Fig. 3.5 differs from that of Fig. 3.4 in that rod OA is an elastic element with rigidity c. As mentioned above, the regime of pure rolling is described by conditions (3.40)-(3.42). The equations of kinetostatics are as follows

$$hm\ddot{y} - mg(l-q) + M + 2lR_1 = 0, \quad m\ddot{y} - cq = 0,$$
$$rcq - M = 0, \quad y - x + q = 0, \quad R_1 + R_2 - mg = 0, \qquad (3.64)$$

where q denotes the elastic displacement of point A from the equilibrium position, y and x denote the coordinates of points A and 0, correspondingly. As one can see from the second and third equations, the inertia force $m\ddot{y}$ is equal to the elastic force cq which is also equal to Mr^{-1}.

Resolving eq. (3.64) for $\ddot{y}, R_1$ and R_2 yields the equation of motion

$$m\ddot{y} = Mr^{-1} \qquad (3.65)$$

and expressions for the reactions forces

$$R_1 = \frac{mgl - (ch + cr + mg)q}{2l} = \frac{lmrg - (h + r + mg/c)M}{2lr},$$
$$R_2 = \frac{mgl + (ch + cr + mg)q}{2l} = \frac{lmrg + (h + r + mg/c)M}{2lr}. \qquad (3.66)$$

In what follows we take argument q instead of M, since they are proportional to each other $(M = rcq)$.

For values of R_1 and R_2 from eq. (3.46) the condition for positiveness of the reaction forces (3.41) reduces respectively to the following ones

$$-q_1 < q < q_1, \quad q_1 = \frac{mgl}{c(h+r) + mg}. \qquad (3.67)$$

The conditions (3.42) for absence of slip are reduced to the following form

$$q_2^- < q < q_2^+; \quad \text{for} \quad h < H_*, \tag{3.68}$$

$$q < q_2^-, \quad \text{and} \quad q < q_2^+ \quad \text{for} \quad h > H_*, \tag{3.69}$$

where

$$H_* = h_* - \frac{mg}{c}, \quad q_2^\pm = \frac{\pm \mu mgl}{2cl \pm \mu(ch + cr + mg)}.$$

Quantities q_1 and q_2^+ are positive and bounded for all values of $h > 0$, whilst q_2^- is unbounded at $h = H_*$, that is

$$q_2^- \begin{cases} < 0 & h < H_*, \\ = \pm \infty & h = H_* \pm 0, \\ > q_2^+ > 0 & h > H_*. \end{cases} \tag{3.70}$$

As mentioned above the value of h_* given by eq. (3.47) is the critical height, which being exceeded results in paradoxes in the rigid body model. As shown in Subsection 3.3.5 the paradoxes are absent in the elastic model for $h < H_*$ and present for $h > H_*$. Moreover, according to the formula for H_* we have $\lim_{c \to \infty} H_* = h_*$.

Let us consider the case of $h < H_*$. Pure rolling occurs when the two conditions (3.67) and (3.68) are simultaneously satisfied. Then

$$\max(-q_1, q_2^-) < q < \min(q_1, q_2^+), \tag{3.71}$$

$$\min(q_1, q_2^+) = q_2^+,$$

$$\max(-q_1, q_2^-) = \begin{cases} q_2^- & h \leq H_1, \\ -q_1 & H_1 \leq h < H_*, \end{cases} \tag{3.72}$$

$$H_1 = \frac{l}{\mu} - r - \frac{mg}{c} = h_1 - \frac{mg}{c}. \tag{3.73}$$

Here H_1 and h_1 are those values of h for which $q_2^- = -q_1$ and $M_2^- = -M_1$, respectively.

Substituting relationships (3.72) and (3.73) into eq. (3.71) we obtain, for the case of $h < H_*$, the condition of pure rolling in the form

$$\begin{aligned} q_2^- < q < q_2^+, & \quad h \leq H_1, \\ -q_1 < q < q_2^+, & \quad H_1 \leq h < H_* \end{aligned} \tag{3.74}$$

or

$$\frac{-\mu mgl}{2cl - \mu(ch + cr + mg)} < q < \frac{\mu mgl}{2cl + \mu(ch + cr + mg)}, \quad h \leq H_1,$$

$$\frac{-mgl}{c(h+r) + mg} < q < \frac{\mu mgl}{2cl + \mu(ch + cr + mg)}, \quad H_1 \leq h < H_*. \quad (3.75)$$

If coordinate q takes one of the limiting values $q_2^\pm$, then condition (3.72) does not hold and rolling of the driving wheel is not accompanied by slip with the velocity sign $\varepsilon_2 = \text{sign}(\dot{x} - wr) = \mp 1$. As will become clear from the below reasoning, q remains constant $q = q_2^\pm$ for $\varepsilon_2 = \mp 1$ and $h < H_*$.

The value of the reaction force R_2 is equal to zero and thus, the contact of wheel 2 with the rail weakens for $q \leq -q_1$. Reaction force R_1 is always positive for $h < H_*$ since (by virtue of the inequality $q \leq q_2^+ < q_1$) coordinate q can not reach those values of q_1, for which $R_1 = 0$.

3.3.5 Rolling with account of compliance

This regime is described by equations

$$hm\ddot{y} - mg(l - q) - \varepsilon_1\varepsilon_2\mu r R_1 + 2lR_1 = 0, \quad m\ddot{y} + \varepsilon_1\varepsilon_2\mu R_1 = 0,$$
$$cq + \varepsilon_1\varepsilon_2\mu R_1 = 0, \quad R_1 + R_2 - mg = 0, \quad (3.76)$$

where $\varepsilon_1 = \text{sign}\, R_1$, $\varepsilon_2 = \text{sign}(\dot{x} - wr)$. By solving eq. (3.76) one can obtain differential equations of motion and the expression for the reaction forces

$$[cl - \varepsilon_1\varepsilon_2\mu(mg + cr + ch)]\ddot{y} = -\varepsilon_1\varepsilon_2\mu cgl,$$
$$[2cl - \varepsilon_1\varepsilon_2\mu(mg + cr + ch)]R_1 = mgcl. \quad (3.77)$$

It follows from the third equation in (3.76) and the second equation in (3.77) that in the regime of rolling the coordinate q and the torque M remain constant for $\mu = \text{const}$. Then $\ddot{x} = \ddot{y}$.

Applying the transformation of system eq. (3.56) to eq. (3.77) we arrive at the condition of absence of paradoxes in the following form

$$h < H_* = \frac{2cl - \mu(mg + cr)}{\mu c} = h_* - \frac{mg}{c}. \quad (3.78)$$

When this inequality holds we have $\varepsilon_1 = 1$ and the solution of system (3.77) is unique.

Provided that condition (3.78) does not hold. i.e. if $h \geq H_*$, then the solution of (3.77) is not unique for $\varepsilon_2 = -1$ (or $q = q_2^+$) and does not exists for $\varepsilon_2 = 1$ ($q = q_2^-$). In such paradoxical regimes, the motion experiences jumps which is a topic of Chapter 4.

3.3.6 Speed-up with account of compliance

The two previous subsections are concerned with the analysis of two sorts of motion, namely, pure rolling and slip of the driving wheel. In practical applications one sort of motion can change into the other and one needs matching solutions for obtaining the complete law of motion. As an example, let us consider speed-up in the positive direction for $h < H_*$, i.e. in the case of absence of Painlevé's paradoxes. The initial conditions are as follows

$$\omega = \text{const} > 0, \quad y(0) = \dot{y}(0) = 0, \quad q(0) = \dot{q}(0) = 0. \tag{3.79}$$

The first of these conditions takes place under a flat characteristic of the motor. The other conditions indicate that motion begins from the equilibrium position when the shifting moment M vanishes. The moment can later increase or decrease. However it is always proportional to displacement q. Because of the continuity of $q(t)$, the condition of rolling without slip, eq. (3.74), will be fulfilled. Hence

$$\dot{x} = \omega r, \quad \ddot{x} = 0, \quad t \geq 0. \tag{3.80}$$

Inserting eq. (3.80) into eq. (3.64) we obtain the following equation

$$m\ddot{q} + cq = 0. \tag{3.81}$$

Under initial conditions (3.79) the law of pure rolling for $t \geq 0$ can be written as follows

$$\begin{aligned} q &= \omega r \lambda^{-1} \sin \lambda t, \quad \lambda = (c/m)^{1/2}, \quad x = \omega r t, \\ y &= \omega r (t - \lambda^{-1} \sin \lambda t), \quad \dot{y} = \omega r (1 - \cos \lambda t). \end{aligned} \tag{3.82}$$

Let us find the condition of remaining inside this regime of motion and the condition of transition to slip. It is evident that the law of pure motion (3.82) must meet condition (3.74) or (3.75). Hence

$$\begin{aligned} q_2^- &< q = \omega r \lambda^{-1} \sin \lambda t < q_2^+, \quad h \leq H_1, \\ -q_1 &< q = \omega r \lambda^{-1} \sin \lambda t < q_2^+, \quad H \leq h < H_*. \end{aligned} \tag{3.83}$$

To ensure fulfillment of condition (3.83) at any time instant t it is necessary and sufficient that the amplitude $\omega r \lambda^{-1}$ is less than all limiting absolute values $|q_1|, |q_2^+|$ and $|q_2^-|$. By virtue of expressions (3.67) and (3.68) we obtain the following condition of maintaining rolling without slip in the case of q_1 and $q_2^\pm$

$$\omega r \lambda^{-1} < \min(|q_1|, |q_2^+|, |q_2^-|) = q_2^+,$$

$$\dot{x} = \omega r < \lambda q_2^+ = \frac{\mu \lambda m g}{2cl + \mu(ch + cr + mg)}. \tag{3.84}$$

Thus, in contrast to the rigid body model, in the case of tangential compliance and a flat motor characteristic pure rolling is feasible if angular velocity ω does not exceed $\lambda q_2^+/r$.

Let us now the situation when condition (3.84) is not fulfilled. Then, as time progresses value of q increases to q_2^+ and pure rolling develops into slip with the velocity sign $\varepsilon_2 = \text{sign}(\dot{x} - \omega r) = -1$. At the time instant of transition t_1 $q(t_1) = q_2^+$, hence, due to eq. (3.82)

$$\sin \lambda t_1 = \frac{\lambda q_2^+}{r\omega}, \quad \cos \lambda t_1 = \frac{[\omega^2 r^2 - (\lambda q_2^+)^2]^{1/2}}{\omega r},$$

$$t_1 = \frac{1}{\lambda} \arcsin \frac{\lambda q_2^+}{\omega r}. \qquad (3.85)$$

Let us now determine the law of motion under slip, i.e. when $t \geq t_1$. In as much as the paradoxes are absent and slip corresponds to the right limit of condition (3.74) then, as established in the previous subsection, the signs of the reaction forces and the slip velocity are $\varepsilon_1 = 1$ and $\varepsilon_2 = -1$, respectively. In this case, taking into account the equality $\ddot{x} = \ddot{y}$ for slip and expression (3.68) for q_2^+, as well as the first relationship in eq. (3.77) we obtain the following differential equation

$$\ddot{x} = \ddot{y} = \frac{\mu c g l}{2cl + \mu(mg + cr + ch)} = \lambda^2 q_2^+. \qquad (3.86)$$

The initial condition is established by reasoning of continuity of coordinates $x(t)$, $y(t)$ and velocity $\dot{y}$ at time instant t_1. Due to eq. (3.82) and (3.85) we have

$$x(t_1) = \omega r t_1, \quad y(t_1) = \omega r t_1 + q_2^+,$$
$$\dot{x}(t_1) = \dot{y}(t_1) = \omega r - \sqrt{\omega^2 r^2 - (\lambda q_2^+)^2}. \qquad (3.87)$$

Using eqs. (3.80) and (3.87) one can see that velocity $\dot{x}$ of the stacker base $0_1 0_2$ decreases abruptly to the value of $[\omega^2 r^2 - (\lambda q_2^+)^2]^{1/2}$ at time instant t_1.

Using differential equation (3.86) and initial condition (3.87) we can set the law of motion under slip in the following form

$$\dot{x} = \dot{y} = \lambda^2 q_2^+(t - t_1) + \omega r - [\omega^2 r^2 - (\lambda q_2^+)^2]^{1/2},$$
$$x(t) = \frac{\lambda^2}{2} q_2^+(t - t_1)^2 + \left\{\omega r - [\omega^2 r^2 - (\lambda q_2^+)^2]^{1/2}\right\}(t - t_1) + \omega r t_1,$$
$$y(t) = x(t) + q_2^+, \quad q(t) = q_2^+, \quad t > t_1. \qquad (3.88)$$

According to this law, velocity $\dot{x}$ continuously increases. Hence, the slip motion changes into pure rolling at time instant t_2 when its velocity becomes equal to the circumferential velocity of wheel 1

$$\dot{x}(t_2) = \lambda^2 q_2^+(t_2 - t_1) + \omega r - [\omega^2 r^2 - (\lambda q_2^+)^2]^{1/2} = \omega r,$$

3. Accounting for dry friction in mechanisms

or

$$t_2 = \frac{\left[\omega^2 r^2 - (\lambda q_2^+)^2\right]^{1/2}}{\lambda^2 q_2^+} + t_1. \tag{3.89}$$

If $t \geq t_2$ we obtain relationships (3.80) and (3.81) once again

$$\dot{x} = \omega r, \quad \ddot{x} = 0, \quad m\ddot{q} + cq = 0 \quad t \geq t_2. \tag{3.90}$$

Therefore, the initial values of coordinates x, q and velocity $\dot{q}$ can be obtained by means of eq. (3.88) and (3.89). The result is

$$q(t_2) = q_2^+, \quad \dot{q}_2(t_2) = 0,$$
$$x(t_2) = -\frac{1}{2} \frac{\omega^2 r^2 - (\lambda q_2^+)^2}{\lambda^2 q_2^+} + \omega r t_2. \tag{3.91}$$

The solutions of eq. (3.90) have the form

$$q = q_2^+ \cos \lambda(t - t_2), \quad x(t_2) = -\frac{1}{2} \frac{\omega^2 r^2 - (\lambda q_2^+)^2}{\lambda^2 q_2^+} + \omega r t, \quad t \geq t_2. \tag{3.92}$$

As one can see condition (3.83) is always fulfilled for $t \geq t_2$, thus the regime of pure rolling with the constant velocity $\dot{x} = \omega r$ is sustained. The particle A of mass m executes harmonic oscillations with respect to the stacker base $O_1 O_2$ with amplitude q_2^+.

In the case when condition (3.84) does not hold, the above analysis shows that rolling of wheel 1 of the stacker is firstly pure rolling ($0 \leq t < t_1$), then it is accompanied by slip ($t_1 \leq t < t_2$) and finally becomes pure rolling ($t \geq t_2$). Unifying eqs. (3.82), (3.88) and (3.92) we arrive at the law of motion in the form

$$x(t) = \begin{cases} \omega r t & 0 \leq t \leq t_1, \\ \omega r t_1 + \{\omega r - [\omega^2 r^2 - (\lambda q_2^+)^2]^{1/2}\}(t - t_1) \\ \quad + \frac{1}{2} \lambda^2 q_2^+ (t - t_1)^2 & t_1 \leq t \leq t_2, \\ -\frac{1}{2} \frac{\omega^2 r^2 - (\lambda q_2^+)^2}{\lambda^2 q_2^+} + \omega r t & t \geq t_2, \end{cases}$$
$$\tag{3.93}$$

$$q(t) = \begin{cases} \omega r \lambda^{-1} \sin \lambda t & 0 \leq t \leq t_1, \\ q_2^+ & t_1 \leq t \leq t_2, \\ q_2^+ \cos \lambda(t - t_2) & t \geq t_2, \end{cases} \tag{3.94}$$

where t_1 and t_2 are determined from eqs. (3.85) and (3.89).

Finally we calculate the normal reaction forces R_1 and R_2. Under the condition of pure rolling (3.84) they can be found by means of formulae (3.66) and expression (3.82) for $q(t)$:

$$R_1 = \frac{mgl - (ch + cr + mg)wr\lambda^{-1} \sin \lambda t}{2l},$$

$$R_2 = \frac{mgl + (ch + cr + mg)wr\lambda^{-1} \sin \lambda t}{2l}. \quad (3.95)$$

When condition (3.84) does not hold, then for $0 \leq t < t_1$ and $t > t_2$ the values of R_1 and R_2 are determined by eq. (3.66) in which q is given by eq. (3.94) whilst for $t_1 \leq t < t_2$ the reaction forces are found by means of the second expression in eq. (3.77) under the condition $\varepsilon_1 = 1, \varepsilon_2 = -1$. The result is as follows

$$R_1 = \begin{cases} [mgl - (ch + cr + mg)wr\lambda^{-1} \sin \lambda t](2l)^{-1} & 0 \leq t \leq t_1, \\ [2cl + \mu(ch + cr + mg)^{-1}]cmgl & t_1 \leq t \leq t_2, \\ \dfrac{2l + \mu(ch + cr + mg)[1 - \cos \lambda(t - t_2)]}{4cl + 2\mu(ch + cr + mg)} mg & t \geq t_2, \end{cases} \quad (3.96)$$

$$R_2 = \begin{cases} [mgl + (ch + cr + mg)wr\lambda^{-1} \sin \lambda t](2l)^{-1} & 0 \leq t \leq t_1, \\ \dfrac{cl + \mu(ch + cr + mg)}{2cl + \mu(ch + cr + mg)} mg & t_1 \leq t \leq t_2, \\ \dfrac{2cl + \mu(ch + cr + mg)[1 + \cos \lambda(t - t_2)]}{4cl + 2\mu(ch + cr + mg)} mg & t \geq t_2. \end{cases} \quad (3.97)$$

By using eqs. (3.96) and (3.97) with the values of t_1 and t_2 determined from eqs. (3.85) and (3.89) one can easily prove equalities $R_i(t_k - 0) = R_i(t_k + 0)$ $(i, k = 1, 2)$. Hence, in contrast to the rigid body model the reaction forces are not continuous at the time instants of transition from one type of motion to the other.

3.3.7 Numerical example

Let

$$\begin{aligned} m &= 2000 \text{ kg}, \quad 2l = 2 \text{ m}, \quad h = 6 \text{ m}, \quad r = 0,2 \text{ m}, \\ \mu &= 0,3, \quad w = 1,5 \text{ s}^{-1}, \quad v = 0,3 \text{ ms}^{-1}. \end{aligned} \quad (3.98)$$

The limiting values for the conditions of pure rolling (3.51) and (3.52) are determined as follows

$$M_1 = \frac{0,22 \times 000 \times 9,81}{6+0,2} = 633 \text{ N} \cdot \text{m},$$

$$M_2^+ = \frac{1 \times 0,3 \times 0,2 \times 2000 \times 9,81}{2+0,3 \times 6,2} = 305 \text{ N} \cdot \text{m},$$

$$M_2^- = \frac{-1 \times 0,3 \times 0,2 \times 2000 \times 9,81}{2-0,3 \times 6,2} = 8409 \text{ N} \cdot \text{m}. \quad (3.99)$$

Due to eq. (3.47) the critical value h_* is

$$h_* = \frac{2}{0,3} - 0,2 = 6,46 \text{ m}. \quad (3.100)$$

Equations (3.98) and (3.100) yields that $h = 6,00$ m$< h_*$. Hence, for the rigid body model, i.e. for $c = \infty$, Painlevé's paradoxes are absent.

According to eq. (3.50), the height h_1 is given by

$$h_1 = \frac{1}{0,3} - 0,2 = 3,13 \text{ m}. \quad (3.101)$$

As h is greater than h_1 and less than h_* ($h_1 < h < h_*$), the condition of pure rolling is given by the second expression in eq. (3.51). Taking into account eq. (3.99) we obtain

$$-633 \text{ N} \cdot \text{m} < M < 305 \text{ N} \cdot \text{m}. \quad (3.102)$$

When this condition is met, the forces of normal reactions acting on the wheel from the rail can be calculated with the help of formulae (3.44), to yield

$$R_1 = 9810 - 15,5 \cdot M; \quad R_2 = 9810 + 15,5 \cdot M. \quad (3.103)$$

Provided that the shifting moment reaches the lower limit (- 633 N·m) the reaction force R_2 becomes zero and the contact between wheel 2 and the rail weakens. Such a situation is not desirable in practice. If the torque reaches the right limiting value (305 N·m), the driving wheel slips. According to condition (3.54) the slip velocity is negative

$$\varepsilon_2 = \text{sign}(\dot{x} - \omega r) = -1.$$

Then by using eq. (3.59) one can obtain the law of slip and the values of the reaction forces

$$\dot{x}(t) = 0,762t + \dot{x}(0) \text{ m} \cdot \text{s}^{-1},$$
$$R_1 = 5083 \text{ N} \cdot \text{m}; \quad R_2 = 14537 \text{ N} \cdot \text{m}, \quad (3.104)$$

where $\dot{x}(0)$ is determined by the initial condition.

Let us consider the speed-up in the positive direction, i.e. $\dot{x}(t) > 0$. The initial conditions for this regime of motion are prescribed in the form (3.62) whilst the law of motion and the reactions forces are determined by eq. (3.63). We have

$$\dot{x} = \begin{cases} 0,762t & \text{for } t < t_1, \\ 0,030 & \text{for } t \geq t_1, \end{cases}$$

$$R_1 = \begin{cases} 5083 & \text{for } t < t_1, \\ 9810 & \text{for } t \geq t_1, \end{cases}$$

$$R_2 = \begin{cases} 14537 & \text{for } t < t_1, \\ 9810 & \text{for } t \geq t_1, \end{cases} \quad (3.105)$$

where $t_1 = 0,81$ s denotes the duration of speed-up.

It should be mentioned that in the case of $h_1 < h < h_*$ for a flat motor characteristic and the rigid body model during the speed-up in the negative direction (to the left) the driving torque becomes smaller than the left limit (3.102), that is $M < -633$ N. This leads to weaking of the contact between wheel 2 and the rail. In other words, for a motor with a flat characteristic the height h should be smaller than h_1, i.e. $h < h_1$.

The above numerical calculations are related to the case of the rigid rod OA. Let us study now the influence of compliance of the stacker on its dynamic properties. As shown above, the paradoxes are not observed for the elastic model if $h < H_*$. Using eq. (3.69) we can reduce this inequality to the following one

$$c > \frac{mg}{h_* - h}.$$

Inserting the numerical values from eqs. (3.98) and (3.100) into this inequality yields

$$c > c_* = 42652 \text{ N} \cdot \text{m}^{-1}. \quad (3.106)$$

This is the condition for absence of paradoxes which is understood as a restriction imposed on rigidity c for prescribed value h. When this condition is fulfilled we have $h > H_*$ and paradoxes appear.

Let us assume for definiteness that $c = 6 \cdot 10^4$ N·m^{-1}. Then calculations using eqs. (3.67), (3.73) and (3.101) yields

$$H_* = 6,133, \ H_1 = 2,80, \ q_1 = 0,050, \ q_2^+ = 0,025, \ q_2^- = -2,315 \text{ m}. \quad (3.107)$$

One can see from eqs. (3.98), (3.100) and (3.107) that $H_1 < h < H_*$. The condition for pure rolling (3.74) takes the form

$$-0,050 \text{ m} < q < 0,025 \text{ m}. \quad (3.108)$$

94   3. Accounting for dry friction in mechanisms

For $q = -0,050$ m the contact between the driving wheel and the rail weakens. If $q = 0,025$ m one observes slip of the driving wheel with a negative sign of the slip velocity $\varepsilon_2 = \text{sign}(\dot{x} - wr)$ where $\varepsilon_1 = 1$. Taking into account these values of ε_1 and ε_2, as well as the other values of the parameters in eq. (3.77), leads to the law of motion and the reaction forces

$$\dot{x}(t) = 0,745t + \dot{x}(0), \quad R_1 = 4966 \text{ N}, \quad R_2 = 14654 \text{ N}, \qquad (3.109)$$

where $\dot{x}(0)$ is given in the form of the initial condition.

Let us apply the developed strategy to speed-up in the positive direction. In this case the initial condition is given by eq. (3.80) where $wr = 1,5 \times 0,2 = 0,3$ ms^{-1}. In addition to this

$$\lambda = \sqrt{\frac{60 \times 10^3}{2 \times 10^3}} = 5,477 \text{ s}^{-1}. \qquad (3.110)$$

It follows from eqs. (3.107) and (3.110) that

$$\lambda q_2^+ = 0,136 < wr = 0,3$$

which means that the condition for rolling without slip (3.84) does not hold. Thus, speed-up consists of three stages: pure rolling, slip and pure rolling.

The time instants t_1 and t_2 can be calculated by means of eqs. (3.85) and (3.89), to give

$$t_1 = 0,085 \text{ s}, \quad t_2 = 0,445 \text{ s}. \qquad (3.111)$$

Under conditions (3.107)-(3.111) the equations for speed-up, eqs. (3.93) and (3.94) take the form

$$x(t) = \begin{cases} 0,3t & t \leq 0,085, \\ 0,027 + 0,0325(t - 0,085) + 0,372(t - 0,085)^2 & 0,085 \leq t \leq 0,445, \\ -0,048 + 0,3t & t \geq 0,445, \end{cases}$$

$$q(t) = \begin{cases} 0,055 \sin 5,477t & t \leq 0,085, \\ 0,025 & 0,085 \leq t \leq 0,445, \\ 0,025 \cos 5,477(t - 0,445) & t \geq 0,445. \end{cases}$$

3.4   Epicyclic mechanism with cylindric teeth of the involute gearing

Let r_1 and r_2 denote respectively the radii of the pitch circle of the fixed wheel 1 and the moving wheel 2, see Fig. 3.6. Furthermore, $a_i = r_i \cos \alpha$ ($i =$

3.4 Epicyclic mechanism with cylindric teeth of the involute gearing

1,2) denote the radii of the base circles of the wheels, α designates the gearing angle, m and $I = \frac{1}{2}mr_2^2$ are the mass and the moment of inertia about axis 0_2 of the wheel Π, respectively. The carrier $0_1 0_2$ is assumed to be massless, M_1 and M_2 denote the torques acting on the carrier and wheel Π, respectively, and μ is the coefficient of friction between the teeth. Finally, the gearing is provided by a single pair of teeth.

Required is the following: to derive the differential equation of motion and the expression for the reaction force in the form of eq. (2.59), to establish the possibility of paradoxical situations, to determine the relationships between the external torques M_1 and M_2 at rest, in the case of transition to motion and under the stationary motion and to express the efficiency in terms of the instantaneous value of the angle of rotation and the coefficient of friction.

3.4.1 Differential equation of motion, equations for the reaction force and the conditions for paradoxes.

In order to derive these equations, one needs the derivatives of the position vector of the point of contact of the teeth, the coefficients for the kinetic energy and the generalised forces. Let us begin by constructing equations for the tooth profile.

Let $0_1 xy$ and $0_2 \zeta \eta$ denote a fixed coordinate system and a system moving together with wheel Π, respectively, $\mathbf{i}_s$ and $\mathbf{i}'_s$ ($s = 1, 2$) denote the unit base vectors of these systems. The axes $0_1 x$ and $0_2 \zeta$ pass through the origins I_1 and I_2 of the involutes of the corresponding surface. According to the definition, see [63], [156], the profile of a tooth is described by the involute of the base circles. Then the position vector of the contact point T in these coordinate systems are respectively described by the equations

$$\overline{0_1 T} = \mathbf{r} = \mathbf{i}_1 a_1 (\cos u + u \sin u) - \mathbf{i}_2 a_1 (\sin u - u \cos u),$$
$$\overline{0_2 T} = \boldsymbol{\rho} = -\mathbf{i}'_1 a_2 (\cos p + p \sin p) + \mathbf{i}'_2 a_2 (\sin p - p \cos p). \quad (3.112)$$

Here

$$u = (\angle I_1 0_1 A_1) = \theta + \alpha, \quad p = (\angle I_2 0_2 A_2) = \frac{r_1 + r_2}{r_2} \tan \alpha - \frac{r_1}{r_2}(\theta + \alpha),$$
$$\mathbf{i}'_1 = \mathbf{i}'_1 \sin \varphi - \mathbf{i}_2 \cos \varphi, \quad \mathbf{i}'_2 = \mathbf{i}'_1 \cos \varphi + \mathbf{i}_2 \sin \varphi,$$
$$\varphi = \pi/2 - \theta + p - \alpha, \quad (3.113)$$

where φ denotes the angle between axes $0_1 x$ and $0_2 \eta$, and θ is the angle of rotation of the carrier. Quantity θ is taken as the generalised coordinate.

Under the process of motion of the epicyclic mechanism, the pitch circle of wheel Π (of radius r_2) rolls without slip on pitch circle 1, whereas rolling of the tooth of wheel Π on the tooth of wheel 1 is accompanied by slip. The quantity $d\mathbf{r}_T^0/dq$ appearing in eqs. (2.13) and (2.15) is the derivative of $\mathbf{r}_T^0$

96   3. Accounting for dry friction in mechanisms

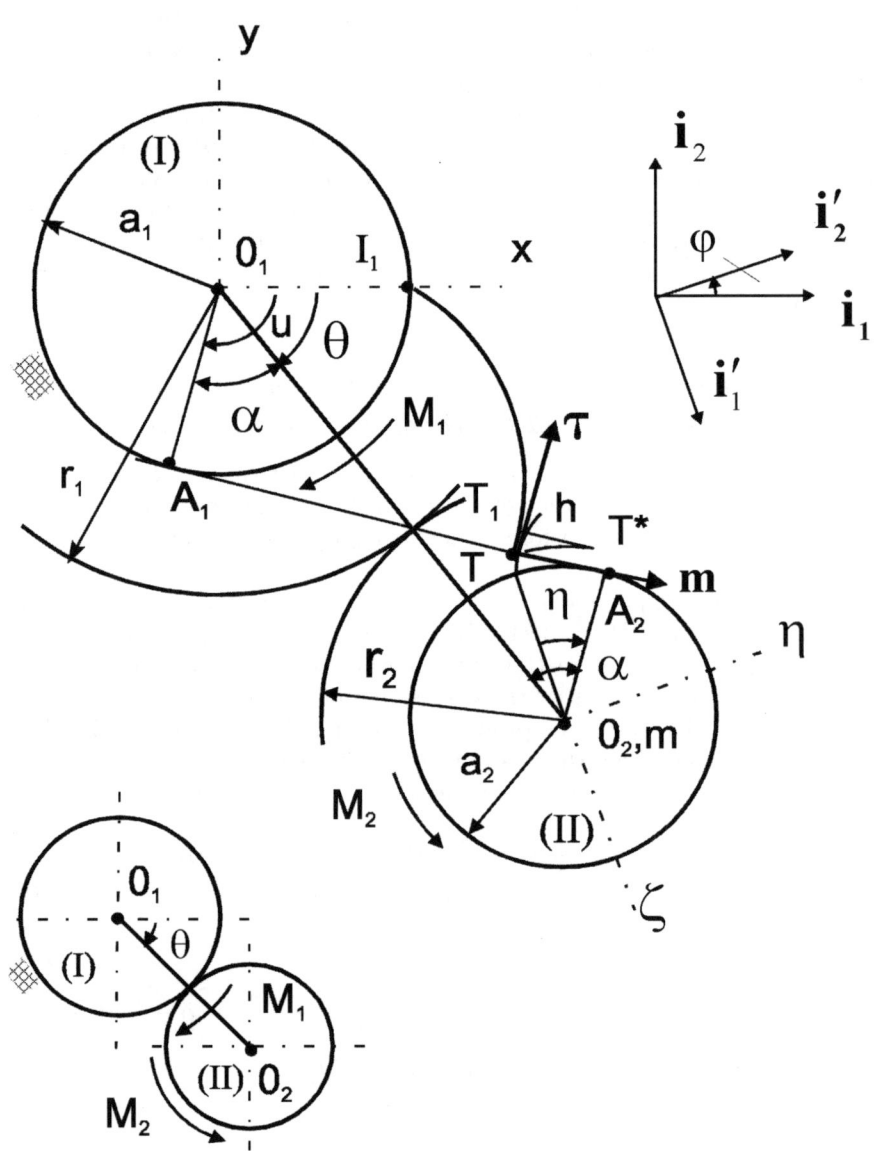

FIGURE 3.6.

3.4 Epicyclic mechanism with cylindric teeth of the involute gearing

with respect to θ, where $\mathbf{r}_T^0$ designates the position vector of the point fixed on Π which is the contact point at the considered time instant. The slip velocity $\mathbf{v}_T^0$ is the velocity of motion of this point. By virtue of eq. (2.26)

$$\mathbf{v}_T^0 = \mathbf{v}^0 + \boldsymbol{\omega} \times \boldsymbol{\rho} = \frac{d\mathbf{r}_T^0}{d\theta}\dot{\theta}. \tag{3.114}$$

Here $\mathbf{v}_0$ is the velocity of centre 0_2 and $\boldsymbol{\omega}$ is the angular velocity of wheel Π. The position vector of centre 0_2 is given by

$$\mathbf{r}_0 = (r_1 + r_2)(\mathbf{i}_1 \cos\theta - \mathbf{i}_2 \sin\theta), \tag{3.115}$$

hence,

$$\mathbf{v}_0 = \frac{d\mathbf{r}_0}{dt} = -(r_1 + r_2)(\mathbf{i}_1 \sin\theta + \mathbf{i}_2 \cos\theta)\dot{\theta}. \tag{3.116}$$

The angular velocity $\boldsymbol{\omega}$ can be determined by taking into account that the contact point of the pitch circles is an instantaneous centre of velocities. Then we have

$$\boldsymbol{\omega} = -\mathbf{i}_3 \frac{r_1 + r_2}{r_2}\dot{\theta}; \quad (\mathbf{i}_3' = \mathbf{i}_1' \times \mathbf{i}_2' = \mathbf{i}_1 \times \mathbf{i}_2). \tag{3.117}$$

The second term in the expression for the slip velocity is calculated by means of eqs. (3.112) and (3.117), to give

$$\boldsymbol{\omega} \times \boldsymbol{\rho} = a_2 \dot{\theta} \frac{r_1 + r_2}{r_2}\left[\mathbf{i}'(\sin p - \cos p) + \mathbf{i}'(\cos p + p\sin p)\right].$$

Some elementary transformations yield

$$\boldsymbol{\omega} \times \boldsymbol{\rho} = a_2 \dot{\theta} \frac{r_1 + r_2}{r_2}\{\mathbf{i}'[\sin(\theta + \alpha) - p\cos(\theta + \alpha)] + \\ \mathbf{i}'[(\cos(\theta + \alpha) + p\sin(\theta + \alpha)]\}. \tag{3.118}$$

Inserting (3.114) into the obtained expressions for $\mathbf{v}_0, \boldsymbol{\omega} \times \boldsymbol{\rho}$, eqs. (3.116) and (3.118), we arrive at the following formulae for the slip velocity $\mathbf{v}_T^0$ and the derivative $d\mathbf{r}_T^0/dq$

$$\mathbf{v}_T^0 = (r_1 + r_2)\dot{\theta}(\sin\alpha - p\cos\alpha)[\mathbf{i}_1 \cos(\theta + \alpha) - \mathbf{i}_2 \sin(\theta + \alpha)],$$

$$\frac{d\mathbf{r}_T^0}{d\theta} = (r_1 + r_2)(\sin\alpha - p\cos\alpha)[\mathbf{i}_1 \cos(\theta + \alpha) - \mathbf{i}_2 \sin(\theta + \alpha)],$$

$$\left|\frac{d\mathbf{r}_T^0}{d\theta}\right| = (r_1 + r_2)|\sin\alpha - p\cos\alpha|. \tag{3.119}$$

Let us note in passing that expressions (3.114) are a particular case of formulae (2.31) and (2.32) for the epicyclic mechanism.

98   3. Accounting for dry friction in mechanisms

Proceeding to consider the system with removed contact we note that the unit vector of the tangent $\boldsymbol{\tau}$ and the normal $\mathbf{m}$ to the tooth surface at the contact point T can be expressed as follows

$$\begin{aligned}\boldsymbol{\tau} &= -\mathbf{i}_1 \cos(\theta + \alpha) + \mathbf{i}_2 \sin(\theta + \alpha),\\ \mathbf{m} &= \mathbf{i}_1 \sin(\theta + \alpha) + \mathbf{i}_2 \cos(\theta + \alpha).\end{aligned} \qquad (3.120)$$

Let wheel II have a small additional clockwise rotation. The contact point T fixed on the wheel moves to position T^* where

$$\mathbf{r}_T^* = \mathbf{r}_T^0 + |\overline{TT^1}|\boldsymbol{\tau} + h\mathbf{m}. \qquad (3.121)$$

As one can see from Fig. 3.6 and eq. (2.33), the trajectory of the additional displacement TT^* is, strictly speaking, not orthogonal to the tooth surface. Since

$$|\overline{TT^1}| = h\cot(\angle T_1TT^*) = h\cot(\angle A_2T0_2) = ph$$

we obtain with the help of eq. (2.33) that

$$\mathbf{r}_T^* = \mathbf{r}_T^0 + ph\boldsymbol{\tau} + h\mathbf{m}. \qquad (3.122)$$

This yields the following expression for the derivative $(\partial \mathbf{r}_T^*/\partial h)_0$

$$\begin{aligned}\left(\frac{\partial \mathbf{r}_T^*}{\partial h}\right)_0 &= p\boldsymbol{\tau} + \mathbf{m} \qquad (3.123)\\ &= \mathbf{i}_1[\sin(\theta+\alpha) - p\cos(\theta+\alpha)] + \mathbf{i}_2[\cos(\theta+\alpha) - p\sin(\theta+\alpha)].\end{aligned}$$

Let us recall that in the case of a non-orthogonal additional displacement the equation of motion contains coefficient λ given by eq. (2.31). Substituting $d\mathbf{r}_T^0/dq$ and $(\partial \mathbf{r}_T^*/\partial h)_0$ from eqs. (2.31) and (2.35) into the expression for λ we have

$$\lambda = \frac{-p}{(r_1+r_2)(\sin\alpha - p\cos\alpha)}. \qquad (3.124)$$

Let us determine the kinetic energy of the system with the removed contact. Because of the additional displacement, the angle of rotation of wheel II gains the following increment

$$\varepsilon = \frac{|\overline{TT^*}|}{|\rho|} = \frac{h}{a_2}. \qquad (3.125)$$

The kinetic energy of the system is then given by

$$\begin{aligned}T &= \frac{1}{2}mv_0^2 + \frac{1}{4}mr_2^2(\omega + \dot{\varepsilon})^2\\ &= \frac{3}{4}m(r_1+r_2)^2\dot{\theta}^2 + \frac{mr^2(r_1+r_2)}{2a_2}\dot{\theta}\dot{h} + \frac{mr_2^2}{4a_2^2}\dot{h}^2. \qquad (3.126)\end{aligned}$$

3.4 Epicyclic mechanism with cylindric teeth of the involute gearing

Here v_0 and ω are given by eqs. (3.116) and (3.117). Using eq. (3.126) we obtain

$$A = A_{11} = \frac{3}{2}m(r_1 + r_2)^2, \quad A_{12} = \frac{mr_2(r_1 + r_2)}{2a_2},$$

$$\frac{\partial A_{11}}{\partial \theta} = \frac{\partial A_{12}}{\partial \theta} = \frac{\partial A_{11}}{\partial h} = 0. \tag{3.127}$$

It is clear that the generalised active force Q_1 corresponding to coordinate θ is as follows

$$Q_1 = M_1 - \frac{r_1 + r_2}{r_2} M_2. \tag{3.128}$$

The generalised active force Q_2^* is determined under the additional rotation (3.125) and has the form

$$Q_2^* = -M_2 \frac{d\varepsilon}{dh} = -\frac{M_2}{a_2}. \tag{3.129}$$

Here a minus sign implies that the positive direction of torque M_2 is prescribed as being opposite to the direction of the additional rotation. If the additional displacement were prescribed as being orthogonal to the tooth surface, the corresponding generalised force Q_2 would be calculated by means of eqs. (2.37), (3.124), (3.128) and (3.129) and have the form

$$Q_2 = Q_2^* - \lambda Q_1 = \frac{1}{\sin\alpha - p\cos\alpha}\left(\frac{pM_1}{r_1 + r_2} - \frac{M_2}{r_2}\tan\alpha\right). \tag{3.130}$$

Now we can apply formulae (2.60)-(2.63). Considering the parameters from eqs. (3.119), (3.123), (3.124) and (3.128)-(3.130) we obtain the following

$$L = \frac{1}{3}(\tan\alpha + 2p)\,\text{sign}(\tan\alpha - p), \quad R_0 = \frac{M_1}{3(a_1 + a_2)} + \frac{2M_2}{3a_2},$$

$$H = 0, \quad G = \frac{2}{3m(r_1 + r_2)}\left[M_1 - \frac{r_1 + r_2}{r_2}M_2 + \right.$$

$$\left. \varepsilon_1\varepsilon_2\mu(pM_1 - \frac{r_1 + r_2}{a_2}\sin\alpha M_2)\,\text{sign}(\tan\alpha - p)\right]. \tag{3.131}$$

For the sake of convenience we replace the generalised coordinate θ by coordinate p with the help of relationship (3.113). Using eq. (2.59) we arrive at the differential equation of motion and the expression for the reaction force in the form

$$\left[1+\varepsilon_1\varepsilon_2\varepsilon_4\frac{\mu}{3}(\tan\alpha+2p)\right]\frac{a_2}{a_1}\ddot{p}=\frac{-2}{3m(r_1+r_2)}$$
$$\left[M_1-\frac{r_1+r_2}{r_2}M_2+\varepsilon_1\varepsilon_2\varepsilon_4\mu(pM_1-\frac{r_1+r_2}{r_2}\tan\alpha M_2)\right]$$
$$\left[1+\varepsilon_1\varepsilon_2\varepsilon_4\frac{\mu}{3}(\tan\alpha+2p)\right]R=M_1[3(a_1+a_2)]^{-1}+2M_2(3a_2)^{-1},$$
(3.132)

where $\varepsilon_1=\text{sign}\,R$, $\varepsilon_2=\text{sign}\,\dot{\theta}=-\text{sign}\,\dot{p}$, $\varepsilon_4=\text{sign}\,(\tan\alpha-p)$. The values of ε_2 and ε_4 are determined by the initial conditions, whilst the value of ε_1 remains unknown and must be determined by means of the criterion for Painlevé's paradoxes.

By Theorem 1 the solution of system (3.132) and thus ε_1 exists and is unique if and only if

$$\mu|L|=\frac{\mu}{3}(\tan\alpha+2p)<1.$$

This inequality holds for any value of p when

$$\mu<\mu_1=\frac{3}{\tan\alpha+2p_{max}}. \qquad (3.133)$$

It is shown below that in the regime of absence of Painlevé's paradoxes the transition from the state of rest to a motion is possible only if $\mu<\mu_2$, where $\mu_2<\mu_1$. Therefore, for the system under consideration, the condition of transition to motion is stronger than that of the absence of paradoxes. The forthcoming analysis is carried out under the assumption that inequality (3.133) holds. Then $\varepsilon_1=\varepsilon_0=\text{sign}\,R_0$. In particular, for $M_1>0, M_2>0$ we have $\varepsilon_1=1$. The value of the reaction is determined and the law of motion is obtained by quadratures.

3.4.2 Relationships between the torques at rest and in the transition to motion

These relationships are determined by conditions (2.75) and (2.76) of Theorem 2 under which the quantities $d\mathbf{r}_T^0/dq$, Q_1 and Q_2 are given by eqs. (3.119), (3.128) and (3.130). The system begins to move from rest in the direction

$$\varepsilon_2=\text{sign}\left(M_1-\frac{r_1+r_2}{r_2}M_2\right) \qquad (3.134)$$

if

$$\left|M_1-\frac{r_1+r_2}{r_2}M_2\right|>\mu\left|pM_1-\frac{r_1+r_2}{r_2}\tan\alpha M_2\right|, \qquad (3.135)$$

3.4 Epicyclic mechanism with cylindric teeth of the involute gearing 101

otherwise it remains at rest.

Let us transform the absolute values on the both sides of inequality (3.135). By virtue of eq. (3.134), for motion in the positive direction we have

$$\varepsilon_2 = 1 \quad \text{or} \quad M_1 > \frac{r_1 + r_2}{r_2} M_2. \tag{3.136}$$

Then eq. (3.135) yields

$$M_1 > \frac{r_1 + r_2}{r_2} \frac{1 + \mu \tan \alpha}{1 + \mu p} M_2 \quad \text{for} \quad M_1 \leq \frac{r_1 + r_2}{r_2} \frac{\tan \alpha}{p} M_2, \tag{3.137}$$

$$M_1 > \frac{r_1 + r_2}{r_2} \frac{1 - \mu \tan \alpha}{1 - \mu p} \quad \text{or} \quad M_1 \geq \frac{r_1 + r_2}{r_2} \frac{\tan \alpha}{p} M_2. \tag{3.138}$$

Solving eqs. (3.136) and (3.137) results in the following relationships

$$\frac{r_1 + r_2}{r_2} \frac{1 + \mu \tan \alpha}{1 + \mu p} M_2 < M_1 < \frac{r_1 + r_2}{r_1} \frac{\tan \alpha}{p} M_2 \quad \text{for} \quad p \leq \tan \alpha. \tag{3.139}$$

For $p > \tan \alpha$ inequalities (3.136) and (3.137) can not be satisfied simultaneously. In order to solve eqs. (3.136) and (3.138) it is necessary to compare the right hand sides of these inequalities. One can see that

$$\frac{\tan \alpha}{p} \geq 1 \geq \frac{1 - \mu \tan \alpha}{1 - \mu p} \quad \text{for} \quad p \leq \tan \alpha,$$

$$\frac{1 - \mu \tan \alpha}{1 - \mu p} \geq 1 \geq \frac{\tan \alpha}{p} \quad \text{for} \quad \tan \alpha \leq p < \mu^{-1}.$$

Hence inequalities (3.136) and (3.138) are simultaneously valid if

$$M_1 > \frac{r_1 + r_2}{r_2} \cdot \frac{\tan \alpha}{p} M_2 \quad \text{for} \quad p \leq \tan \alpha, \tag{3.140}$$

$$M_1 > \frac{r_1 + r_2}{r_2} \cdot \frac{1 - \mu \tan \alpha}{1 - \mu p} M_2 \quad \text{for} \quad \tan \alpha \leq p < \mu^{-1}. \tag{3.141}$$

Inequalities (3.139)-(3.141) can be set in the form of the following condition for transition in the positive direction

$$\begin{aligned} M_1 &> \frac{r_1 + r_2}{r_2} \frac{1 + \mu \tan \alpha}{1 + \mu p} M_2 \quad \text{for} \quad p \leq \tan \alpha, \\ M_1 &> \frac{r_1 + r_2}{r_2} \frac{1 - \mu \tan \alpha}{1 - \mu p} M_2 \quad \text{for} \quad p \geq \tan \alpha. \end{aligned} \tag{3.142}$$

The condition for transition to motion in the negative direction is obtained by analogy and is given by

$$M_1 < \frac{r_1 + r_2}{r_2} \frac{1 - \mu \tan \alpha}{1 - \mu p} M_2 \quad \text{for} \quad p \leq \tan \alpha,$$
$$M_1 > \frac{r_1 + r_2}{r_2} \frac{1 + \mu \tan \alpha}{1 + \mu p} M_2 \quad \text{for} \quad p \geq \tan \alpha. \quad (3.143)$$

The condition for maintaining the initial state of rest is obtained when both conditions (3.142) and (3.143) are not fulfilled, i.e. when

$$\frac{r_1 + r_2}{r_2} \frac{1 - \mu \tan \alpha}{1 - \mu p} M_2 \leq M_1 \leq \frac{r_1 + r_2}{r_2} \frac{1 + \mu \tan \alpha}{1 + \mu p} M_2 \quad \text{for} \quad p \leq \tan \alpha,$$
$$\frac{r_1 + r_2}{r_2} \frac{1 + \mu \tan \alpha}{1 + \mu p} M_2 \leq M_1 \leq \frac{r_1 + r_2}{r_2} \frac{1 - \mu \tan \alpha}{1 - \mu p} M_2 \quad \text{for} \quad p \geq \tan \alpha.$$
$$(3.144)$$

Along with the condition of absence of the paradoxes (3.133) let us notice another restriction imposed on p and μ. Let the value M_2 be finite. Then $M_1 \to +\infty$ for $\mu p \to 1 - 0$ due to the second condition in eq. (3.142) while $M_1 \to -\infty$ for $\mu p \to 1 + 0$ due to the first condition in eq. (3.143). This means that the transfer of motion is feasible only for $\mu p \neq 1$. On the other hand, as it is known from the theory of mechanisms and machines [63], [156], the lower limit $p_{\min}$ of the polar angle p is less than μ^{-1}. Hence, the condition of transfer $\mu p \neq 1$ reduces to the following one

$$p < \mu^{-1} \quad \text{or} \quad \mu < \mu_2 = 1/p_{\max}. \quad (3.145)$$

Indeed, if we assumed $\mu > \mu_2$, then there would exist a time instant in the motion when $\mu p = 1$ and thus the transfer would be impossible.

Condition (3.145) is stronger than condition (3.133) since

$$\mu_1 - \mu_2 = \frac{p_{\max} - \tan \alpha}{(\tan \alpha + 2p_{\max})p_{\max}} > 0.$$

Hence, for the considered epicyclic mechanism Painlevé's paradoxes do not appear when the condition for transfer holds.

Finally, by using notation $\varepsilon_4 = \text{sign}(\tan \alpha - p)$ and condition for maintaining the state of rest $\varepsilon_3 = -\varepsilon_2$ and the condition of the transition to motion $\varepsilon_3 = \varepsilon_2 = \pm 1$ introduced in Section 2.4, we can represent relationships (3.142)-(3.144) in the following form

$$\frac{r_2}{r_1 + r_2} M_1 < \frac{1 - \varepsilon_4 \tan \alpha}{1 - \varepsilon_4 \mu p} M_2, \quad \varepsilon_2 = \varepsilon_3 = -1,$$
$$\frac{1 - \varepsilon_4 \mu \tan \alpha}{1 - \varepsilon_4 \mu p} M_2 \leq \frac{r_2}{r_1 + r_2} M_1 \leq \frac{1 + \varepsilon_4 \mu \tan \alpha}{1 + \varepsilon_4 \mu p} M_2, \quad \varepsilon_2 = -\varepsilon_3,$$
$$\frac{1 + \varepsilon_4 \mu \tan \alpha}{1 + \varepsilon_4 \mu p} M_2 < \frac{r_2}{r_1 + r_2} M_1, \quad \varepsilon_2 = \varepsilon_3 = 1. \quad (3.146)$$

This is a particular form of Theorem 2 for the considered mechanism.

3.4.3 Regime of uniform motion

Under a uniform motion the relationships between M_1 and M_2 are determined by substituting $\ddot{p} = 0$, $\varepsilon_1 = 1$ into the first equation in (3.127). One obtains the first and third expressions in (3.146) where the inequality signs are replaced by equality signs

$$\frac{r_2}{r_1 + r_2} M_1 = \frac{1 + \varepsilon_2 \varepsilon_4 \mu \tan \alpha}{1 + \varepsilon_2 \varepsilon_4 \mu p} M_2. \qquad (3.147)$$

Substituting this result into the second equation in eq. (3.132) and taking into account that $\varepsilon_1 = \text{sign } R_0$ in the case of absence of paradoxes, we arrive at the following formula for the normal reaction force

$$R = \frac{M_2}{a_2(1 + \varepsilon_2 \varepsilon_4 \mu p)} = \frac{M_1}{(a_1 + a_2)(1 + \varepsilon_2 \varepsilon_4 \mu \tan \alpha)}. \qquad (3.148)$$

It is now easy to express the instantaneous value of the efficiency in terms of the generalised coordinate p. It is known, see e.g. [63], [156], that if M_1 is a driving torque then the motion takes place in the positive direction $\varepsilon_2 = 1$ and the efficiency is given by

$$\eta_+ = \frac{r_1 + r_2}{r_2} \frac{M_2}{M_1}.$$

If M_2 is a driving torque, then $\varepsilon_2 = -1$ and the efficiency is

$$\eta_- = \frac{r_1 + r_2}{r_2} \frac{M_1}{M_2}.$$

By means of eq. (3.147) we arrive at the following equations for the efficiency of epicyclic mechanism

$$\eta_+ = \frac{1 + \varepsilon_4 \mu p}{1 + \varepsilon_4 \mu \tan \alpha}, \quad \eta_- = \frac{1 - \varepsilon_4 \mu \tan \alpha}{1 - \varepsilon_4 \mu p}. \qquad (3.149)$$

As follows from these formulae one can see that in particular the efficiency is equal to unity ($\eta_+ = \eta_- = 1$) at the gearing pole $p = \tan \alpha$. Indeed, as expression (3.119) suggests the slip velocity $\mathbf{v}_T^0$ vanishes for $p = \tan \alpha$ and the systems becomes an ideal one for a single time instant. This fact is well known in the theory of mechanisms and machines, see [63], [156].

3.5 Gear transmission with immovable rotation axes

Let r_1 and r_2 denote respectively the radii of the pitch circle of wheels 1 and 2, see Fig. 3.7. The centres of the wheels are assumed to be fixed in space. Furthermore, $a_i = r_i \cos\alpha$ $(i = 1, 2)$ denotes the radii of the base circles of the wheels, α designates the gearing angle, J_1 and J_2 are the moments of inertia about their rotation axes O_1 and O_2, and M_1 and M_2 are the external torques acting on wheels 1 and 2, respectively. The gearing is provided by a single pair of teeth.

3.5.1 Differential equations of motion and the condition for absence of paradoxes

Figure 3.7 shows that if the moving axes O_1xy and $O_2\xi\zeta$ are related to wheels 1 and 2, respectively, then we arrive at the model of the previous problem. In this case the quantities $d\mathbf{r}_T^0/d\theta$ and $\mathbf{v}_T^0$ given by eq. (3.119) are the local derivatives described in the O_1xy axes and the position-vector of the contact point fixed on wheel 2. Formula (3.117) provides us with an expression for the angular velocity of wheel 2 with respect to system O_1xy. In accordance with the rule of Subsection 3.4 for the positive direction of rotation of the carrier, in the system considered the angular velocity of rotation $\dot\theta$ of wheel 1 is viewed as being positive if the wheel rotates counterclockwise. The same rule is also true for torque M_1.

Thus, the kinematic expressions (3.112)-(3.124) derived for the epicyclic mechanism can be applied directly to the case of the moving axes of the wheel rotation. In order to construct the differential equation of motion and the expression for the reaction force in the form of eq. (2.59) it is necessary to obtain the kinetic energy and the generalised active forces.

Similar to the case of the epicyclic mechanism, under an additional rotation of wheel 2 through angle ε the redundant coordinate h is determined as follows

$$h = \varepsilon a_2. \qquad (3.150)$$

The kinetic energy of the system is then given by

$$T = \frac{1}{2}J_1\dot\theta^2 + \frac{1}{2}J_2\left(\frac{r_2}{r_2}\dot\theta + \dot\theta\right)^2 = \frac{1}{2}J_1\dot\theta^2 + \frac{1}{2}J_2\left(\frac{r_1^2}{r_2^2}\dot\theta^2 + \frac{2r_1}{r_2 a_2}\dot\theta\dot h + \frac{\dot h^2}{a_2^2}\right), \qquad (3.151)$$

thus

$$A_{11} = J_1 + \frac{r_1^2}{r_2^2}J_2, \quad A_{12}^* = \frac{r_1 J_2}{r_2^2 \cos\alpha}. \qquad (3.152)$$

The generalised forces Q_1 and Q_2^* are

$$Q_1 = M_1 - \frac{r_1}{r_2}M_2, \quad Q_2^* = -\frac{M_2}{a_2}. \qquad (3.153)$$

3.5 Gear transmission with immovable rotation axes

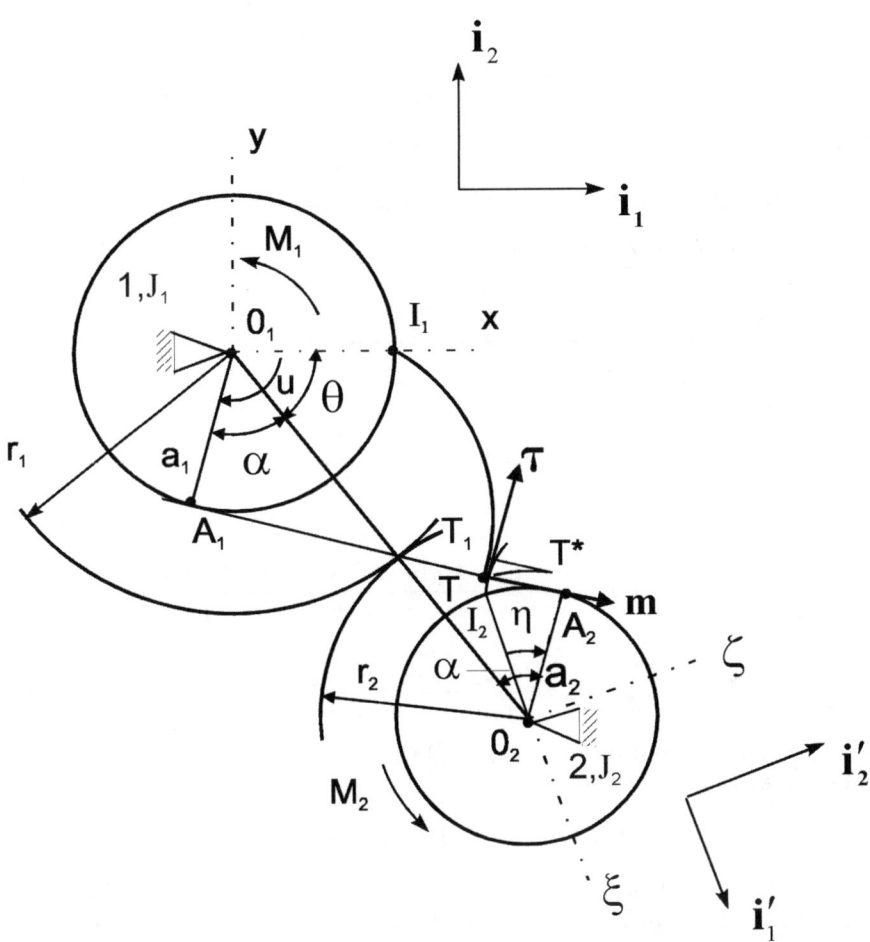

FIGURE 3.7.

The generalised force Q_2, which would appear under an additional displacement along the normal to the surface of the tooth, is calculated by means of eq. (2.22) and taking into account eqs. (2.86) and the following equality

$$Q_2 = Q_2^* - \lambda Q_1 = -\frac{M_2}{a_2} + \frac{p(r_2 M_1 - r_1 M_2)}{r_2(r_1 + r_2)(\sin\alpha - \cos\alpha)}.$$

Further, in order to calculate L, H, G, and R_0 by means of eqs. (2.60)-(2.63) it is necessary to use the kinematic relationships (3.112)-(3.124) and expressions (2.3) and (2.4). The result is as follows

$$L = \frac{r_2(r_1 + r_2)J_1 \tan\alpha + r_1(r_1 J_2 - r_2 J_1)u}{r_1^2 J_2 + r_2^2 J_1} \operatorname{sign}(\tan\alpha - p) \quad (3.154)$$

$$= \frac{r_1(r_1 + r_2)J_2 \tan\alpha + r_2(r_2 J_1 - r_1 J_2)p}{r_1^2 J_2 + r_2^2 J_1} \operatorname{sign}(\tan\alpha - p),$$

$$H = 0, \quad (3.155)$$

$$G = \frac{r_2^2}{r_2^2 J_1 + r_1^2 J_2} \times \quad (3.156)$$

$$\left[M_1 - \frac{r_1}{r_2} M_2 + \varepsilon_1 \varepsilon_2 \mu \frac{(M_1 + M_2)pr_2 \cos\alpha - M_2(r_1 + r_2)\sin\alpha}{r_2 \cos\alpha \operatorname{sign}(\tan\alpha - p)} \right],$$

$$R_0 = \frac{r_1 J_2 M_1 + r_2 J_1 M_2}{(r_1^2 J_2 + r_2^2 J_1)\cos\alpha}. \quad (3.157)$$

The differential equation of motion and the expression for the reaction force have the form of eq. (2.59), i.e.

$$(1 + \varepsilon_1 \varepsilon_2 \mu L)\ddot\theta = G, \qquad (1 + \varepsilon_1 \varepsilon_2 \mu L)R = R_0. \quad (3.158)$$

In addition to this, by using eq. (3.154) the condition for the existence and uniqueness of $\mu|L| < 1$ takes the form

$$-1 < \mu \frac{r_1(r_1 + r_2)J_2 \tan\alpha + r_2(r_2 J_1 - r_1 J_2)p}{r_1^2 J_2 + r_1^2 J_1} < 1. \quad (3.159)$$

3.5.2 Regime of uniform motion

Let condition (3.159) be satisfied. Then the relationship between the torques M_1 and M_2 in the regime of uniform motion can be established by substituting $\ddot\theta = 0$ and $\varepsilon_1 = \varepsilon_0 = \operatorname{sign} R_0$ into the first of the expressions in eq. (3.158). The result is

$$\frac{M_1}{M_2} = \frac{r_1 + \varepsilon_0 \varepsilon_2 \mu[(r_1 + r_2)\tan\alpha - r_2 p]}{r_2(1 + \varepsilon_0 \varepsilon_2 \varepsilon_4 \mu p)} = \frac{r_1(1 + \varepsilon_0 \varepsilon_2 \varepsilon_4 \mu u)}{r_2(1 + \varepsilon_0 \varepsilon_2 \varepsilon_4 \mu p)}, \quad (3.160)$$

where $\varepsilon_4 = \text{sign}(\tan\alpha - p)$.

Let us use eq. (3.160) to obtain the restriction imposed on u and p under which the motion for $M_1 \neq 0$ and $M_2 \neq 0$ is feasible. Let $\varepsilon_0\varepsilon_2\varepsilon_4 = -1$ which, for any values of ε_0 and ε_4, can always be achieved by a choice of an appropriate direction of motion, i.e. by means of an appropriate sign of velocity $\varepsilon_2 = \text{sign}\,\dot\theta$. The numerator of the right fraction vanishes as $\mu u \to 1$ whereas its denominator vanishes as $\mu p \to 1$. Hence, in order to ensure that neither M_1 nor M_2 becomes unbounded for finite values of the other quantity it is necessary and sufficient that $\mu p \neq 1$ and $\mu u \neq 1$. Since $\mu p_{\min} \ll 1$, $\mu u_{\min} \ll 1$ and the polar angles p and u vary continuously, the above conditions reduce to the following

$$\mu p_{\max} < 1, \quad \mu u_{\max} < 1, \quad p_{\min} > \frac{\mu(r_1+r_2)\tan\alpha - r_1}{\mu r_2}. \tag{3.161}$$

In addition to this because $p_{\max} > \tan\alpha$ and $u_{\max} > \tan\alpha$, [156], eq. (3.161) yields the following inequality

$$\mu \tan\alpha < 1. \tag{3.162}$$

The condition for absence of paradoxes (3.159) and the condition of gearing (3.161) are already at our disposal. Let us prove that condition (3.159) is automatically fulfilled when condition (3.161) holds, i.e. (3.159) follows from (3.161). To this end, we rewrite (3.159) in the form

$$\frac{-r_1^2 J_2 - r_2^2 J_1 - \mu r_1(r_1+r_2)J_2\tan\alpha}{\mu r_2(r_2 J_1 - r_1 J_2)} < p <$$
$$\frac{r_1^2 J_2 + r_2^2 J_1 - \mu r_1(r_1+r_2)J_2\tan\alpha}{\mu r_2(r_2 J_1 - r_1 J_2)} \quad \text{for} \quad r_2 J_1 > r_1 J_2, \tag{3.163}$$

$$\frac{r_1^2 J_2 + r_2^2 J_1 - \mu r_1(r_1+r_2)J_2\tan\alpha}{\mu r_2(r_2 J_1 - r_1 J_2)} < p <$$
$$\frac{-r_1^2 J_2 - r_2^2 J_1 - \mu r_1(r_1+r_2)J_2\tan\alpha}{\mu r_2(r_2 J_1 - r_1 J_2)} \quad \text{for} \quad r_2 J_1 < r_1 J_2, \tag{3.164}$$

$$\mu \tan\alpha < 1 \quad \text{for} \quad r_2 J_1 = r_1 J_2 \tag{3.165}$$

and take into account the relationships

$$p_{min} > 0, \quad p_{max} < \frac{r_1+r_2}{r_2}\tan\alpha \tag{3.166}$$

which are easily obtained with the help of eq. (3.113) and Fig. 3.6. One can immediately see that condition (3.165) is fulfilled due to eq. (3.162).

108   3. Accounting for dry friction in mechanisms

The left hand side of eq. (3.163) is negative, whereas, due to eq. (3.166), angle p is always positive during the motion. For this reason, the left inequality (3.163) is satisfied. In order to prove the right inequality in (3.163) we equate its right hand side to $p_{\max}$ and take into account eqs. (3.162) and (3.166), i.e.

$$p_{\max} - \frac{r_1^2 J_2 + r_2^2 J_1 - \mu r_1 (r_1 + r_2) J_2 \tan \alpha}{\mu r_2 (r_2 J_1 - r_1 J_2)}$$
$$\leq \frac{\mu r_2 (r_1 + r_2) J_1 \tan \alpha - r_1^2 J_2 - r_2^2 J_1}{\mu r_2 (r_2 J_1 - r_1 J_2)}$$
$$< \frac{r_2 (r_1 + r_2) J_1 - r_1^2 J_2 - r_2^2 J_1}{\mu r_2 (r_2 J_1 - r_1 J_2)} < \frac{r_1^2 J_2 - r_1^2 J_2}{\mu r_2 (r_2 J_1 - r_1 J_2)} = 0$$

which is required.

To prove the left inequality in (3.164), we make use of eqs. (3.161) and (3.162) to write down the following inequality

$$p_{\min} - \frac{r_1^2 J_2 + r_2^2 J_1 - \mu r_1 (r_1 + r_2) J_2 \tan \alpha}{\mu r_2 (r_2 J_1 - r_1 J_2)} >$$
$$\frac{r_2 (r_1 + r_2) J_1 (\mu \tan \alpha - 1)}{\mu r_2 (r_2 J_1 - r_1 J_2)} > 0.$$

One can see that this inequality holds for all $p > p_{\min}$.

The first inequality (3.164) also holds by virtue of the relationship

$$p_{\max} + \frac{r_1^2 J_2 + r_2^2 J_1 + \mu r_1 (r_1 + r_2) J_2 \tan \alpha}{\mu r_2 (r_2 J_1 - r_1 J_2)} <$$
$$\frac{r_1 r_2 J_2 - 2 r_2^2 J_1 - \mu r_1 (r_1 + r_2) J_2 \tan \alpha}{\mu r_2 (r_1 J_2 - r_2 J_1)} <$$
$$-\frac{r_2^2 J_1 + \mu r_1 (r_1 + r_2) J_2 \tan \alpha}{\mu r_2 (r_1 J_2 - r_2 J_1)} < 0.$$

Thus, when the condition of gearing (3.161) holds the condition of absence of paradoxes, eq. (3.159) or eq. (3.164), is also fulfilled.

Let us proceed to determination of the normal reaction in the regime of uniform motion. Substituting the relationship between the moments (3.160) into eq. (3.157) we obtain the following expression for R_0 in terms of M_2

$$R_0 = \frac{1 + \varepsilon_0 \varepsilon_2 \mu L}{a_2 (1 + \varepsilon_0 \varepsilon_2 \mu p)} M_2 . \tag{3.167}$$

Let $M_2 > 0$. Then, by virtue of the condition for absence of paradoxes $\mu |L| < 1$ and the condition of gearing (3.161), we have

$$\varepsilon_1 = \varepsilon_0 = \operatorname{sign} R_0 = 1 . \tag{3.168}$$

Inserting R_0 and ε_1 given by eqs. (3.167) and (3.168), into the second relationship (3.158) yields the value of the normal reaction in the regime of uniform motion

$$R = \frac{M_2}{a_2(1+\varepsilon_2\varepsilon_4\mu p)} = \frac{M_1}{a_1(1+\varepsilon_2\varepsilon_4\mu u)}. \qquad (3.169)$$

This expression coincides with the formula derived by Kolchin in [63].

Following [63] we can write the instantaneous loss factor due to tooth friction in the form

$$\alpha_\mu = \frac{\mu R |\mathbf{v}_T^0|}{M_1 \dot\theta},$$

where $\mathbf{v}_T^0$ and R are taken from eqs. (3.119) and (3.169). The result is as follows

$$\alpha_\mu = \frac{\varepsilon_4\mu(r_1+r_2)(\sin\alpha - p\cos\alpha)}{a_1(1+\varepsilon_2\varepsilon_4\mu u)}$$
$$= \frac{\varepsilon_4\mu(r_1+r_2)(\sin\alpha - p\cos\alpha)}{a_1 + \varepsilon_2\varepsilon_4\mu[(r_1+r_2)\sin\alpha - r_2 p\cos\alpha]}. \qquad (3.170)$$

Thus the efficiency is given by

$$\eta = 1 - \alpha_\mu = \frac{1+\varepsilon_2\varepsilon_4\mu p}{1+\varepsilon_2\varepsilon_4\mu u}. \qquad (3.171)$$

Clearly, the same expression for the efficiency is obtained if one substitutes relationships (3.122) and (3.168) into the following formula

$$\eta = \frac{r_1 M_2}{r_2 M_1}.$$

Let us notice in passing that $u = p = \tan\alpha$, $\eta = 1$ at the pole of gearing.

3.5.3 Transition from the state of rest to motion

The condition for maintaining the state of rest and the condition for transition to motion are given by eqs. (3.113) and (2.76), where the quantities $d\mathbf{r}_T^0/d\theta, Q_1$ and Q_2 are taken from eqs. (3.119), (3.153) and (3.154). The system begins to move in the direction

$$\varepsilon_2 = \mathrm{sign}\,(M_1 - \frac{r_1}{r_2}M_2) \qquad (3.172)$$

provided that

$$\left| M_1 - \frac{r_1}{r_2}M_2 \right| > \mu \left| \frac{(M_1+M_2)pr_2\cos\alpha - M_2(r_1+r_2)\sin\alpha}{\varepsilon_4 r_2 \cos\alpha} \right|, \qquad (3.173)$$

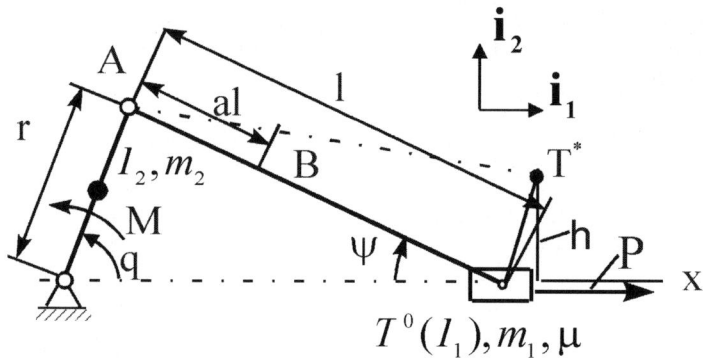

FIGURE 3.8.

otherwise it remains at rest. Solving the system of equations (3.172) and (3.173) allows us to obtain the following result

$$\frac{r_2}{r_1}M_1 < \frac{1-\varepsilon_4\mu u}{1-\varepsilon_4\mu p}M_2, \quad \varepsilon_2 = \varepsilon_3 = -1,$$

$$\frac{1-\varepsilon_4\mu u}{1-\varepsilon_4\mu p}M_2 \leq \frac{r_2}{r_1}M_1 \leq \frac{1+\varepsilon_4\mu u}{1+\varepsilon_4\mu p}M_2, \quad \varepsilon_2 = -\varepsilon_3,$$

$$\frac{1+\varepsilon_4\mu u}{1+\varepsilon_4\mu p}M_2 < \frac{r_2}{r_1}M_1, \quad \varepsilon_2 = \varepsilon_3 = 1. \qquad (3.174)$$

The first and third expressions represent conditions for transition to motion into the negative ($\varepsilon_2 = \varepsilon_3 = -1$) and positive ($\varepsilon_2 = \varepsilon_3 = 1$) directions, respectively. The second relationship expresses the condition for staying at rest.

3.6 Crank mechanism

The schematic of this mechanism is shown in Fig. 3.8. It consists of slider T (particle I_1) of mass m_1, a crank OA of length r whose mass m_2 is uniformly distributed over the length and a massless coupler AI_1 of length l ($l > r$). The crank and the slider are subjected to a torque M and a tangential force P, respectively.

3.6.1 Equation of motion and reaction force

The angle between the crank OA and axis Ox in the initial position is taken as the generalised coordinate q. The position vectors of the slider T

3.6 Crank mechanism

and the centre of mass I_2 of the crank can be expressed as follows

$$\mathbf{r}_T^0 = \mathbf{r}_1^0 = \left(r\cos q + \sqrt{l^2 - r^2\sin^2 q}\right)\mathbf{i}_1, \quad \mathbf{r}_2^0 = \frac{r}{2}(\cos q\,\mathbf{i}_1 + \sin\mathbf{i}_2). \tag{3.175}$$

Under an additional displacement of the slider T along the circle with centre A from point T^0 to position T^* we have

$$\mathbf{r}_T^* = \mathbf{r}_1^* = \left[r\cos q + \sqrt{l^2 - (r\sin q - h)^2}\right]\mathbf{i}_1 + h\,\mathbf{i}_2, \quad \mathbf{r}_2^* = \mathbf{r}_2^0. \tag{3.176}$$

Let us recall that a superscript $*$ implies the case when the additional displacement is not prescribed along the normal to the guide.

Differentiating eqs. (3.175) and (3.176) with respect to q and h yields the following expressions

$$\frac{d\mathbf{r}_T^0}{dq} = -r\sin q\left(1 + \frac{r\cos q}{l\cos\Psi}\right)\mathbf{i}_1, \quad \left(\frac{\partial\mathbf{r}_T^*}{\partial h}\right)_0 = \frac{r\sin q}{l\cos\Psi}\mathbf{i}_1 + \mathbf{i}_2,$$

$$\frac{d\mathbf{r}_2^0}{dq} = -\frac{r}{2}(\mathbf{i}_1\sin q - \mathbf{i}_2\cos q), \quad \left(\frac{\partial\mathbf{r}_2^*}{\partial h}\right)_0 = 0, \tag{3.177}$$

where Ψ denotes angle $\angle OT^0 A$. It is clear that

$$l\cos\Psi = \sqrt{l^2 - r^2\sin^2 q}. \tag{3.178}$$

The coefficients of equations of the type of (2.13) are determined by using the expressions of Section 2.1 together with formulae for the derivatives of vectors (2.13) and (2.13). The result is as follows

$$\lambda = -\frac{1}{l\cos\Psi + r\cos q}, \tag{3.179}$$

$$Q_1 = M - Pr\sin q\left(1 + \frac{r\cos q}{l\cos\Psi}\right), \quad Q_2 = \frac{Pr\sin q}{l\cos\Psi}, \tag{3.180}$$

$$S_1 = -\varepsilon_1\varepsilon_\mu\frac{r|\sin q|}{l\cos\Psi}(l\cos\Psi + r\cos q)P, \quad S_2^* = \left(1 - \varepsilon_1\varepsilon_\mu\frac{r|\sin q|}{l\cos\Psi}\right)P, \tag{3.181}$$

$$\begin{aligned} A &= m_1 r^2 \sin^2 q(l\cos\Psi + r\cos q)^2 l^{-2}\cos^{-2}\Psi + m_2 r^2/3, \\ A_{12}^{*0} &= -m_1 r^2 \sin^2 q(l\cos\Psi + r\cos q)l^{-2}\cos^{-2}\Psi, \end{aligned} \tag{3.182}$$

112   3. Accounting for dry friction in mechanisms

$$\frac{dA}{dq} = \frac{2m_1 r^2 \sin^2 q}{l^2 \cos^2 \Psi}(l\cos\Psi + r\cos q)\left[l^3 \cos^3 \Psi \cos q - rl^2(\cos^2 q - \cos^2 \Psi \sin q)\right],$$

$$\left(\frac{\partial A_{12}^*}{\partial q}\right)_0 = -\frac{2m_1 r^2 l^2 \sin q \cos q}{l^4 \cos^4 \Psi}(l\cos\Psi + r\cos q),$$

$$\frac{\partial A_{12}^{*0}}{\partial q} = -\frac{m_1 r^2 \sin q \cos q}{l^4 \cos^4 \Psi}(l\cos\Psi + r\cos q)(2l^2 - r^2 \sin^2 q) -$$
$$-\frac{m_1 r^2 \sin^2 q}{l^4 \cos^4 \Psi}[l^3 \cos^3 \Psi + (r^2 - l^2)r\sin q]. \quad (3.183)$$

In order to obtain expressions for coefficients L, H, G and R_0 of eq. (2.59) it is necessary to insert expressions for A, A_{12}^* and their derivatives into eqs. (2.60)-(2.63). As the expressions obtained are very cumbersome we restrict ourselves to a single expression for L because this expression is needed for the forthcoming analysis

$$L = \frac{m_2 r \cos\Psi |\sin q|}{3m_1(l\cos\Psi + r\cos q)^2 \sin^2 q + m_2 l^2 \cos^2 \Psi}, \quad (3.184)$$

where $l\cos\Psi = \sqrt{l^2 - r^2 \sin^2 q}$. According to eq. (3.184) L is a positive periodic function of the generalised coordinate q with period 2π. At points $k\pi$ and $(2k+1)\pi/2$ it has the following values

$$L = \begin{cases} 0 & q = k\pi, \\ \dfrac{rm_2}{\sqrt{l^2 + r^2}(3m_1 + m_2)} & q = (2k+1)\dfrac{\pi}{2}. \end{cases} \quad (3.185)$$

Figure 3.9 displays $L(q)$ for the case of $m_2 = 2m_1$, $l = 1, 5r$.

3.6.2 Condition for complete absence of paradoxes

The paradoxical situations of non-existence and non-uniqueness of solution do not occur for all values of the generalised coordinate q if

$$\mu L_{\max} < 1. \quad (3.186)$$

On the other hand, eqs. (3.184) and (3.185) suggest that as l increases (under condition $l > r$ and $q \neq k\pi$) the value of $L_{\max}$ decreases continuously and tends to zero for $l \gg r$. Given the coefficient of friction μ, inequality (3.186) holds if

$$l > l_1, \quad (3.187)$$

3.6 Crank mechanism   113

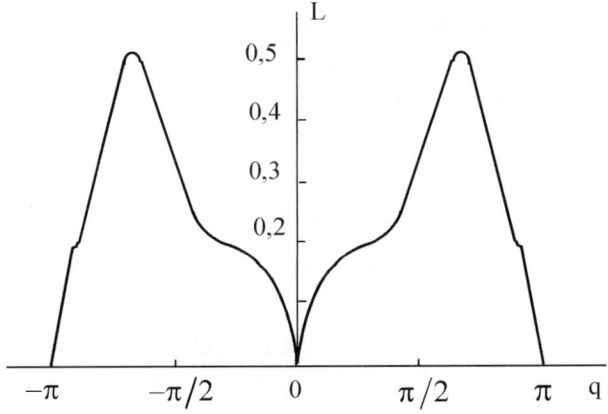

FIGURE 3.9.

where l_1 is the root of equation $\mu L_{\max} - 1 = 0$. For example, for $m_1 = m_2$, $\mu = 1$ we have $l_1 = 1, 1r$. Let us conditionally refer to l_1 as the critical length of the coupler.

Expression (3.184) allows one to judge the influence of the mass distribution on the possibility of paradoxical situations. For example, for $m_2 = 0$ the influence coefficient corresponding to L is zero. For this reason, the condition of absence of paradoxes (3.186) is fulfilled for any value of the friction coefficient μ. As ratio m_2/m_1 grows, the values of L and $L_{\max}$ increase and tend to the following values

$$L_* = \frac{r|\sin q|}{\sqrt{l^2 - r^2 \sin^2 q}}, \quad L_{*\max} = \frac{r}{\sqrt{l^2 - r^2}}. \tag{3.188}$$

Thus, for $m_2 \gg m_1$ the condition of complete absence of paradoxes (3.186) takes the form of the limiting condition

$$\frac{\mu r}{\sqrt{l^2 - r^2}} - 1 < 0 \tag{3.189}$$

which is equivalent to the following requirement

$$l > l_* = r\sqrt{\mu^2 + 1}. \tag{3.190}$$

This discussion suggests that as the mass ratio m_2/m_1 increases, the critical length of the coupler l_1 increases and tends to the limit l_*. For instance, for $\mu = 1$, $m_2 = m_1$ we have $l_1 = 0, 77 l_*$, whereas for $\mu = 1$, $m_2 = 2m_1$ we have $l_1 = 0, 90 l_*$.

As has been frequently mentioned above, when the conditions of absence of paradoxes (3.187) (or (3.190) in the case $m_2 \gg m_1$) is met the reaction force is unique and the law of motion is uniquely determined by quadratures.

3.6.3 The property of self-braking in the case of no paradoxes

Determination of the points of self-braking and the points of debraking was carried out by means of Theorem 4. An arbitrary point on crank OA located at a distance ρ from the centre O is characterised by the following expressions for the position-vector and its derivatives

$$\mathbf{r}^* = \mathbf{r}^0 = \rho\left(\cos q\,\mathbf{i}_1 + \sin q\,\mathbf{i}_2\right),$$

$$\frac{d\mathbf{r}^0}{dq} = -\rho\left(\sin q\,\mathbf{i}_1 - \cos q\,\mathbf{i}_2\right), \quad \frac{\partial \mathbf{r}^*}{\partial h} = 0,$$

$$\left(\frac{\partial \mathbf{r}^*}{\partial h}\right)_0 = -\lambda \frac{d\mathbf{r}^0}{dq} = \frac{\rho}{l\cos\Psi + r\cos q}\left(\mathbf{i}_1 \sin q - \mathbf{i}_2 \cos q\right). \tag{3.191}$$

Inserting $d\mathbf{r}_T^0/dq$, $d\mathbf{r}^0/dq$ and $(\partial \mathbf{r}^*/\partial h)_0$ due to eqs. (3.177) and (3.191) into eq. (2.94) we derive the following condition for self-braking

$$|\gamma| = \frac{l\cos\Psi}{\mu r |\sin q|} = \frac{\sqrt{l^2 - r^2 \sin^2 q}}{\mu r |\sin q|} \leq 1. \tag{3.192}$$

Hence,

$$l \leq r\sqrt{1+\mu^2}|\sin q| = l_*|\sin q|. \tag{3.193}$$

Condition (3.193) together with the condition for absence of paradoxes (3.187) suggests that if the length of the coupler satisfies the inequality

$$l_1 \leq l \leq l_* = \sqrt{1+\mu^2}\,r, \tag{3.194}$$

then all points of crank OA are points of self-braking for those values of q for which

$$|\sin q| \geq \frac{l}{r\sqrt{1+\mu^2}} = \frac{l}{l_*} \tag{3.195}$$

and are points of debraking for the remaining values of q.

If $l > l_*$, then the points on OA are points of debraking for any value of q. In this case the condition of absence of paradoxes (3.187) is also fulfilled for all q, since $l_* > l_1$.

In order to establish the property of self-braking for the coupler, let us consider an arbitrary point B located such that distance $AB = \alpha l$ ($0 \leq \alpha \leq 1$). Under an additional displacement of the slider along the mentioned circle its position vector can be expressed in the form

$$\mathbf{r}_3^* = [r\cos q + \alpha\sqrt{l^2 - (r\sin q - h)^2}]\mathbf{i}_1 + [(1-\alpha)r\sin q + \alpha h]\mathbf{i}_2. \tag{3.196}$$

Hence,

$$\frac{d\mathbf{r}_3^0}{dq} = -\frac{r\sin q}{l\cos\Psi}(l\cos\Psi + \alpha r\cos q)\mathbf{i}_1 + (1-\alpha)r\cos q\,\mathbf{i}_2, \tag{3.197}$$

$$\frac{\partial \mathbf{r}_3^0}{\partial h} = \frac{\alpha r \sin q}{l \cos \Psi} \mathbf{i}_1 + \alpha \mathbf{i}_2 . \qquad (3.198)$$

The value of $(\partial \mathbf{r}_3/\partial h)_0$ can be obtained by formula (2.18) with account of eqs. (3.179), (3.197) and (3.198). The result is

$$\left(\frac{\partial \mathbf{r}_3^0}{\partial h} \right)_0 = \frac{r \sin q}{l \cos \Psi} \left[\alpha - \frac{l \cos \Psi + \alpha \cos q}{l \cos \Psi + r \cos q} \right] \mathbf{i}_1 + \left[\alpha - \frac{(1-\alpha) r \cos q}{l \cos \Psi + r \cos q} \right] \mathbf{i}_2 . \qquad (3.199)$$

According to Theorem 4, the necessary condition for self-braking of point B of the coupler is that vectors $d\mathbf{r}_3^0/dq$ and $(\partial \mathbf{r}_3/\partial h)_0$ are collinear. By virtue of eqs. (3.197) and (3.199) it is easy to prove that this takes place only in the case when $\alpha = 0$, i.e. when points B and A coincide. Indeed, equating the vector product $d\mathbf{r}_3^0/dq \times (\partial \mathbf{r}_3/\partial h)_0$ to zero we can write

$$(l \cos \Psi + \alpha r \cos q) \left[\alpha + \frac{(1-\alpha) r \cos q}{l \cos \Psi + r \cos q} \right] +$$
$$(1-\alpha) r \cos q \left[\alpha - \frac{l \cos \Psi + \alpha r \cos q}{l \cos \Psi + r \cos q} \right] = 0 .$$

Hence,

$$\alpha \left(\sqrt{l^2 - r^2 \sin^2 q} + r \cos q \right) = 0 . \qquad (3.200)$$

Since the quantity in the parentheses in eq. (3.200) is always positive we obtain that $\alpha = 0$. Therefore, all points of the coupler, except possibly point A, are points of debraking. We notice that the property of debraking of point A was determined above earlier in the study of the crank.

3.7 Link mechanism of a planing machine

The carriage of a planing machine is set into translatory motion by means of a link mechanism, see Fig. 3.10. The mass of the carriage is denoted by m, the centre of mass of the crank coincides with the centre of rotation O and its moment of inertia about O is J. A torque M and a load $\mathbf{F} = -P\mathbf{i}_1$ are applied to the crank and the carriage, respectively. Coulomb friction is assumed to act between the slider T and the link CB, whereas any friction between the remaining joints is neglected. The distances are as follows: $OT = r$, $OC = a (a > r)$, $OA = b$, $CB = l$.

3.7.1 Differential equations of motion and the expression for the reaction force

The position vectors of point $C(\mathbf{r}_c^0)$, the slider $T(\mathbf{r}_1^0)$ and the centre of mass I of the carriage $(\mathbf{r}_2^0)$ are given with respect to a fixed coordinate system

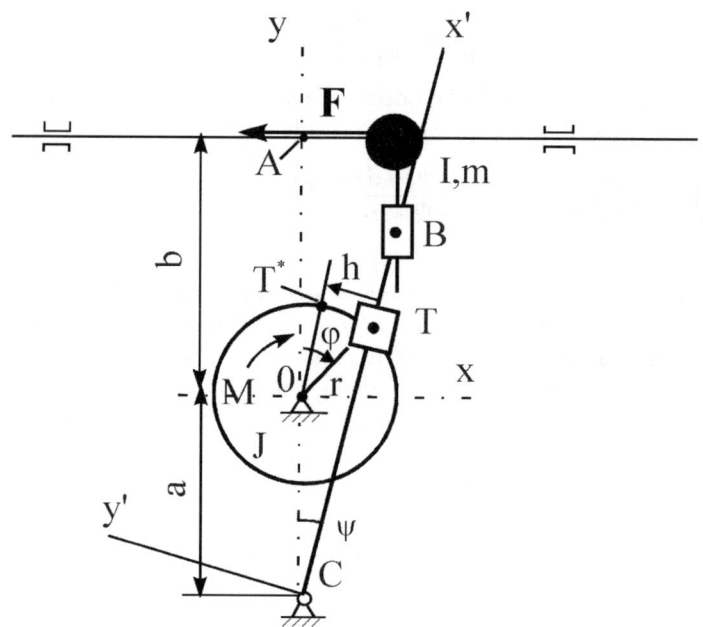

FIGURE 3.10.

Oxy in the following form

$$\mathbf{r}_c^0 = -a\mathbf{i}_2, \quad \mathbf{r}_1^0 = \mathbf{i}_1 r \sin\varphi + \mathbf{i}_2 r \cos\varphi,$$
$$\mathbf{r}_2^0 = \mathbf{i}_1 lr \sin\varphi (a^2 + r^2 + 2ar\cos\varphi)^{-1/2} + b\mathbf{i}_2. \quad (3.201)$$

Let a moving coordinate system $Cx'y'$, whose axis Cx' is coincident with CB, be related to link CB. In this coordinate system, the position vector of the slider is given by

$$\mathbf{r}_T^0 = \sqrt{a^2 + r^2 + 2ar\cos\varphi}\,\mathbf{i}'_1 = \eta_0\,\mathbf{i}_1, \quad (3.202)$$

where $\eta_0 = \sqrt{a^2 + r^2 + 2ar\cos\varphi}$ is the distance CT.

Removing mentally the contact constraint, we fix the link and rotate the crank counterclockwise. In the moving coordinate system, the position vector of the slider takes the form

$$\begin{aligned}\mathbf{r}_T^* &= \left[a(a + r\cos\varphi)\eta_0^{-1} + \sqrt{r^2 - (h - ar\eta_0^{-1}\sin\varphi)^2}\right]\mathbf{i}'_1 + h\mathbf{i}'_1 \\ &= \eta\mathbf{i}'_1 + h\mathbf{i}'_2.\end{aligned} \quad (3.203)$$

In the fixed coordinate system, the position vectors of points T^* and I are determined as follows

$$\begin{aligned}\mathbf{r}_1^* &= \mathbf{r}_1^0 + (\eta - \eta_0)(\mathbf{i}_1 \sin\Psi + \mathbf{i}_2 \sin\Psi) - h(\mathbf{i}_1 \cos\Psi + \mathbf{i}'_2 \sin\Psi), \\ \mathbf{r}_2^* &= \mathbf{r}_2^0,\end{aligned} \quad (3.204)$$

where η is the projection of vector $\mathbf{r}_T^*$ on axis Cx' and Ψ denotes the angle between axis Cx' and Cy determined by the following relationship

$$\sin \Psi = \frac{r \sin \varphi}{\eta_0}, \quad \cos \Psi = \frac{a + r \cos \varphi}{\eta_0}. \quad (3.205)$$

Equations (3.201)-(3.204) yield the following expressions for the derivatives

$$\frac{d\mathbf{r}_T^0}{d\varphi} = -\frac{ar \sin \varphi}{\eta_0} \mathbf{i}'_1, \quad \left(\frac{\partial \mathbf{r}_T^*}{\partial h}\right)_0 = -\frac{a \sin \varphi}{a \cos \varphi + r} \mathbf{i}'_1 + \mathbf{i}'_2, \quad (3.206)$$

$$\frac{d\mathbf{r}_1^0}{d\varphi} = r(\mathbf{i}_1 \cos \varphi - \mathbf{i}_2 \sin \varphi), \quad \frac{d^2 \mathbf{r}_1^0}{d\varphi^2} = -r(\mathbf{i}_1 \sin \varphi + \mathbf{i}_1 \cos \varphi),$$

$$\left(\frac{\partial \mathbf{r}_1^*}{\partial h}\right)_0 = -\frac{\eta_0}{a \cos \varphi + r} (\mathbf{i}_1 \cos \varphi - \mathbf{i}_2 \sin \varphi), \quad (3.207)$$

$$\frac{d\mathbf{r}_2^0}{d\varphi} = lr\eta_0^{-3}(\eta_0^2 \cos \varphi + ar \sin^2 \varphi) \mathbf{i}_1, \quad \left(\frac{\partial \mathbf{r}_2^*}{\partial h}\right)_0 = 0,$$

$$\frac{d^2 \mathbf{r}_2^0}{d\varphi^2} = lr\eta_0^{-5} \left[3ar(\eta_0^2 \cos \varphi + ar \sin^2 \varphi) - \eta_0^4\right] \mathbf{i}_1. \quad (3.208)$$

Inserting eq.(3.206) into eqs. (2.8) and (2.18) yields

$$\lambda = -\eta_0 r^{-1}(a \cos \varphi + r)^{-1}, \quad (3.209)$$

$$\left(\frac{\partial \mathbf{r}_1}{\partial h}\right)_0 = 0, \quad \left(\frac{\partial \mathbf{r}_2}{\partial h}\right)_0 = l\eta_0^{-2} \frac{\eta_0^2 \cos \varphi + ar \sin^2 \varphi}{a \cos \varphi + r} \mathbf{i}_1. \quad (3.210)$$

Let us determine the coefficients in the expression for the kinetic energy. To this end, we denote the angle between the crank OT^* and the vertical Oy under the removed constraint as Φ. The kinetic energy of the system is given by

$$T = \frac{1}{2} J \dot{\Phi}^2 + \frac{1}{2} m \left|\frac{d\mathbf{r}_2^0}{d\varphi}\right|^2 \dot{\varphi}^2.$$

Here $\dot{\Phi}$ denotes the angular velocity of the crank. Its square is as follows

$$\dot{\Phi}^2 = \frac{1}{r^2} \left|\frac{\partial \mathbf{r}_1^*}{\partial \varphi} \dot{\varphi} + \frac{\partial \mathbf{r}_1^*}{\partial h} \dot{h}\right|^2.$$

Thus

$$T = \frac{1}{2} \left(\frac{J}{r^2} \left|\frac{\partial \mathbf{r}_1^*}{\partial \varphi}\right|^2 + \frac{1}{2} m \left|\frac{d\mathbf{r}_2^0}{d\varphi}\right|^2\right) \dot{\varphi}^2 + \frac{J}{r^2} \frac{\partial \mathbf{r}_1^*}{\partial \varphi} \cdot \frac{\partial \mathbf{r}_1^*}{\partial h} \dot{\varphi} \dot{h} + \frac{1}{2} \frac{J}{r^2} \left|\frac{\partial \mathbf{r}_1^*}{\partial h} \dot{h}\right|^2.$$

118   3. Accounting for dry friction in mechanisms

Taking into account eqs. (3.206)-(3.208), we express the coefficients of the kinetic energy in the following form

$$A_{11}^{*0} = A = J + mr^2 l^2 \eta_0^{-6}(\eta_0^2 \cos\varphi + ar\sin^2\varphi)^2,$$
$$A_{12}^{*0} = -\frac{\eta_0 J}{r(a\cos\varphi + r)}. \qquad (3.211)$$

Now we can determine the coefficients of the left hand side of eq. (2.13)

$$\frac{1}{2}\frac{dA}{d\varphi} = \frac{ml^2 r^2(\eta^2 \cos\varphi + ar\sin^2\varphi)}{\eta_0^8}\left[3ar\left(\eta_0^2\cos\varphi + ar\sin^2\varphi\right) - \eta_0^4\right]\sin\varphi, \qquad (3.212)$$

$$\left(\frac{\partial A_{12}^*}{\partial \varphi} - \frac{1}{2}\frac{\partial A_{11}^*}{\partial \varphi}\right)_0 = \left(\frac{J}{r^2}\frac{d\mathbf{r}_1^0}{d\varphi}\cdot\frac{\partial^2 \mathbf{r}_1^*}{\partial\varphi\partial h} - \frac{J}{r^2}\frac{d\mathbf{r}_1^0}{d\varphi}\cdot\frac{\partial^2 \mathbf{r}_1^*}{\partial\varphi\partial h}\right)_0 = 0. \qquad (3.213)$$

In order to determine the generalised forces it is sufficient to substitute the vector derivatives (2.7) and (2.8) into eq. (2.19). The result is as follows

$$Q_1 = M - \frac{lr}{\eta_0^3}(\eta_0^2\cos\varphi + ar\sin^2\varphi)P, \quad Q_2^* = -\frac{\eta_0 M}{r(a\cos\varphi + r)}. \qquad (3.214)$$

Thus, all coefficients of the system of equations in the form of (2.13) are derived. All information for calculation of the coefficients of system (2.59) is also at our disposal. For example, substituting the expressions due to eqs. (3.206)-(3.214) into eqs. (2.60)-(2.63) we obtain

$$L = (1 - \frac{J\eta_0^6}{\nu})\frac{a|\sin\varphi|}{a\cos\varphi + r}, \qquad (3.215)$$

$$H = (1 + \varepsilon_1\varepsilon_2\frac{\mu a|\sin\varphi|}{a\cos\varphi + r})\frac{ml^2 r^2 x(3ar\chi - \eta_0^4)\sin\varphi}{\eta_0^2 \nu}, \qquad (3.216)$$

$$G = \frac{\eta_0^6}{\nu}[M - \frac{lr\chi}{\eta_0^3}(1 + \varepsilon_1\varepsilon_2\frac{\mu a|\sin\varphi|}{a\cos\varphi + r})P], \qquad (3.217)$$

$$R_0 = \frac{\eta_0 M}{r(a\cos\varphi + r)} - \frac{J}{r\eta_0(a\cos\varphi + r)\nu}[\eta_0^5(\eta_0^3 M - lr\chi P) - ml^2 r^2 \chi\dot\varphi^2]. \qquad (3.218)$$

Here the new notation

$$\chi = \eta_0^2\cos\varphi + ar\sin^2\varphi, \quad \nu = J\eta_0^6 + ml^2 r^2 \chi^2$$

is adopted. Finally, inserting eqs. (3.215)-(3.218) into (2.59) we arrive at the following differential equation of motion

$$\left[1+\varepsilon_1\varepsilon_2\mu\left(1-\frac{J\eta_0^6}{\nu}\right)\frac{a|\sin\varphi|}{a\cos\varphi+r}\right]\ddot{\varphi} + \left[1+\varepsilon_1\varepsilon_2\frac{\mu a|\sin\varphi|}{a\cos\varphi+r}\right] \times$$
$$\frac{ml^2r^2\chi(3a\chi-\eta_0^4)\sin\varphi}{\eta_0^2\nu}\dot{\varphi}^2 - \frac{\eta_0^6}{\nu}\left[M - \frac{lr\chi}{\eta_0^3}\left(1+\varepsilon_1\varepsilon_2\frac{\mu a|\sin\varphi|}{a\cos\varphi+r}\right)P\right] = 0, \quad (3.219)$$

and the expression for the reaction force

$$\left[1+\varepsilon_1\varepsilon_2\mu\left(1-\frac{J\eta_0^6}{\nu}\right)\frac{a|\sin\varphi|}{a\cos\varphi+r}\right]R - \frac{\eta_0 M}{r(a\cos\varphi+r)} +$$
$$\frac{J}{r\eta_0(a\cos\varphi+r)\nu}\left[\eta_0^5(\eta_0^3 M - lr\chi P) - ml^2r^2\chi\dot{\varphi}^2\right] = 0. \quad (3.220)$$

3.7.2 Feasibility of Painlevé's paradoxes

Let us first clarify the feasibility of the paradoxes in the particular case of $J=0$. For $J=0$ it follows from eq. (3.215) that

$$L = \frac{a|\sin\varphi|}{a\cos\varphi+r}. \quad (3.221)$$

Quantity L has the following poles

$$\varphi_\pm = \pi \pm \arccos(r/a). \quad (3.222)$$

Within the period $[0,2\pi]$ this function is equal to zero when $\varphi = 0, \pi, 2\pi$, positive in $(0,\varphi_-)$ and $(\varphi_+, 2\pi)$, negative in (φ_-, φ_+) and unbounded ($\pm\infty$) as $\varphi \to \varphi_- \mp 0$ and $\varphi \to \varphi_+ \pm 0$. The condition for the paradoxes

$$\mu|L| > 1 \quad \text{or} \quad \mu a|\sin\varphi| > |a\cos\varphi + r| \quad (3.223)$$

is necessarily fulfilled in certain vicinities of the points $\varphi_\pm$. The boundary of the region of the paradoxes is described by the equation

$$\mu|L| - 1 = \frac{\mu a|\sin\varphi|}{|a\cos\varphi+r|} - 1 = 0.$$

The roots are given by

$$\cos\varphi = \frac{-r \pm \mu\sqrt{(a^2(\mu^2+1)-r^2)}}{a(1+\mu^2)}$$

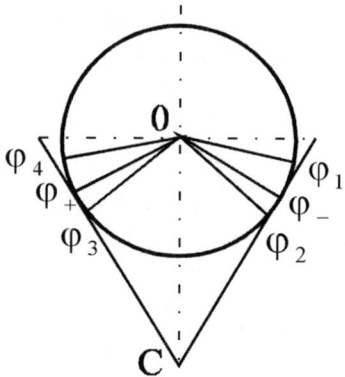

FIGURE 3.11.

or

$$\begin{aligned}
\varphi_1 &= \pi - \arccos \frac{r - \mu\sqrt{a^2(\mu^2 + 1) - r^2}}{a(\mu^2 + 1)}, \\
\varphi_2 &= \pi - \arccos \frac{r + \mu\sqrt{a^2(\mu^2 + 1) - r^2}}{a(\mu^2 + 1)}, \\
\varphi_3 &= \pi + \arccos \frac{r + \mu\sqrt{a^2(\mu^2 + 1) - r^2}}{a(\mu^2 + 1)}, \\
\varphi_4 &= \pi + \arccos \frac{r - \mu\sqrt{a^2(\mu^2 + 1) - r^2}}{a(\mu^2 + 1)}.
\end{aligned} \qquad (3.224)$$

One can see from eqs. (3.222)-(3.224) and Fig. 3.11 that the intervals of the paradoxes (φ_1, φ_2) and (φ_3, φ_4) are located symmetrically about axis Oy and include points φ_- and φ_+, respectively. If $\dot\varphi > 0$, the solution of the dynamic problem (3.220) is not unique in (φ_1, φ_-) and (φ_+, φ_4), and does not exist in $(\varphi_-, \varphi_2), (\varphi_3, \varphi_+)$. When the velocity changes its sign $\varepsilon_2 = \mathrm{sign}\,\dot\varphi$ the paradox type changes as well, that is, the non-uniqueness becomes the non-existence and vice versa.

The greater the friction coefficient μ, the broader the intervals of the paradoxes. In the limiting case we have

$$\begin{aligned}
&\varphi_1 = 0, \quad \varphi_2 = \varphi_3 = \pi, \quad \varphi_4 = 2\pi \quad \text{for} \quad \mu \to \infty, \\
&\varphi_1 = \varphi_2 = \varphi_-, \quad \varphi_3 = \varphi_4 = \varphi_+ \quad \text{for} \quad \mu = 0.
\end{aligned} \qquad (3.225)$$

This means that for very large values of μ the region of the paradoxes actually occupies the whole period $(0, 2\pi)$, whereas for small values of μ this region reduces to the two points φ_- and φ_+.

Let us determine now the influence of the moment of inertia of the crank J on the feasibility of the paradoxes. To this aim, we rewrite expression

(3.215) in an explicit form

$$L = \frac{a|\sin\varphi|}{a\cos\varphi + r} \times \qquad (3.226)$$

$$\frac{ml^2r^2[(a^2 + r^2 + 2ar\cos\varphi)\cos\varphi + ar\sin^2\varphi]^2}{J(a^2 + r^2 + 2ar\cos\varphi)^3 + ml^2r^2[(a^2 + r^2 + 2ar\cos\varphi)\cos\varphi + ar\sin^2\varphi]^2}.$$

It is easy to prove that the following equation

$$\chi(\varphi) \equiv (a^2 + r^2 + 2ar\cos\varphi)\cos\varphi + ar\sin^2\varphi = 0$$

has roots of the type (3.222). Therefore, quantity L is an undeterminate form of the sort $0/0$ at points $(\varphi_\mp)$.

On the other hand, when φ tends to $\varphi_\mp$ we have

$$\lim_{\cos\varphi \to -r/a} L = \frac{ml^2r^4(a\cos\varphi + r)\cos^2\varphi a|\sin\varphi|}{J(a^2 + r^2 + 2ar\cos\varphi)^3} = 0. \qquad (3.227)$$

For $\varphi \neq \varphi_\mp$ the sign of L coincides with the sign of $a\cos\varphi + r$. This means that the influence coefficient L is equal to zero at points $\varphi = 0, \pi, 2\pi, \varphi_\pm$, positive in $(0, \varphi_-)$ and $(\varphi_+, 2\pi)$ and negative in (φ_-, φ_+). Thus, in contrast to the case $J = 0$, in the general case $(J \neq 0)$ the influence coefficient L is a continuous bounded periodic function of coordinate φ. When J grows, quantity L decreases and, in turn, the region of the paradoxes narrows, see Figs. 3.12 and 3.13.

3.7.3 The property of self-braking

Let us consider first the property of self-braking of the points of the crank OT in the case of no paradoxes. The equations of straight lines $\Pi^1_\pm$ forming the stagnation angle Ψ of point T of crank OT can be derived by inserting the values of $d\mathbf{r}^0_1/d\varphi$ and $\partial\mathbf{r}_1/\partial h$, eqs. (2.100) and (2.103), into eq. (2.99) The result is

$$\Pi^1_\pm : (\boldsymbol{\rho}_1 - \mathbf{r}^0_1) \cdot (\mathbf{i}_2 r\cos\varphi - \mathbf{i}_2\sin\varphi) = 0. \qquad (3.228)$$

As one can see, these straight lines are coincident and pass through O and T. Moreover, inasmuch as vector $d\mathbf{r}_1/d\varphi$ is perpendicular to these lines the stagnation angle Ψ is equal to zero and the shift angle Ψ is equal to π, see Theorem 5.

It is evident that expressions (3.207) and (3.210) for quantities $d\mathbf{r}^0_1/d\varphi$ and $(\partial\mathbf{r}_1/\partial h)_0$ and, thus, eq. (3.228) are valid not only for point T of the crank but for any of its points provided that r is understood to be the distance from the centre O. For this reason, all points of the crank are points of debraking with zero stagnation angle.

Let us determine the property of self-braking of the points of the carrier in the case of no paradoxes. The derivatives of the position-vector of any

122   3. Accounting for dry friction in mechanisms

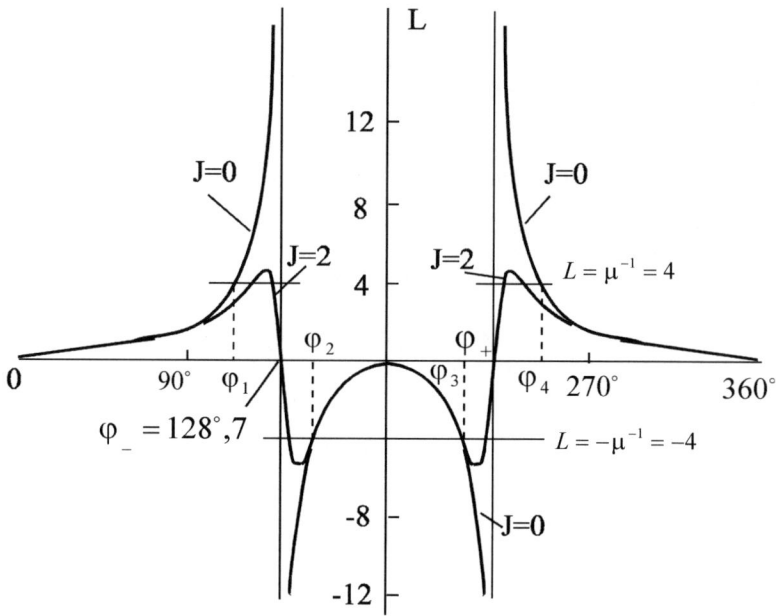

FIGURE 3.12.

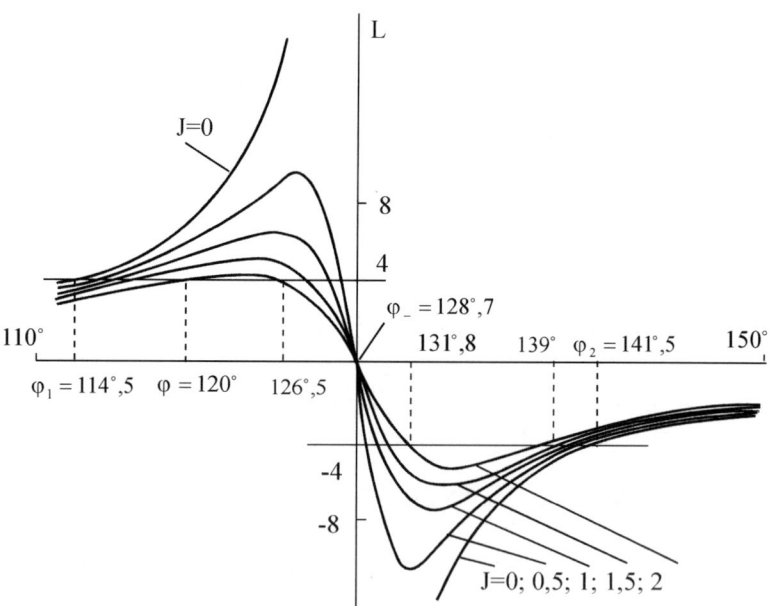

FIGURE 3.13.

point of the carrier with respect to φ and h are respectively equal to $d\mathbf{r}_2^0/d\varphi$ and $(\partial \mathbf{r}_2/\partial h)_0$ determined by eqs. (3.208) and (3.210). They are seen to be collinear. Inserting these expressions into eq. (2.94) yields

$$|\gamma| = \frac{a\cos\varphi + r}{\mu a |\sin\varphi|}. \qquad (3.229)$$

Using Theorem 4, the considered points are points of self-braking if

$$\mu a |\sin\varphi| \geq |a\cos\varphi + r|. \qquad (3.230)$$

This condition coincides with condition (3.223). Thus, the region of self-braking of the points of the carrier is coincident with the region of the paradoxes of the mechanism for $J = 0$.

When condition (3.230) does not hold true, i.e. in the case in which

$$\mu a |\sin\varphi| < |a\cos\varphi + r|, \qquad (3.231)$$

all points of the support are points of debraking. As vectors $d\mathbf{r}_2^0/d\varphi$ and $(\partial \mathbf{r}_2/\partial h)_0$ are collinear and satisfy condition (2.104), the stagnation angle Ψ is equal to zero. This means that the system can be shifted by applying any force to the carrier (provided that there is no force acting at other points).

Let us establish now the property of self-braking in the paradoxical case $\mu|L| > 1$. We notice that $(\partial \mathbf{r}_2/\partial h)_0 = 0$ and $d\mathbf{r}_2^0/d\varphi \neq 0$ for the points of the crank and thus the second relationship in eq. (2.115) holds. As for the points of the carrier (for example point A), the second condition in eq. (2.115) is equivalent to inequality (3.231), which, as follows from eq. (3.226) is satisfied for $\mu|L| > 1$. With the help of Theorem 6 we conclude that in the paradoxical case all points of the crank and the carrier are simultaneously points of self-braking and debraking.

3.7.4 Numerical example

Let $l = 1$ m, $r = 0,25$ m, $a = 0,4, m = 500$ kg, $\mu = 0,25$ and the moment of inertia take the following values: $J = 0, 0,5, 1, 1,5, 2$ kg·m². For these numerical values expression (3.226) takes the form

$$L = \frac{31,25(0,1 + 0,2225\cos\varphi + 0,1\cos^2\varphi)^2}{31,25(0,1 + 0,2225\cos\varphi + 0,1\cos^2\varphi)^2 + J(0,2225 + 0,2\cos\varphi)^3} \times \frac{0,4|\sin\varphi|}{0,4\cos\varphi + 0,25}. \qquad (3.232)$$

A plot of $L(\varphi)$ is shown in Fig. 3.13 for φ within the interval $(110^0, 150^0)$, i.e. from values of φ near the pole φ_-, for four taken values of J. The region of paradoxes is given by intersections of the horizontal lines $L = \pm\mu^{-1} = \pm 4$. For $J = 0$ this region consists of two intervals

$$\varphi_1 = 114,5^0, \quad \varphi_2 = 141,5^0 \quad \text{and} \quad \varphi_3 = 218,5^0, \quad \varphi_4 = 245,5^0. \quad (3.233)$$

For $J \neq 0$ this region is split into four intervals, for instance, for $J = 1$ these intervals are as follows

$$(118^0, 126,5^0), \ (130^0, 140,5^0), \ (219,5^0, 230^0), \ (233,5^0, 242^0). \quad (3.234)$$

As eqs. (3.233) and (3.234), as well as Fig. 3.13 suggest, the region of paradoxes narrows with the growth of J.

4

Systems with many degrees of freedom and a single frictional pair. Solving Painlevé's paradoxes

The chapter is concerned with deriving equations for the class of systems with many degrees of freedom and a single frictional pair, and solving three problems of Painlevé's paradoxes, namely, the criteria of paradoxes, the origin of their appearance and the true motions. The derivation of the equations and determination of the criterion for the paradoxes are a generalisation of the procedures suggested in Sections 2.1 and 2.3. Understanding the reason for the paradoxes and the true motions under the paradoxical situations is carried out by means of a limiting process in which an elastic contact joint is made absolutely rigid, see [84], [86].

The theoretical results obtained are applied to the further analysis of the Painlevé-Klein scheme, as well as for an elliptic pendulum and the Zhukovsky-Froude pendulum. Additionally, the condition for instability of a stationary cutting regime is proved by generalising the solution to the metal cutting, cf. [83].

4.1 Lagrange's equations with a removed constraint

We consider a system of N particles having n degrees of freedom and subjected to stationary holonomic two-sided constraints. Let all of these constraints, except a single contact constraint with Coulomb friction described in Chapter 2, be ideal. In contrast to the case of a single degree of freedom, in the case for systems with many degrees of freedom the slider can move in any direction on the surface U of the counterpart body. For this reason, while removing the contact constraint a locus of the slider is a certain three-dimensional region about guide U.

The position vectors of the particles $\mathbf{r}_1^0, ... \mathbf{r}_N^0$ and the slider $\mathbf{r}_T^0$ are functions of n independent generalised coordinates $q_1, ..., q_n$. Hence,

$$\mathbf{r}_i^0 = \mathbf{r}_i^0(q_1, ..., q_n), \quad \mathbf{r}_T^0 = \mathbf{r}_T^0(q_1, ..., q_n),$$

$$\mathbf{v}_i^0 = \sum_{k=1}^n \left(\frac{\partial \mathbf{r}_i^0}{\partial q_k}\right) \dot{q}_k, \quad \mathbf{v}_T^0 = \sum_{k=1}^n \left(\frac{\partial \mathbf{r}_T^0}{\partial q_k}\right) \dot{q}_k, \qquad (4.1)$$

where $\mathbf{v}_i^0$ and $\mathbf{v}_T^0$ denote the velocity of motion of the $i-th$ particle and the slipping velocity of the slider along the guide, respectively. In the case of slip with rolling the quantity $\mathbf{v}_T^0$ is given by eq. (2.31), where the Gaussian coordinates u^α, $p^\alpha (\alpha = 1, 2)$ are functions of the generalised coordinates $q_1, ..., q_n$.

Let us assume that in the case of the removed contact constraint the slider moves from position T^0 to a certain position T^*, see Fig. 2.1. The projection of $T^0 T^*$ onto the normal U at point T^0 is taken as the redundant coordinate, i.e.

$$h = (\mathbf{r}_T^* - \mathbf{r}_T^0) \cdot \mathbf{m}, \qquad (4.2)$$

where $\mathbf{m}$ is the unit vector of the normal. In what follows, this coordinate is subjected to the following condition

$$h = \dot{h} = \ddot{h} = 0. \qquad (4.3)$$

Since

$$\mathbf{r}_T^* = \mathbf{r}_T^*(q_1, ..., q_n, h),$$
$$\mathbf{r}_T^*(q_1, ..., q_n, 0) \equiv \mathbf{r}_T^0(q_1, ..., q_n) \qquad (4.4)$$

we have

$$\left(\frac{\partial \mathbf{r}_T^*}{\partial q_k}\right)_0 = \left(\frac{\partial \mathbf{r}_T^0}{\partial q_k}\right) \quad (k = 1, ...n). \qquad (4.5)$$

The total constraint force of the removed contact force is given by eq. (2.10)

$$\mathbf{R}_\sigma = \left(-\varepsilon_1 \mu \frac{\mathbf{v}_T^0}{|\mathbf{v}_T^0|} + \mathbf{m}\right) R, \qquad (4.6)$$

where R is the normal reaction force and $\varepsilon_1 = \text{sign } R$. This reaction force is related to the generalised coordinates $q_1, ..., q_n$ and the redundant coordinate h as follows

$$S_j = \mathbf{R}_\sigma \cdot \left(\frac{\partial \mathbf{r}_T^*}{\partial q_j}\right)_0 \quad (j = 1, ..., n), \quad S^* = \mathbf{R}_\sigma \cdot \left(\frac{\partial \mathbf{r}_T^*}{\partial h}\right)_0.$$

Taking into account eqs. (4.5) and (4.6) as well as the relationships

$$\mathbf{m} \cdot \left(\frac{\partial \mathbf{r}_T^0}{\partial q_j}\right) = 0, \quad \frac{\mathbf{v}_T^0}{|\mathbf{v}_T^0|} \cdot \left(\frac{\partial \mathbf{r}_T^0}{\partial q_j}\right) = \left(\frac{\partial v_T^0}{\partial \dot{q}_j}\right), \quad \frac{\mathbf{v}_T^0}{|\mathbf{v}_T^0|} \cdot \left(\frac{\partial \mathbf{r}_T^*}{\partial h}\right)_0 = \left(\frac{\partial v_T^*}{\partial \dot{h}}\right)_0,$$

$$\mathbf{m} \cdot \left(\frac{\partial \mathbf{r}_T^*}{\partial h}\right)_0 = \lim_{T^* \to T^0} \frac{(\mathbf{r}_T^* - \mathbf{r}_T^0) \cdot \mathbf{m}}{(\mathbf{r}_T^* - \mathbf{r}_T^0) \cdot \mathbf{m}} = 1 \qquad (4.7)$$

we obtain

$$S_j = -\varepsilon_1 \mu \frac{\partial v_T^0}{\partial \dot{q}_j} R, \quad S^* = \left(1 - \varepsilon_1 \mu \frac{\partial v_T^*}{\partial \dot{h}}\right)_0 R. \qquad (4.8)$$

By using the method of constraint removal, [103], we can cast Lagrange's equations in the form

$$\sum_{k=1}^{n} A_{ks}^0 \ddot{q}_k + \sum_{k=1}^{n} \sum_{m=1}^{n} (\Gamma_{km,s})_0 \, \dot{q}_k \dot{q}_m = Q_s - \varepsilon_1 \mu \frac{\partial v_T^0}{\partial \dot{q}_s} R, \quad (s = 1, ..., n)$$
(4.9)

$$\sum_{k=1}^{n} A_{k,n+1}^{*0} \ddot{q}_k + \sum_{k=1}^{n} \sum_{m=1}^{n} (\Gamma_{km,n+1}^*)_0 \, \dot{q}_k \dot{q}_m = Q^* + \left(1 - \varepsilon_1 \mu \frac{\partial v_T^*}{\partial \dot{h}}\right)_0 R,$$
(4.10)

where Q_s and Q^* denote the generalised active forces corresponding to coordinates q_s and h respectively, $A_{ks}^0 (k = 1, ..., n)$ are the coefficients of the kinetic energy of the system before the constraints are removed and $A_{ij}^* (i, j = 1, ..., n+1)$ are those after the constraints have been removed. Christoffel's symbols of the first kind are given by

$$(\Gamma_{km,s})_0 = \frac{1}{2}\left(\frac{\partial A_{ks}^0}{\partial q_m} + \frac{\partial A_{ms}^0}{\partial q_k} - \frac{\partial A_{km}^0}{\partial q_s}\right),$$

$$(\Gamma_{km,n+1}^*)_0 = \frac{1}{2}\left(\frac{\partial A_{k\,n+1}^*}{\partial q_m} + \frac{\partial A_{m\,n+1}^*}{\partial q_k} - \frac{\partial A_{km}^*}{\partial h}\right)_0. \qquad (4.11)$$

In the particular case in which the trajectory of the additional displacement of the slider is orthogonal to the velocity vector $((\partial \mathbf{r}_T/\partial h)_0 \perp \mathbf{v}_T^0)$ we use the general notation but omit subscript $*$. Then

$$\left(\frac{\partial v_T}{\partial \dot{h}}\right)_0 = \frac{\mathbf{v}_T^0}{|\mathbf{v}_T^0|} \cdot \left(\frac{\partial \mathbf{r}_T}{\partial h}\right)_0 = 0, \quad S = R. \qquad (4.12)$$

Hence, Lagrange's equations are as follows

$$\sum_{k=1}^{n} A_{ks}^0 \ddot{q}_k + \sum_{k=1}^{n} \sum_{m=1}^{n} (\Gamma_{km,s})_0 \, \dot{q}_k \dot{q}_m = Q_s - \varepsilon_1 \mu \frac{\partial v_T^0}{\partial \dot{q}_s} R, \quad (s = 1, ..., n),$$

$$\sum_{k=1}^{n} A_{k\,n+1}^0 \ddot{q}_k + \sum_{k=1}^{n} \sum_{m=1}^{n} (\Gamma_{km,n+1})_0 \, \dot{q}_k \dot{q}_m = Q + R. \qquad (4.13)$$

128   4. Systems with many degrees of freedom and a single frictional pair

It goes without saying that systems (4.10) and (4.13) are equivalent. More precisely, the $(n+1)-th$ equation in (4.10) is a linear combination of the equations in (4.13), as was stated in Chapter 2.

4.2 Equation for the constraint force, differential equation of motion and the criterion of paradoxes

4.2.1 Determination of the constraint force and acceleration

The goal is to resolve systems (4.10) and (4.13) for R and $\ddot{q}$. Let system (4.13) be set in the form

$$DX \equiv \begin{bmatrix} A^0_{11} & A^0_{21} & \cdots & A^0_{n1} & \varepsilon_1\mu\dfrac{\partial v^0_T}{\partial \dot{q}_1} \\ \cdots & \cdots & \cdots & \cdots & \cdots \\ A^0_{12} & A^0_{2n} & \cdots & A_{nn} & \varepsilon_1\mu\dfrac{\partial v^0_T}{\partial \dot{q}_n} \\ A^0_{1n+1} & A^0_{2n+1} & \cdots & A_{nn+1} & -1 \end{bmatrix} \begin{bmatrix} \ddot{q}_1 \\ \cdots \\ \ddot{q}_n \\ R \end{bmatrix}$$

$$= \begin{bmatrix} Q_1 & -\sum_{k,l}\Gamma_{kl,1} & \ddot{q}_k\dot{q}_l \\ \cdots & \cdots & \cdots \\ Q_n & -\sum_{k,l}\Gamma_{kl,n} & \ddot{q}_k\dot{q}_l \\ Q & -\sum_{k,l}\Gamma_{kl,n+1} & \ddot{q}_k\dot{q}_l \end{bmatrix}, \qquad (4.14)$$

where X denotes a column of variables $\ddot{q}_1, ..., \ddot{q}_n, R$ and D is the matrix of coefficients of X.

The algebraic adjuncts of the elements $\varepsilon_1\mu(\partial v^0_T/\partial \dot{q}_k)$ and (-1) of matrix D are correspondingly

$$-A_k = -\begin{bmatrix} A^0_{11} & \cdots & A^0_{n1} \\ \cdots & \cdots & \cdots \\ A^0_{1k-1} & \cdots & A^0_{n\,k-1} \\ A_{1n+1} & \cdots & A_{nn+1} \\ A_{1k+1} & \cdots & A_{nk+1} \\ \cdots & \cdots & \cdots \\ A^0_{1n} & \cdots & A^0_{nn} \end{bmatrix}, \qquad (4.15)$$

$$A = \begin{bmatrix} A^0_{11} & \cdots & A^0_{n1} \\ \cdots & \cdots & \cdots \\ A^0_{1n} & \cdots & A^0_{nn} \end{bmatrix}. \qquad (4.16)$$

4.2 Constraint force, equation of motion and the criterion of paradoxes

As one can see, A_k is obtained from A by replacing elements $A^0_{1k}, ..., A^0_{nk}$ of the $k-th$ row by the elements $A^0_{1n+1}, ..., A^0_{n\,n+1}$.

Along with D we consider the matrix of quadratic form of the generalised velocities of the system with removed constraint

$$a = \begin{bmatrix} A^0_{11} & \cdots & A^0_{n1} & A^0_{n+11} \\ \cdots & \cdots & \cdots & \cdots \\ A^0_{1n} & \cdots & A^0_{nn} & A^0_{n+1n} \\ A^0_{1,n+1} & \cdots & A^0_{n,n+1} & A^0_{n+1,n+1} \end{bmatrix}. \tag{4.17}$$

It is clear that the algebraic adjuncts of the elements $A^0_{n+1,1}, ..., A^0_{n+1,n+1}$ of this matrix are respectively $-A_1, ..., -A_n$ and A. Thus, the elements of the $(n+1)-th$ row of the inverse matrix a^{-1} are

$$A^{k\,n+1} = \frac{-A_k}{\det a}, \quad (k=1,...,n), \qquad A^{n+1\,n+1} = \frac{A}{\det a}. \tag{4.18}$$

Expanding the determinant of matrix D in terms of its last column and taking into account eqs. (4.15)-(4.18) we obtain the following result

$$\det D = -A(1 + \varepsilon_1 \mu \mathsf{L}), \tag{4.19}$$

where

$$\mathsf{L} = \sum_{k=1}^{n} \frac{A_k}{A} \frac{\partial v^0_T}{\partial \dot{q}_k} = -\sum_{k=1}^{n} \frac{A^{k\,n+1}}{A^{n+1\,n+1}} \frac{\partial v^0_T}{\partial \dot{q}_k}. \tag{4.20}$$

For the sake of convenience of resolving eq. (4.20) by means of Cramer's rule, we introduce the following notation

$$R_0 = \frac{1}{A} \begin{bmatrix} A^0_{11} & \cdots & A^0_{n1} & \sum \Gamma_{k,l,1} \dot{q}_k \dot{q}_l - Q_1 \\ \cdots & \cdots & \cdots & \cdots \\ A^0_{1n} & \cdots & A^0_{nn} & \sum \Gamma_{k,l,h} \dot{q}_k \dot{q}_l - Q_n \\ A^0_{1\,n+1} & \cdots & A^0_{n\,n+1} & \sum \Gamma_{k,l,n+1} \dot{q}_k \dot{q}_l - Q \end{bmatrix}$$

$$= \frac{1}{A^{n+1\,n+1}} \left[\sum_{k=1}^{n} \sum_{l=1}^{n} \Gamma^{n+1}_{k,l} \dot{q}_k \dot{q}_l - \sum_{k=1}^{n} A^{kn+1} Q_k - A^{n+1\,n+1} Q \right], \tag{4.21}$$

$$F_s = \frac{1}{A} \begin{bmatrix} A^0_{11} & \cdots & A^0_{s-11} & \sum_{k,l=1}^{n} \Gamma_{k,l,r} \dot{q}_k \dot{q}_l & A^0_{s+1,1} \cdots \varepsilon_1 \mu \frac{\partial v^0_T}{\partial \dot{q}_1} \\ \cdots & \cdots & \cdots & \cdots & \cdots \\ A^0_{1n} & \cdots & A^0_{s-1n} & \sum \Gamma_{k,l,n} \dot{q}_k \dot{q}_l & A^0_{s+1,n} \cdots \varepsilon_1 \mu \frac{\partial v^0_T}{\partial \dot{q}_n} \\ A^0_{1\,n+1} & \cdots & A^0_{s-1\,n+1} & \sum \Gamma_{k,l,n+1} \dot{q}_k \dot{q}_l & A^0_{s+1,\,n+1} \cdots -1 \end{bmatrix}, \tag{4.22}$$

130   4. Systems with many degrees of freedom and a single frictional pair

$$G_s = -\frac{1}{A}\begin{bmatrix} A^0_{11} & \cdots & A^0_{s-11} & Q_l A^0_{s+11} & \cdots & \varepsilon_1\mu\dfrac{\partial v^0_T}{\partial \dot{q}_1} \\ \cdots & \cdots & \cdots & \cdots & \cdots & \cdots \\ A^0_{1\,n+1} & \cdots & A^0_{s-1\,n+1} & QA^0_{s+1\,n+1} & \cdots & -1 \end{bmatrix},$$
(4.23)

where γ^l_{kt} denotes Christoffel's symbols of the second kind

$$\Gamma^l_{k,t} = \sum_{s=1}^{n+1} A^{sl}\gamma_{kt,s}, \quad (l,k,t = 1,...,n+1).$$

The value of AR_0 is obtained from $\det D$ by replacing the elements of the $(n+1)-th$ column by the elements of the right hand side of eq. (4.14), where R_0 denotes the normal reaction force of the contact constraint in the ideal case ($\mu = 0$). The value of $A(F_s + G_s)$ is obtained from $\det D$ by a similar replacement of the elements of the $s-th$ column. Besides, F_s and G_s contain terms depending correspondingly on velocities and active forces.

If we consider eq. (4.10) instead of (4.13), then instead of eq. (4.20) we obtain the following expression for L

$$\mathsf{L} = \sum_{k=1}^{n} \frac{A^*_k}{A}\frac{\partial v^0_T}{\partial \dot{q}_k} - \left(\frac{\partial v^*_T}{\partial h}\right)_0,$$
(4.24)

where A^*_k is determined by formula (4.15) in which $A^0_{s\,n+1}$ is replaced by $A^{*0}_{s\,n+1}$. In addition to this, R_0, F_s and G_s are also given by determinants (4.21)-(4.23) in which the elements $A^0_{s\,n+1}, (\gamma_{kl,n+1})_0$ and (-1) are replaced by the elements $A^{*0}_{s\,n+1}, (\gamma_{kl,n+1})^*_0$ and $\varepsilon_1\mu(\partial v^*_T/\partial h)_0 - 1$, respectively.

Finally, resolving eq. (4.14) and accounting for eqs. (4.19)-(4.24) we arrive at the differential equations of motion and the equations for the reaction force

$$(1 + \varepsilon_1\mu\mathsf{L})\ddot{q}_s = F_s + G_s \quad (s = 1,...,n),$$
(4.25)

$$(1 + \varepsilon_1\mu\mathsf{L})R = R_0.$$
(4.26)

As will be shown below, this form of notation is convenient for the proof of the paradoxes' conditions and the true motion.

Equations (4.25) and (4.26) resemble relationships (2.59) derived for a system with one degree of freedom and a single frictional pair. Moreover, while comparing eqs. (4.18)-(4.23) with eqs. (2.60)-(2.63) one can see that eqs. (4.25) and (4.26) become identical to those in eq. (2.59) if $n = 1$. Indeed, in this case formulae (4.20), (4.22) and (4.23) reduce to the following

$$\mathsf{L} = \varepsilon_2\frac{A_{12}}{A}\left|\frac{d\mathbf{r}^0_T}{dq}\right| = \varepsilon_2 L, \quad (\varepsilon_2 = \operatorname{sign}\dot{q}), \quad F_1 = -H\dot{q}^2, \quad G_1 = G.$$

Formula (4.21) for the normal reaction force R_0 of the ideal system takes the form of eq. (2.63).

The product μL characterises the change in the normal reaction force caused by Coulomb friction. For this reason, L is referred to as the index of influence of the contact constraint.

4.2.2 Criterion of Painlevé's paradoxes

The sign of the reaction force $\varepsilon_1 = \text{sign } R$ can be determined with the help of eq. (4.26). Provided that for some values of $q_1, ..., q_n$ and $\dot{q}_1, ..., \dot{q}_n$ the sign of ε_1 can not be determined or there exist simultaneously two signs of $\varepsilon_1 = \pm 1$, then for these values of the coordinates and velocities, the dynamical problem has either no solution or the solution is non-unique.

Theorem 7. If

$$\mu |L| < 1, \tag{4.27}$$

then

$$\varepsilon_1 = \varepsilon_0 = \text{sign } R_0 \tag{4.28}$$

and thus the solution of problem (4.25)-(4.26) exists and is unique. However if

$$\mu |L| > 1, \tag{4.29}$$

then

$$\varepsilon_1 = \begin{cases} \pm 1 & \text{for} \quad \varepsilon_0 \text{ sign } L = 1 \\ \pm 1 \sqrt{-1} = \pm i & \text{for} \quad \varepsilon_0 \text{ sign } L = -1 \end{cases} \tag{4.30}$$

and the solution is not unique for $\varepsilon_0 \text{ sign } L = 1$ and does not exist for $\varepsilon_0 \text{ sign } L = -1$.

Proof. It follows from eq. (4.26) that

$$\varepsilon_1 \text{sign}(1 + \varepsilon_1 \mu L) = \varepsilon_0. \tag{4.31}$$

Then under condition (4.27) we obtain $\varepsilon_1 = \varepsilon_0$.

When condition (4.29) is met, we have $\text{sign}(1 + \varepsilon_1 \mu L) = \varepsilon_1 \text{ sign } L$. Then eq. (4.31) takes the form $\varepsilon_1^2 = \varepsilon_0 \text{ sign } L$ which is equivalent to relationship (4.30).

Remark 1. Since $L = L(q_1, ...q_n, \dot{q}_1, .., \dot{q}_n)$ and $R_0 = R_0(q_1, ...q_n, \dot{q}_1, .., \dot{q}_n)$, then conditions (4.29) and (4.30) determine the region of paradoxes whose border in phase space $(q_1, q_2, ...q_n, \dot{q}_1, .., \dot{q}_n)$ is given by the following equation

$$\mu |L(q_1, ...q_n, \dot{q}_1, .., \dot{q}_n)| - 1 = 0.$$

Remark 2. By virtue of Theorem 7 the coefficient $(1+\varepsilon_1\mu\mathsf{L})$ in front of $\ddot{q}_s$ and R in eqs. (4.25) and (4.26) can be cast as follows

$$1+\varepsilon_1\mu\mathsf{L} = 1\pm\mathsf{L} \quad \text{for} \quad \begin{cases} \varepsilon_0\,\text{sign}\,\mathsf{L} = \pm 1 & \text{for} \quad \mu|\mathsf{L}| < 1, \\ \varepsilon_0\,\text{sign}\,\mathsf{L} = 1 & \text{for} \quad \mu|\mathsf{L}| > 1. \end{cases}$$

Hence, on the border of the region of paradoxes, we have

$$\lim_{\mu|\mathsf{L}|\to 1-0}(R/R_0)\begin{cases} = 1/2 & \text{for} \quad \varepsilon_0\,\text{sign}\,\mathsf{L} = 1, \\ = \infty & \text{for} \quad \varepsilon_0\,\text{sign}\,\mathsf{L} = -1, \end{cases}$$

$$\lim_{\mu|\mathsf{L}|\to 1+0}(R/R_0)\begin{cases} = 1/2 \text{ or } \infty & \text{for} \quad \varepsilon_0\,\text{sign}\,\mathsf{L} = 1, \\ \text{does not exist} & \text{for} \quad \varepsilon_0\,\text{sign}\,\mathsf{L} = -1. \end{cases}$$

As one can see, the left limit is determined uniquely, however it tends to infinity at $\varepsilon_0\,\text{sign}\,\mathsf{L} = -1$ whilst the right limit is either non-unique or does not exist. For this reason, the boundary of the region is considered as belonging to this region.

4.3 Determination of the true motion

The non-existence and non-uniqueness of the solution of the dynamical problem contradicts the main principle of mechanics which states that the motion exists and is unique, [110], [117]. This means that eqs. (4.25) and (4.26) are not correct in the region of paradoxes because they do not correctly describe the dynamics of the system in this region. This incorrectness is clearly a consequence of one of the assumptions made for the frictional contact interaction of the bodies.

The first assumption is that the equations do not account for the gap between the slider and the two parallel planes which ensures a two-sided contact constraint. When the gap is taken into account we have two different systems with one-sided constraints instead of one system with a two-sided constraint. As the examples of the inhomogeneous disc, stacker etc. show, paradoxes occur even in this case. With this in view, the gap can not be used to explain the considered phenomenon.

The second assumption is an idealisation which suggests that the contacting bodies are absolutely rigid. Many researchers are of the opinion that this contradiction can be removed by introducing elastic deformations in the contact zone. This viewpoint has been illustrated time and again for many examples, see [1], [27], [94], [125].

On the other hand, as mentioned earlier, the question of plausibility of the "position of elastic deformation" remains open because of the absence of a general investigation. As for the question of the true motion, there is no consensus of opinion on how to establish these motions, see [49], [117],

[118]. An attempt to make up for this deficiency is undertaken in what follows.

By using general equations (4.10) or (4.13) one can easily prove that no paradox occurs if the normal reaction force R is viscous-elastic. Indeed, let R be uniquely expressed in terms of the generalised coordinates and velocities

$$R = f(q_1, ..., q_n, \dot{q}_1, ..., \dot{q}_n).$$

Then the first n equations (4.10) or (4.13) take the form

$$\sum_{k=1}^{n} A_{ks}^0 \ddot{q}_k + \sum_{k=1}^{n}\sum_{m=1}^{n} \Gamma_{k,m,s} \dot{q}_k \dot{q}_m = Q_s - \mu \frac{\partial v_T^0}{\partial \dot{q}_s} |f(q_1, ..., q_n, \dot{q}_1, ..., \dot{q}_n)|$$

$$(s = 1, ..., n).$$

Therefore, according to given $q_1, ..., q_n, \dot{q}_1, ..., \dot{q}_n$ the values of the reaction force R and accelerations $\ddot{q}_1, ..., \ddot{q}_n$ are determined uniquely, that is no paradox occurs.

This confirms the validity of the above-mentioned opinion for the origin of the considered phenomena but does not exhaustively solve the problem. It is necessary to know what motion should be ascribed to the system in any paradoxical situation. To this aim let us consider the problem of a limiting process from an elastic contact to an absolutely rigid one. The viscosity will be neglected as it causes only attenuation rather than influencing the results of the study.

4.3.1 Limiting process

Let a system with Coulomb friction admit paradoxes. In order to understand them we consider a new system which is obtained from the old system by replacing the rigid contact joint by an elastic one. The law of motion derived for such an elastic system under a limiting process of changing the rigidity to infinity is taken as being the law of motion for the rigid system.

We assume that the result of the elastic deformation is that the slider moves along the guide by a small value

$$h = -R/c, \qquad (4.32)$$

where c denotes an equivalent rigidity. For the elastic system under consideration the position vectors $\mathbf{r}_1, ..., \mathbf{r}_n$ and $\mathbf{r}_T$ of the particles and the slider, respectively, are functions not only of coordinates $q_1, ..., q_n$ but also displacement h. The kinetic energy of this system can be expressed in the form

$$T = \frac{1}{2}\sum_{s=1}^{n}\sum_{k=1}^{n} A_{ks} \dot{q}_s \dot{q}_k + \sum_{s=1}^{n} A_{s,n+1} \dot{q}_s \dot{h} + \frac{1}{2} A_{n+1,n+1} \dot{h}^2. \qquad (4.33)$$

134   4. Systems with many degrees of freedom and a single frictional pair

The generalised reaction forces are calculated according to eqs. (4.8) and (4.12)

$$S_j = \varepsilon_1 \mu \frac{\partial v_T}{\partial \dot{q}_j} ch, \quad (j=1,...,n), \quad S = -ch. \tag{4.34}$$

Coefficients A_{ik} and the slip velocity v_T depend on $q_1,...,q_n$ and h. Since h is small we can set

$$A_{ik} = A_{ik}^0, \quad \frac{\partial v_T}{\partial \dot{q}_j} = \frac{\partial v_T^0}{\partial \dot{q}_j}. \tag{4.35}$$

By virtue of relationship (4.35) matrix $a = (A_{sk})$ of the quadratic form (4.33) is coincident with matrix (4.17). The elements $A^{n+1,s}(s=1,...,n+1)$ of the $(n+1)-th$ row of the inverse a^{-1} are given by eqs. (4.15), (4.16) and (4.18).

Constructing Lagrange's equations and having solved them for the generalised accelerations, we obtain

$$\ddot{h} + (1+\varepsilon_1\mu L)A^{n+1,n+1}ch + \sum_{k=1}^{n}\sum_{l=1}^{n}\Gamma_{kl}^{n+1}\dot{q}_k\dot{q}_l + 2\sum_{k=1}^{n}\Gamma_{k,n+1}^{n+1}\dot{q}_k\dot{h} +$$
$$\Gamma_{n+1,n+1}^{n+1}\dot{h}^2 - \sum_{k=1}^{n}A^{k,n+1}Q_k - A^{n+1,n+1}Q = 0, \tag{4.36}$$

where Γ_{kl}^s denotes Christoffel's symbols of the second kind, and $\varepsilon_1 = \mathrm{sign}\,R = -\mathrm{sign}\,h$. Substituting relationships (4.21) and (4.32) into eq. (4.36) yields the following equation for the reaction force

$$\ddot{R} + 2\sum_{k=1}^{n}\Gamma_{k,n+1}^{n+1}\dot{q}_k\dot{R}^2 + c^{-1}\Gamma_{n+1,n+1}^{n+1}\dot{R}^2 +$$
$$cA^{n+1,n+1}(1+\varepsilon_1\mu L)R = cR_0 A^{n+1,n+1}, \tag{4.37}$$

which differs from the algebraic equation (4.26) by the presence of terms depending on $\dot{R}$ and $\ddot{R}$. Therefore, when we consider the contact joint in Sections 4.1 and 4.2 as being absolutely rigid we restrict ourselves to the case of a stationary or slowly changing value of R.

Let us introduce the non-dimensional reaction force

$$x = R/R_0 = -ch/R_0. \tag{4.38}$$

Then eq. (4.37) takes the form

$$\ddot{x} + (1+\varepsilon_1\mu L)A^{n+1,n+1}cx - A^{n+1,n+1}c = -2\dot{x}\sum_{k=1}^{n}\Gamma_{k,n+1}^{n+1}\dot{q}_k +$$
$$c^{-1}R_0\Gamma_{k,n+1}^{n+1}\dot{x}^2. \tag{4.39}$$

Furthermore, by introducing the non-dimensional time

$$\tau = \sqrt{cA^{n+1,n+1}|1+\varepsilon_1\mu\mathsf{L}|}\,t \tag{4.40}$$

we transform eq. (4.39) to the following form

$$\frac{d^2x}{d\tau^2} + \frac{1+\varepsilon_1\mu\mathsf{L}}{|1+\varepsilon_1\mu\mathsf{L}|}x - \frac{1}{|1+\varepsilon_1\mu\mathsf{L}|} = \gamma f\left(\frac{dx}{d\tau}\right), \tag{4.41}$$

where

$$\gamma = (cA^{n+1,n+1}|1+\varepsilon_1\mu\mathsf{L}|)^{-1/2}\sum_{k=1}^{n}\Gamma_{k,n+1}^{n+1}\dot{q}_k,$$

$$f\frac{dx}{d\tau} = -2\frac{dx}{d\tau} + \frac{\Gamma_{n+1,n+1}^{n+1}R_0(A^{n+1,n+1}|1+\varepsilon_1\mu\mathsf{L}|)^{1/2}}{\sqrt{c}\sum_{k=1}^{n}\Gamma_{k,n+1}^{n+1}\dot{q}_k}\left(\frac{dx}{d\tau}\right)^2. \tag{4.42}$$

Here γ is a non-dimensional small parameter which tends to zero as rigidity c tends to infinity. Hence, when $c \to \infty$ the perturbation $\gamma f(dx/d\tau)$ can be neglected and then instead of eq. (4.41) we have the following differential equations for the reaction force

$$\frac{d^2x}{d\tau^2} + \frac{1+\varepsilon_1\mu\mathsf{L}}{|1+\varepsilon_1\mu\mathsf{L}|}x - \frac{1}{|1+\varepsilon_1\mu\mathsf{L}|} = 0. \tag{4.43}$$

In this equation the free term $|1+\varepsilon_1\mu\mathsf{L}|^{-1}$ remains unchanged as $c \to \infty$ in a sufficiently large interval of time τ due to the following condition

$$\lim_{c\to\infty}\frac{d}{d\tau}\frac{1}{|1+\varepsilon_1\mu\mathsf{L}|} = \lim_{c\to\infty}\frac{1}{\sqrt{cA^{n+1,n+1}|1+\varepsilon_1\mu\mathsf{L}|}}\frac{d}{dt}\frac{1}{|1+\varepsilon_1\mu\mathsf{L}|} = 0$$

and the coefficient of x is equal to ± 1.

Thus, the problem of limiting process $c \to \infty$ is reduced to the differential equation with constant coefficients (4.43).

When the paradoxes are absent, then $\mu|\mathsf{L}| < 1$. In this case $|1+\varepsilon_1\mu\mathsf{L}| = 1+\varepsilon_1\mu\mathsf{L}$ and it follows from eq. (4.43) that

$$x = r_0\sin(\tau+\varphi_0) + \frac{1}{1+\varepsilon_1\mu\mathsf{L}} = r_0\sin(\omega_0 t+\varphi_0) + \frac{1}{1+\varepsilon_1\mu\mathsf{L}},$$

$$\omega_0 = \sqrt{cA^{n+1,n+1}|1+\varepsilon_1\mu\mathsf{L}|} \quad\text{for}\quad \mu|\mathsf{L}| < 1,$$

where r_0 and q_0 are determined from the initial conditions. The stationary value

$$x = \frac{1}{1+\varepsilon_0\mu\mathsf{L}}, \quad R = \frac{R_0}{1+\varepsilon_0\mu\mathsf{L}}, \quad \varepsilon_1 = \varepsilon_0$$

coincides with the root of the algebraic equation (4.26) of the rigid system for $\mu|\mathsf{L}| < 1$. One can see that outside the region of paradoxes, the stationary value of the reaction force becomes coincident with the value of the reaction force of the rigid system as $c \to \infty$. In other words, in the case of no paradoxes, the assumption of a rigid contact does not affect the dynamic characteristic of the system with friction.

Let us construct the equation for the reaction force (4.43) under paradoxes, i.e. in the case of $\mu|L| > 1$. We distinguish between two cases, namely $\varepsilon_1 \operatorname{sign} L = 1$ and $\varepsilon_1 \operatorname{sign} L = -1$. In the first case $\varepsilon_1 L = |L|$, and the solution is given by

$$x = r\sin(\tau + \varphi) + \frac{1}{1+\mu|L|} = r\sin(\omega t + \varphi) + \frac{1}{1+\mu|L|},$$
$$\omega = \sqrt{cA^{n+1,n+1}(1+\mu|L|)} \quad (\mu|L| > 1,\ \varepsilon_1 \operatorname{sign} L = 1). \quad (4.44)$$

The phase trajectories are ellipses

$$\left(x - \frac{1}{1+\mu|L|}\right)^2 + \frac{\dot{x}^2}{\omega^2} = r^2, \quad (\mu|L| > 1,\ \varepsilon_1 \operatorname{sign} L = 1) \quad (4.45)$$

with the stable centre

$$x^+ = \frac{1}{\mu|L|+1}, \quad \dot{x} = 0, \quad R^+ = \frac{R_0}{\mu|L|+1}. \quad (4.46)$$

In the second case, in which $\varepsilon_1 \operatorname{sign} L = -1$, we have $\varepsilon_1 L = -|L|$, and solution (4.43) can be represented in the form

$$x = r_1 e^\tau + r_2 e^{-\tau} - \frac{1}{\mu|L|-1} = r_1 e^{\lambda t} + r_2 e^{-\lambda t} - \frac{1}{\mu|L|-1},$$
$$\lambda = \sqrt{cA^{n+1,n+1}(\mu|L|-1)} \quad (\mu|L| > 1,\ \varepsilon_1 \operatorname{sign} L = -1), \quad (4.47)$$

where r_1, r_2 are the integration constants. In accordance with eq. (4.45) the phase trajectories are the hyperbolas

$$\left(x + \frac{1}{\mu|L|-1}\right)^2 - \frac{\dot{x}^2}{\lambda} = 4r_1 r_2, \quad (\mu|L| > 1,\ \varepsilon_1 \operatorname{sign} L = -1) \quad (4.48)$$

with the unstable saddle point

$$x^- = \frac{-1}{\mu|L|-1}, \quad \dot{x} = 0, \quad R^- = \frac{-R_0}{\mu|L|-1} \quad (4.49)$$

and the asymptotes

$$\dot{x} = \pm \lambda \left[x + \frac{1}{\mu|L|-1}\right].$$

We considered above the case of $\mu|L| > 1$. The differential equation for the reaction force is derived and its solutions are obtained for two signs $\varepsilon_1 \operatorname{sign} L = \pm 1$. In order to find the true reaction force in any paradoxical situation we will establish the region of each sign in the plane $(x, \dot{x})$ and match the solutions.

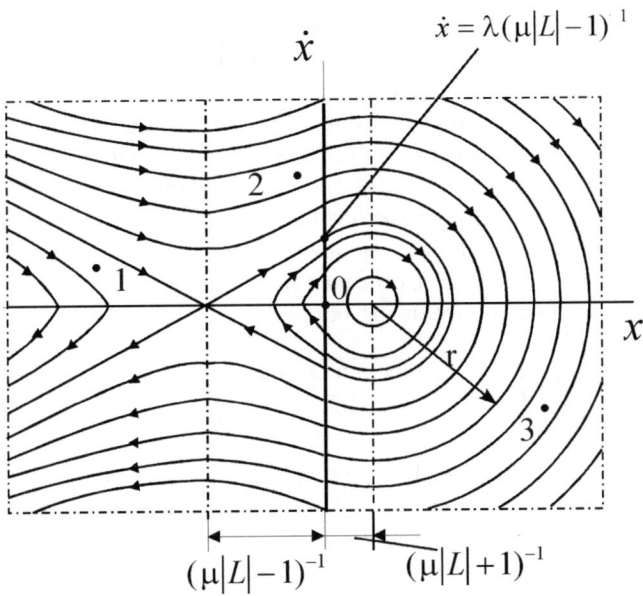

FIGURE 4.1.

4.3.2 True motions under the paradoxes

Let us determine the true laws of change of the reaction force x and the coordinates q_s in the paradoxical situations. We begin with the theorem on non-uniqueness. By Theorem 7

$$\mu|L| > 1, \quad \varepsilon_0 \operatorname{sign} L = 1. \tag{4.50}$$

For $x > 0$ it follows from eqs. (4.38) and (4.50) that $\varepsilon_1 = \varepsilon_0$, $\varepsilon_1 \operatorname{sign} L = 1$. Correspondingly, quantity x obeys the law (4.44). Hence, in the right half-space $(x, \dot{x})$ the ellipses (4.45) are observed.

Since for $x < 0$ and due to eqs. (4.38) and (4.50) we have $\varepsilon_1 = -\varepsilon_0$ and $\varepsilon_1 \operatorname{sign} L = -1$, then in the left half-space $(x, \dot{x})$ quantity x is given by eq. (4.47), and the phase curves are hyperbolas (4.48).

In the case of non-uniqueness of the problem of the dynamics of the rigid system the complete phase portrait is obtained by matching the left hyperbolas with the right ellipses, see Fig. 4.1. Now it becomes clear that the reaction force R and thus the accelerations $\ddot{q}_1, ..., \ddot{q}_n$ are determined uniquely under the limiting process $c \to \infty$. The two roots of eq. (4.26) of the rigid system for $\varepsilon_1 = \pm 1$ are stationary values of the reaction force corresponding to the centre of the ellipses and the saddle point of the hyperbolas. Moreover, depending upon the initial values $x(0)$ and $\dot{x}(0)$ the motion can belong to one of two typical cases: (i) the case in which the representing point $(x, \dot{x})$ moves permanently about the centre of the ellipses $(x^+, 0)$ and (ii) the case in which the value of x becomes negative at a certain time instant and then its absolute value increases rapidly due to exponential law (4.47) where $r_1 < 0$.

In the first case, due to the viscous damping which was not taken into account within the derivation of eq. (4.39), the oscillatory components of quantities x and $\dot{x}$ attenuate and the representing point approaches the centre $(x^+, 0)$. Indeed, if, instead of (4.32), we take $R = -ch - \alpha \dot{h}$, with α denoting the coefficient of the viscous damping, then eq. (4.33) gains the term $(1+\varepsilon_1 \mu \mathsf{L})A^{n+1,n+1}\alpha \dot{h}$. Entering the notation $x = -chR_0^{-1}$ we obtain, instead of eq. (4.43), the following equation

$$\frac{d^2 x}{d\tau^2} + \frac{1+\varepsilon_1 \mu \mathsf{L}}{|1+\varepsilon_1 \mu \mathsf{L}|}\left(\alpha\sqrt{\frac{A^{n+1,n+1}|1+\varepsilon_1 \mu \mathsf{L}|}{c}}\frac{dx}{d\tau} + x\right) - \frac{1}{|1+\varepsilon_1 \mu \mathsf{L}|} = 0.$$

Under the condition $\varepsilon_1 \operatorname{sign} \mathsf{L} = 1$, $\mu|\mathsf{L}| > 1$ this equation describes a damped oscillation in the vicinity of point $x = x^+$. Therefore, in the case under consideration the law of motion eventually becomes equivalent to the original Painlevé's principle $\varepsilon_1 = \varepsilon_0$, $R = R^+$.

The second case takes place, for instance, when the initial representing point is coincident with one of the points 1, 2, 3 in Fig. 4.1. Let us prove that a so-called tangential impact occurs. Resolving eq. (4.10) together with eq. (4.38) we obtain

$$\ddot{q}_s + \varepsilon_1 \mu R_0 x K_s - E_s = 0,$$

$$K_s = \sum_k a^{ks}\frac{v_T}{\dot{q}_k}, \quad E_s = \sum_k a^{ks} Q_k - \sum_{k,l}\Gamma^s_{k,l}\dot{q}_k \dot{q}_l, \qquad (4.51)$$

where a^{ks} are the elements of the matrix which is the inverse of the matrix of quadratic form of velocities of the rigid system. Using eqs. (4.47) and (4.51) yields the change in velocity $\dot{q}_s$ within the time interval Δt

$$\Delta \dot{q}_s = \int_0^{\Delta t} \ddot{q}_s dt \qquad (4.52)$$

$$= \varepsilon_1 \lambda^{-1}\mu R_0 K_s \left[r_1(e^{\lambda \Delta t} - 1) - r_2(e^{-\lambda \Delta t} - 1) + x^- \Delta t\right] + E_s \Delta t.$$

Judging from eq. (4.51), among the generalised coordinates there exists at least one coordinate q_s such that

$$K_s \neq 0, \qquad (4.53)$$

otherwise Coulomb friction would have no influence on the dynamics of the system and the system itself would be ideal.

If we expand the right hand side of eq. (4.52) as a power series, then by accounting for expression (4.47) for λ we can see that the terms are proportional to $(\Delta t/1! + \lambda(\Delta t)^2/2! + \lambda^2(\Delta t)^3/3! + ...)$ and Δt, respectively. On the other hand, λ is proportional to $\sqrt{c}$. Thus, the second term can be neglected under condition (4.53) and $c \to \infty$. As follows from eq. (4.52)

$$\Delta \dot{q}_s = -\lambda^{-1} \mu R_0 r_1 (e^{\lambda \Delta t} - 1) K_s. \tag{4.54}$$

Since $\Delta t > 0$, we have sign $\Delta \dot{q}_s = -\text{sign}(R_0 r_1 K_s)$ and by means of eqs. (4.47) and (4.54) we obtain

$$\Delta t = \lambda^{-1} \ln \left[1 + \frac{\lambda |\Delta \dot{q}_s|}{(\mu |R_0 r_1 K_s|)} \right]. \tag{4.55}$$

Thus, as $\Delta \dot{q}_s$ is bounded we have

$$\lim_{c \to \infty} \Delta t = 0. \tag{4.56}$$

Therefore, for a fixed increment in the generalised velocity $\Delta \dot{q}_s$, the time interval Δt decreases with growth of c and tends to zero. The velocities of the particles $\mathbf{v}_i$ and the slider $\mathbf{v}_T$ also admit jumps due to the condition

$$\Delta \mathbf{v}_i \approx \sum_s \left(\frac{\mathbf{r}_i}{q_s}\right) \Delta \dot{q}_s, \quad \Delta \mathbf{v}_T \approx \sum_s \left(\frac{\mathbf{r}_T}{q_s}\right) \Delta \dot{q}_s. \tag{4.57}$$

Such a discontinuous change in the velocity of the system with friction is referred to as tangential impact, see [49].

The obtained results of the analysis of paradoxical non-uniqueness can be generalised in the form of the following theorem.

Theorem 8. Paradoxical non-uniqueness reflects the situation in which the reaction force R has two stationary values $R^{\pm}$ equal to the roots of equation (4.26) for $\varepsilon_1 = \pm \text{sign L}$. Values R^+ and R^- correspond respectively to the stable centre (4.46) and unstable saddle point (4.49). For some initial values of displacements and velocities $x(0)$ and $\dot{x}(0)$ the motion is about the centre whereas for other initial values quantity x rapidly increases in accordance with eq. (4.47) which results in a tangential impact (4.54)-(4.57), i.e. a discontinuous change in velocity, see Fig. 4.1.

In contrast to the principle suggested by Painlevé [117], [118] and Klein [62], Theorem 8 confirms the feasibility of not only the stable stationary solution for the reaction force in the system but also the unstable non-stationary solutions resulting in tangential impact. With this in mind, the viewpoint of Ivanov [49] on the question of the true motion in the situation of the non-uniqueness does not contradict the Painlevé-Klein principle and even complements it. The present approach to the problem of tangential impact is different from [49] to some extent.

Let us establish the true motion in the situation of non-existence of the solution of the problem of dynamics of the rigid system (4.25)-(4.26). According to Theorem 7, in this case

$$\mu |L| > 1, \quad \varepsilon_0 \text{sign L} = -1. \tag{4.58}$$

140   4. Systems with many degrees of freedom and a single frictional pair

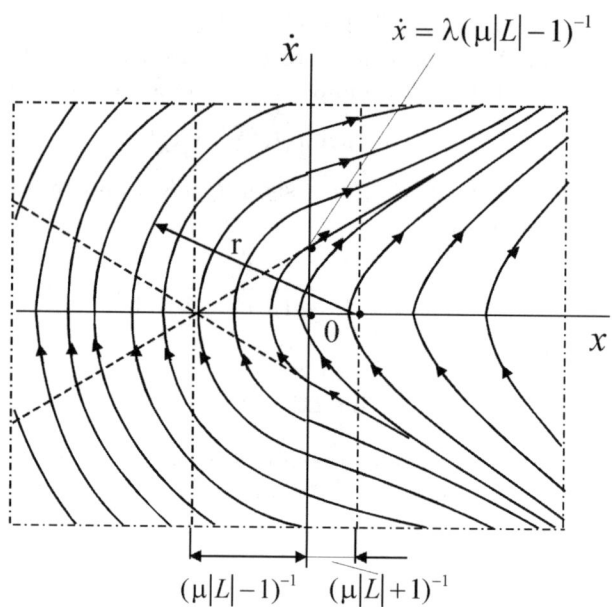

FIGURE 4.2.

Inasmuch as in the right half-space $x > 0$ and $\varepsilon_1 = \varepsilon_0$ we obtain from eq. (4.58) that $\varepsilon_1 \operatorname{sign} L = -1$. In accordance with this condition x has an exponential character, cf. (4.47). Thus, the right hand side of the phase portrait, Fig. 4.2, consists of hyperbolas (4.48).

In the left hand side of the phase plane we have $\varepsilon_1 = -\varepsilon_0$, $\varepsilon_1 \operatorname{sign} L = 1$, and x is described by a sine-function (4.44). Hence, the left hand side of the phase portrait consists of the arches $(x < 0)$ of the ellipses, see (4.45).

As one can see from Fig. 4.2, the centre (4.46) and the saddle point (4.49) lie off the regions of ellipses and hyperbolas, respectively, that is, there exist no stationary solutions $x^{\pm}$. However, in accordance with the prescribed initial values $x(0)$ and $\dot{x}(0)$ the solution exists and is unique. For any $x(0)$ and $\dot{x}(0)$ the representative point reaches eventually the first quarter of the phase space where x increases exponentially, see eq. (4.47) with $r_1 > 0$. Taking into account eq. (4.51) we can again arrive at relationships (4.54)-(4.57), confirming the discontinuous change in the velocity. Thus, we have proved the following theorem.

Theorem 9. Paradoxical non-existence reflects the situation in which the reaction force R has no stationary values. For any initial values $x(0)$ and $\dot{x}(0)$ quantity x eventually becomes positive and increases rapidly due to the exponential law (4.47) which leads to a tangential impact in the form (4.54)-(4.57), see Fig. 4.2.

Corollary to Theorems 8 and 9. Under any tangential impact the motion of the single-degree-of-freedom system is stopped and a dynamic seizure occurs.

Proof. In the case of $n = 1$ eq. (4.52) takes the form

$$A\ddot{q} = -\varepsilon_1\varepsilon_2\mu \left|\frac{d\mathbf{r}_T}{dq}\right| R_0 x - \frac{1}{2}\frac{dA}{dq}\dot{q}^2 + Q_1,$$

$$\varepsilon_1 = \text{sign}(R_0 x), \quad \varepsilon_2 = \text{sign}\,\dot{q}. \qquad (4.59)$$

As mentioned above, in the case of tangential impact (in the case of both non-uniqueness and non-existence) quantity x is given by formula (4.47). It follows from eq. (4.59) that

$$\text{sign}\,\ddot{q} = -\text{sign}\left(\mu\left|\frac{d\mathbf{r}_T}{dq}\right| R_0 r\right| e^{\lambda t}\right)\text{sign}\,\dot{q} = -\text{sign}\,\dot{q}. \qquad (4.60)$$

Relationship (4.60) shows that the sign of acceleration $\ddot{q}$ is opposite to the sign of velocity $\dot{q}$ and thus the absolute value of the latter will decrease until the motion is stopped. The increment in velocity from time instant $t = 0$ to the instant when the stop takes place is given by $\triangle \dot{q} = -\dot{q}(0)$. Besides, for $n = 1$

$$\frac{\partial v_T}{\partial \dot{q}} = \left|\frac{d\mathbf{r}_T}{dq}\right|\text{sign}\,\dot{q}.$$

Hence, by virtue of eqs. (4.55) and (4.56)

$$\dot{q}(\triangle t) = 0, \quad \triangle t = \frac{1}{\lambda}\ln\left(1 + \left|\frac{\lambda A \dot{q}(0)}{\mu R_0 r(d\mathbf{r}_T/dq)}\right|\right), \quad \lim_{c\to\infty}\triangle t = 0 \qquad (4.61)$$

and an instantaneous stop occurs.

Thus, for single-degree-of-freedom systems the dynamic seizure is inevitable in the paradoxical situations of non-existence and may occur for some initial values of the reaction force and its time-derivative. The corollary is proved.

Theorems 8 and 9 and their corollary enables us to judge the true motion of all mechanisms studied in Chapter 3 and Sections 4.5 and 4.6 without repeating the procedure of accounting for the elastic deformation of the contact. Nonetheless, in order to convince ourselves of the validity of the approach, this procedure is repeated for the Painlevé-Klein scheme in Section 4.4 once again. As it is shown below, we arrive at the same results which are predicted by Theorems 8 and 9 and their corollary.

4.4 True motions in the Painlevé-Klein problem in paradoxical situations

The description of the Painlevé-Klein scheme is given in Section 3.2 and in Fig. 3.3. Here for the brevity of notation we take $M_1 = M_2 = 1$.

4.4.1 Equations for the reaction force

Let us derive the differential equation for the reaction force and construct its solution. To this end, in addition to the expression for the coefficients (3.5) and (3.9), it is necessary to determine A_{22}. Let us assume that the slider, due to the contact compliance, moves vertically by the value of $h = -c^{-1}R$. Then

$$A_{22} = \left|\frac{\partial \mathbf{r}_1}{\partial h}\right|^2 + \left|\frac{\partial \mathbf{r}_2}{\partial h}\right|^2 = 1 + \tan^2 \varphi. \tag{4.62}$$

Making use of eqs. (3.9) and (4.62) we can write

$$a = \begin{bmatrix} 2 & \tan \varphi \\ \tan \varphi & 1 + \tan^2 \varphi \end{bmatrix}, \quad a^{-1} = \frac{1}{2 + \tan^2 \varphi} \begin{bmatrix} 1 + \tan^2 \varphi & -\tan \varphi \\ -\tan \varphi & 2 \end{bmatrix},$$

$$\det a = 2 + \tan^2 \varphi, \quad \mathsf{L} = \varepsilon_2 \frac{\tan \varphi}{2}, \quad R_0 = \frac{P_1 - P_2}{2} \tan \varphi, \quad \varepsilon_2 = \operatorname{sign} \dot{q}. \tag{4.63}$$

For these coefficients, the differential equation for the reaction force (4.43) takes the form

$$\frac{d^2 y}{d\tau^2} + \frac{2 + \varepsilon_1 \varepsilon_2 \mu \tan \varphi}{|2 + \varepsilon_1 \varepsilon_2 \mu \tan \varphi|} y - \frac{2}{|2 + \varepsilon_1 \varepsilon_2 \mu \tan \varphi|} = 0, \tag{4.64}$$

where $y = R/R_0$, $\tau = [c|2 + \varepsilon_1 \varepsilon_2 \mu \tan \varphi|/(2 + \tan^2 \varphi)]^{1/2} t$.

Let us construct the solution of eq. (4.64) in the case of no paradoxes. Due to eq. (3.11), the paradoxes do not appear for $M_1 = M_2 = 1$ if

$$\tan \varphi < 2/\mu, \quad \text{or} \quad \varphi < \arctan(2/\mu). \tag{4.65}$$

In this case, eq. (4.64) yields

$$y = r_0 \sin \left[\sqrt{\frac{c(2 + \varepsilon_1 \varepsilon_2 \mu \tan \varphi)}{|2 + \tan^2 \varphi|}} t + \Psi\right] + \frac{2}{2 + \mu \tan \varphi \varepsilon_1 \varepsilon_2}.$$

The stationary solution

$$y = \frac{2}{2 + \varepsilon \mu \tan \varphi}, \quad R = \frac{P_1 - P_2 \tan \varphi}{2 + \varepsilon \mu \tan \varphi}, \quad \varepsilon = \operatorname{sign}[\dot{q}(P_1 - P_2)]$$

coincides with the root of the algebraic equation for the reaction force of the rigid Painlevé-Klein scheme subject to condition (4.65). This confirms the validity of the general conclusion that in the case of no paradoxes there is no need to take account of the compliance of the contact joint.

Let us consider now the solution of eq. (4.64) under the condition for paradoxes

$$\frac{2}{\mu} < \tan \varphi < \infty \quad \text{or} \quad \arctan \frac{2}{\mu} < \frac{\pi}{2}. \tag{4.66}$$

4.4 True motions in the Painlevé-Klein problem in paradoxical situations

If additionally $\varepsilon_1\varepsilon_2 = 1$, then the solution of eq. (4.64) is as follows

$$y = r \sin\left[\sqrt{\frac{c(2 + \mu \tan \varphi)}{2 + \tan^2 \varphi}} t + \Psi\right] + \frac{2}{2 + \mu \tan \varphi} \quad (\varepsilon_1\varepsilon_2 = 1). \quad (4.67)$$

The phase trajectories are ellipses

$$\left(y - \frac{2}{2 + \mu \tan \varphi}\right)^2 + \frac{2 + \tan^2 \varphi}{c(2 + \mu \tan \varphi)} \dot{y}^2 = r^2 \quad (\varepsilon_1\varepsilon_2 = 1) \quad (4.68)$$

with the stable centre

$$y^+ = \frac{2}{2 + \mu \tan \varphi}, \quad \dot{y} = 0, \quad R^+ = \frac{(P_1 - P_2) \tan \varphi}{2 + \mu \tan \varphi}. \quad (4.69)$$

Provided that under condition (4.66) we have $\varepsilon_1\varepsilon_2 = -1$, then the general solution of eq. (4.64) can be cast in the form

$$y = r_1 \exp\sqrt{\frac{c(\mu \tan \varphi - 2)}{2 + \tan^2 \varphi}} t - r_2 \exp\left(-\sqrt{\frac{c(\mu \tan \varphi - 2)}{2 + \tan^2 \varphi}} t\right) - \frac{2}{\mu \tan \varphi - 2}$$

$$(\varepsilon_1\varepsilon_2 = -1). \quad (4.70)$$

The phase trajectories are hyperbolas

$$\left(y - \frac{2}{2 - \mu \tan \varphi}\right)^2 - \frac{2 + \tan^2 \varphi}{c(\mu \tan \varphi - 2)} \dot{y}^2 = 4r_1 r_2 \quad (\varepsilon_1\varepsilon_2 = -1) \quad (4.71)$$

with unstable saddle points

$$y^- = \frac{-2}{\mu \tan \varphi - 2}, \quad \dot{y} = 0, \quad R^- = \frac{-(P_1 - P_2) \tan \varphi}{\mu \tan \varphi - 2} \quad (4.72)$$

and the asymptotes

$$\dot{y} = \pm\sqrt{\frac{c\mu(\tan \varphi - 2)}{2 + \tan^2 \varphi}} \left(y + \frac{2}{\mu \tan \varphi - 2}\right).$$

4.4.2 True motions for the paradoxes

Let us determine the true motions for the paradoxes. As mentioned in Section 4.2 under condition (4.66) non-uniqueness and non-existence take place if $\varepsilon_2 = \text{sign}(P_1 - P_2)$ and if $\varepsilon_2 = -\text{sign}(P_1 - P_2)$, respectively. Let us consider first the non-uniqueness, i.e. the case

$$\arctan\left(\frac{2}{\mu}\right) < \varphi < \frac{\pi}{2}, \quad \varepsilon_2 = \varepsilon_0 = \text{sign}(P_1 - P_2). \quad (4.73)$$

In this case it follows from eq. (4.73) that $\varepsilon_1 = \varepsilon_2 = \text{sign} y$. For this reason, $\varepsilon_1 = \varepsilon_2 = 1$ in the right phase half-space, and the representing point moves along ellipses (4.68). In the left half-space $\varepsilon_1 = \varepsilon_2 = -1$, and the representing point moves along the hyperbolas (4.71). The phase trajectories are obtained by matching the left hyperbolas with the right ellipses, as is shown in Fig. 4.1. Using eqs. (4.67)-(4.71) and Fig. 4.1 we can determine the true value of the reaction force under the prescribed initial values of $y(0)$ and $\dot{y}(0)$. It is necessary to distinguish between two typical cases: (i) the case of the centre of ellipses (4.68) and (ii) the case of the unbounded growth of the absolute value of the reaction force due to eq. (4.70).

In the first case the law of motion is given by formulae (3.16) and (3.17) under the condition $M_1 = M_2 = 1$ and a plus sign in front of μ.

The analysis of motion in the second case can be carried out by means of the equation

$$2\ddot{q} = -\varepsilon_2 \mu |R_0 y| + P_1 + P_2 \,. \tag{4.74}$$

By virtue of eq. (4.70) and in the case of an unbounded growth of y we have

$$\text{sign}\,\ddot{q} = -\varepsilon_2 \,\text{sign}\,|R_0 y| = -\varepsilon_2 = -\,\text{sign}\,\dot{q}\,. \tag{4.75}$$

As one can see, the sign of the acceleration is opposite to the sign of the velocity, hence, the absolute value of the latter decreases until the motion stops. Inserting eq. (4.70) into eq. (4.74) we obtain after integration that

$$\Delta \dot{q} = -\varepsilon_2 \frac{\mu}{2} \sqrt{\frac{2 + \tan^2 \varphi}{c(\mu \tan \varphi - 2)}} |R_0 r_1| \times$$
$$\left[\exp \left(\sqrt{\frac{c(\mu \tan \varphi - 2)}{2 + \tan^2 \varphi}} \Delta t \right) - 1 \right] + o(\Delta t), \tag{4.76}$$

where $o(\Delta t)$ denotes terms of higher order of smallness. On the other hand, an increment in the velocity from the time instant $t = 0$ until the instant of the stop is equal to $\Delta \dot{q} = -\dot{q}(0)$. Then

$$\varepsilon_2 \Delta \dot{q} = -|\dot{q}(0)|, \quad \Delta t = \sqrt{\frac{2 + \tan^2 \varphi}{c(\mu \tan \varphi - 2)}} \times$$
$$\ln \left[1 + \frac{4|\dot{q}(0)|\sqrt{c(\mu \tan \varphi - 2)}}{\mu |r_1||P_1 - P_2| \tan \varphi \sqrt{2 + \tan^2 \varphi}} \right], \quad \lim_{c \to \infty} \Delta t = 0. \tag{4.77}$$

Thus, in the case of non-uniqueness of the solution of the dynamic problem of the rigid Painlevé-Klein scheme, eq. (3.10) describes the process for two stationary values $y^{\pm}$. Under some initial conditions the value of the reaction tends to y^+ whereas under other initial conditions the value of y rapidly increases and the motion stops.

Let us now proceed to consider the situation of non-existence, when

$$\arctan\frac{2}{\mu} < \varphi < \frac{\pi}{2}, \quad \varepsilon_2 = -\varepsilon_0 = -\operatorname{sign}(P_1 - P_2). \tag{4.78}$$

It follows from eq. (4.78) and equality $y = R/R_0$ that $\varepsilon_1 = -\varepsilon_2 \operatorname{sign} y$. Then we have $\varepsilon_1\varepsilon_2 = -1$ for $y > 0$ and $\varepsilon_1\varepsilon_2 = 1$ for $y < 0$. The phase trajectories consist of the arches of ellipses (4.68) in the left half-space $(y, \dot{y})$ and hyperbolas (4.71) in the right half-space, as depicted in Fig. 4.2. In this case there are no stationary solutions and y eventually tends to infinity for any initial conditions. However in this case, relationships (4.75)-(4.77) follows from eqs. (4.70) and (4.74).

Therefore, in the situation of non-existence of the solution to the problem for the rigid Painlevé-Klein scheme, the reaction force grows without bound and the motion soon stops.

As mentioned in Chapter 1 the Painlevé-Klein scheme with an elastic rod M_1M_2 is considered in [26], [125], [128]. The solution in the phase plane is given by Butenin in [26].

4.5 Elliptic pendulum

This pendulum is considered as an example of a system with two degrees of freedom. We can convince ourselves that applying the approach developed in Sections 4.1 and 4.2 allows one to obtain easily the differential equations of motion and the equations for the reaction force, as well as to determine the regions of the non-existence and non-uniqueness of the dynamic problem.

The system considered is shown in Fig. 4.3. Mass M_1 slips on a rough horizontal plane and mass M_2 is connected to mass M_1 by means of a massless rod of length l and moves in the vertical plane. Abscissa x of particle M_1 and angle θ between the rod M_1M_2 and axis Ox are taken as the generalised coordinates.

Removing contact implies that the bodies obtain imaginary vertical translation h. Then

$$\mathbf{r}_T = \mathbf{r}_1 = x\mathbf{i}_1 + h\mathbf{i}_2, \quad \mathbf{r}_2 = (x + l\cos\theta)\mathbf{i}_2 + (h + l\sin\theta)\mathbf{i}_2. \tag{4.79}$$

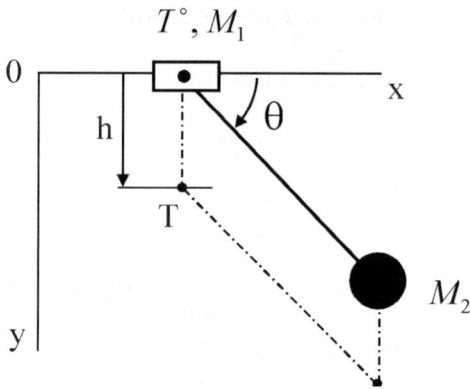

FIGURE 4.3.

As above, T denotes the slider and equality $\mathbf{r}_T = \mathbf{r}_1$ indicates that the role of the slider is played by particle M_1. It follows from (4.79) that

$$\frac{\partial \mathbf{r}_1}{\partial x} = \mathbf{i}_1, \quad \frac{\partial \mathbf{r}_1}{\partial \theta} = 0, \quad \frac{\partial \mathbf{r}_1}{\partial h} = \mathbf{i}_2,$$

$$\frac{\partial \mathbf{r}_2}{\partial x} = \mathbf{i}_1, \quad \frac{\partial \mathbf{r}_2}{\partial \theta} = -l\sin\theta\,\mathbf{i}_1 + l\cos\theta\,\mathbf{i}_2, \quad \frac{\partial \mathbf{r}_2}{\partial h} = \mathbf{i}_2,$$

$$\frac{\partial v_T^0}{\partial \dot{x}} = \operatorname{sign}\dot{x} = \varepsilon_2, \quad \frac{\partial v_T^0}{\partial \theta} = \left(\frac{\partial v_T}{\partial h}\right)_0 = 0. \qquad (4.80)$$

These expressions for the derivatives are used to calculate the coefficients of the kinetic energy

$$A_{11} = \sum_{i=1}^{2} M_i \left|\frac{\partial \mathbf{r}_i}{\partial x}\right|^2 = M_1 + M_2, \quad A_{12} = \sum M_i \frac{\partial \mathbf{r}_i}{\partial x} \cdot \frac{\partial \mathbf{r}_i}{\partial \theta} = -M_2 l \sin\theta,$$

$$A_{22} = \sum_{i=1}^{2} M_i \left|\frac{\partial \mathbf{r}_i}{\partial \theta}\right|^2 = M_2 l^2, \quad A_{13} = \sum M_i \frac{\partial \mathbf{r}_i}{\partial x} \cdot \frac{\partial \mathbf{r}_i}{\partial h} = 0,$$

$$A_{23} = \sum_{i=1}^{2} M_i \frac{\partial \mathbf{r}_i}{\partial \theta} \cdot \frac{\partial \mathbf{r}_i}{\partial h} = M_2 l \cos\theta. \qquad (4.81)$$

Furthermore,

$$\frac{\partial A_{11}}{\partial x} = \frac{\partial A_{11}}{\partial \theta} = \frac{\partial A_{11}}{\partial h} = \frac{\partial A_{22}}{\partial x} = \frac{\partial A_{22}}{\partial \theta} = \frac{\partial A_{22}}{\partial h} = \frac{\partial A_{12}}{\partial x} = 0,$$

$$\frac{\partial A_{12}}{\partial \theta} = -M_2 l \cos\theta, \quad \frac{\partial A_{23}}{\partial \theta} = -M_2 l \sin\theta. \qquad (4.82)$$

4.5 Elliptic pendulum

System of equations (4.13) takes the form

$$(M_1 + M_2)\ddot{x} - M_2 l \sin\theta \ddot{\theta} - M_2 l \cos\theta \dot{\theta}^2 = -\varepsilon_1\varepsilon_2\mu R,$$

$$-\sin\theta \ddot{x} + l\ddot{\theta} = g\cos\theta,$$

$$M_2 l \cos\theta \ddot{\theta} - M_2 l \sin\theta \dot{\theta}^2 = (M_1 + M_2)g + R. \qquad (4.83)$$

Parameters A_1, A_2 and A are calculated by formulae (4.15) and (4.16), for which the coefficients $A_{ik}(i, k = 1, 2, 3)$ are taken from eq. (4.81). The result is

$$A_1 = M_2^2 l^2 \sin\theta \cos\theta, \quad A_2 = M_2(M_1 + M_2)l\cos\theta,$$
$$A = M_2 l^2 (M_1 + M_2 \cos^2\theta). \qquad (4.84)$$

Now we have everything to apply formulae (4.20)-(4.23). Inserting relationships (4.81)-(4.82) and (4.84) into eqs. (4.20) and (4.21) we find the expressions for the index of the influence L and the normal reaction R_0

$$\mathsf{L} = \varepsilon_2 \frac{M_2 \sin\theta \cos\theta}{M_1 + M_2 \cos^2\theta}, \qquad (4.85)$$

$$R_0 = -\frac{M_1(M_1 + M_2)g + M_1 M_2 l \sin\theta \dot{\theta}^2}{M_1 + M_2 \cos^2\theta}. \qquad (4.86)$$

By analogy, coefficients F_s and $G_s (s = 1, 2)$ of equations (4.25) and (4.26) can be calculated by means of eqs. (4.22) and (4.23). Substituting the obtained expressions for L, R_0, F_s and G_s into eqs. (4.25) and (4.26) we obtain the differential equations of motion and the expression for the reaction force

$$(M_1 + M_2 \cos^2\theta + \varepsilon_1\varepsilon_2\mu M_2 \sin\theta \cos\theta)\ddot{x} = M_2 l \cos\theta \dot{\theta}^2 +$$
$$M_2 g \sin\theta \cos\theta + \varepsilon_1\varepsilon_2\mu[M_2 l \sin\theta \dot{\theta}^2 + M_2 g \cos^2\theta - (M_1 + M_2)g],$$

$$(M_1 + M_2 \cos^2\theta + \varepsilon_1\varepsilon_2\mu M_2 \sin\theta \cos\theta)\ddot{\theta} = M_2 \sin\theta \cos\theta \dot{\theta}^2 -$$
$$\frac{M_1 + M_2}{l} g \cos\theta + \varepsilon_1\varepsilon_2\mu[M_2 \sin^2\theta \dot{\theta}^2 + l^{-1}(M_1 + M_2)g \sin\theta],$$

$$(M_1 + M_2 \cos^2\theta + \varepsilon_1\varepsilon_2\mu M_2 \sin\theta \cos\theta)R = -M_1(M_1 + M_2)g -$$
$$M_1 M_2 l \sin\theta \dot{\theta}^2. \qquad (4.87)$$

We proceed now to establish the regions of the paradoxes. By Theorem 7, for index L given by eq. (4.85), paradoxes occur if

$$\frac{\mu M_2 |\sin\theta \cos\theta|}{M_1 + M_2 \cos^2\theta} \geq 1. \qquad (4.88)$$

For values of angle θ within the intervals $(0, \pi/2)$ and $(\pi, 3\pi/2)$ we have $\sin\theta\cos\theta > 0$, thus, condition (4.88) takes the form

$$\frac{\mu M_2 \sin\theta \cos\theta}{M_1 + M_2 \cos^2\theta} \geq 1 \quad \theta \in (0, \pi/2), (\pi, 3\pi/2). \tag{4.89}$$

In the intervals $(\pi/2, \pi)$ and $(3\pi/2, 2\pi)$ we have $\sin\theta\cos\theta < 0$ and condition (4.88) reduces to the following one

$$\frac{-\mu M_2 \sin\theta \cos\theta}{M_1 + M_2 \cos^2\theta} \geq 1 \quad \theta \in (\pi/2, \pi), (3\pi/2, 2\pi). \tag{4.90}$$

When relationships (4.89) and (4.90) are fulfilled the solution of problem (4.87) is non-unique if

$$\varepsilon_2[(M_1 + M_2)g + M_2 \sin\theta l \dot\theta^2] \sin 2\theta < 0 \tag{4.91}$$

and does not exist if

$$\varepsilon_2[(M_1 + M_2)g + M_2 l \sin\theta \dot\theta^2] \sin 2\theta > 0. \tag{4.92}$$

Therefore, the region of non-existence is given by the relationships (4.89), (4.90) and (4.92), whilst the region of non-uniqueness is described by relationships (4.89), (4.90) and (4.91).

4.6 The Zhukovsky-Froude pendulum

This pendulum is depicted in Fig. 4.4. The pin 1 of radius r rotates with angular velocity ω. The journal 2 is massless and a particle M of mass m is attached to the journal at a distance l from the centre O.

In classical mechanics, the Zhukovsky-Froude pendulum is considered as a mechanical system for which the moment of the frictional force is a decreasing function $F(\dot\varphi - \omega)$ of the relative velocity of gliding [99], [141]. Such an approximation of the resistance force is quite sufficient for solving the problem of self-excited oscillations about the equilibrium position. On the other hand, this force does not reflect the influence of the coefficient of Coulomb friction μ and the coordinate φ on the value of the normal reaction force. Hence, it can not be utilised for considering the motion in the whole phase space $(\varphi, \dot\varphi)$ or for determining the feasibility of non-existence of the solution of the dynamic problem.

An exact equation for the reaction force and an exact equation of motion are derived in what follows. By using these equations we establish the condition of non-existence of the solution, find the equilibrium positions, analyse the properties of free vibrations and obtain the condition for joint rotation of the pin and the journal.

4.6 The Zhukovsky-Froude pendulum   149

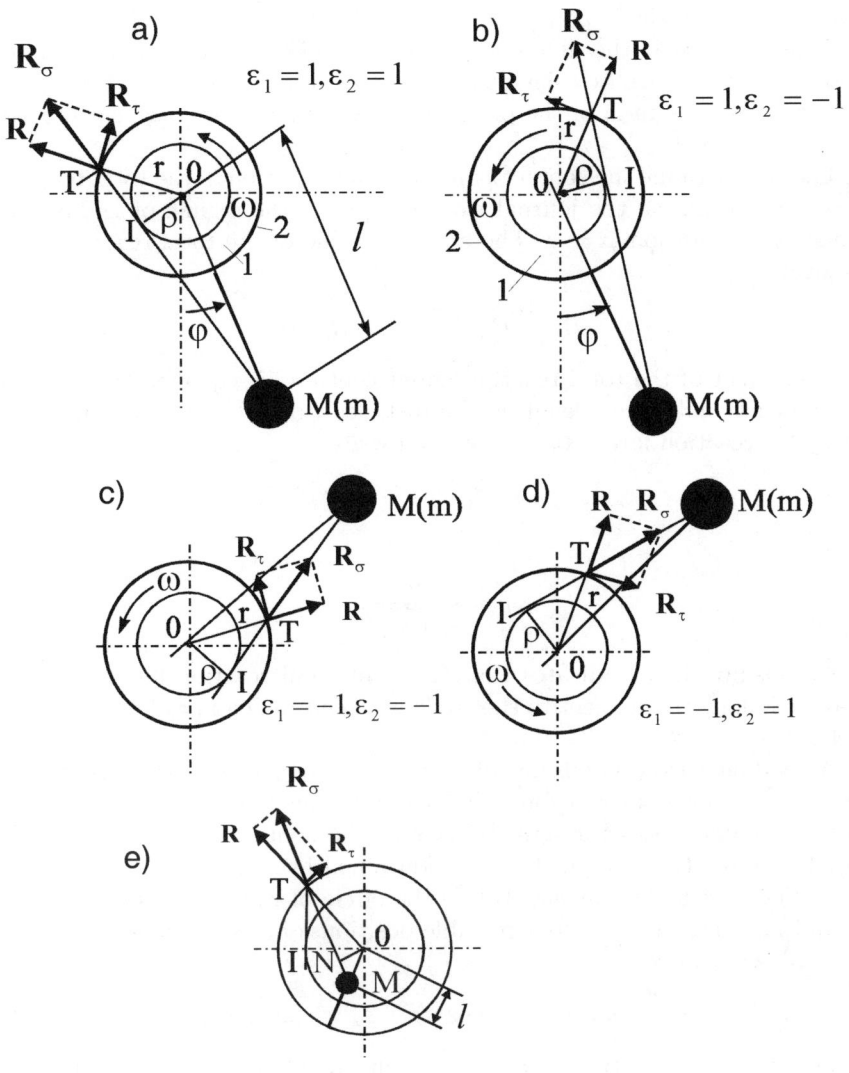

FIGURE 4.4.

4.6.1 Equation for the reaction force and condition for the non-existence of the solution

A rotation of the journal relative to the pin is possible if the radius of the journal is greater than r. The contact between the pin and the journal takes place at a single point which is the point of tangency of two circles. It is assumed that the difference in radii is infinitesimally small and the radii can be taken to be coincident for the dynamic calculation. To put this another way, the radii of the pin and journal coincide and their contact is at a single point.

Let us determine the contact point and the direction of the total reaction force $\mathbf{R}_\sigma$ acting on the journal from the pin in the regime of gliding, i.e. under the condition $\dot\varphi \neq \omega$. The absolute value of the total reaction force is given by

$$R_\sigma = R\sqrt{1+\mu^2}, \quad R > 0. \tag{4.93}$$

The moment of the total reaction about centre O is equal to the moment of the frictional force. Denoting the distance from O to the action line of the joint reaction force $\mathbf{R}_\sigma$ by ρ we can write

$$\rho R\sqrt{1+\mu^2} = r\mu R, \tag{4.94}$$

thus

$$\rho = \frac{\mu r}{\sqrt{1+\mu^2}}. \tag{4.95}$$

Relationship (4.95) indicates that the joint total reaction force is always directed along the tangent to the circle with centre O and radius ρ, see Fig. 4.4.

According to the condition of equilibrium of particle M subjected to the gravity force, inertia force and the reaction force, the action line of the latter must pass through M. Hence, the contact point T, which the reaction force $\mathbf{R}_\sigma$ is applied to, is determined by intersection of the circle of radius r with the tangent MI to the circle of radius ρ. As one can see from Fig. 4.4, there exist four possible positions of point T depending upon the following values

$$\varepsilon_1 = \mathrm{sign}(l\dot\varphi^2 + g\cos\varphi), \quad \varepsilon_2 = \mathrm{sign}(\dot\varphi - \omega). \tag{4.96}$$

Let us construct the equations for the pendulum under gliding. We project all forces, the inertia force included, onto TM and equate their sum to zero. Additionally, equating the sum of the moments of these forces about O to zero yields

$$-\ddot\varphi ml \sin\varepsilon\beta + \dot\varphi^2 ml \cos\beta + mg\cos(\varphi+\varepsilon\beta) - \varepsilon_1 R\sqrt{1+\mu^2} = 0,$$
$$\ddot\varphi ml^2 + mgl\sin\varphi + \varepsilon_2 \mu r R = 0. \tag{4.97}$$

4.6 The Zhukovsky-Froude pendulum

$$\varepsilon = \varepsilon_1 \varepsilon_2 = \text{sign}[(l\dot\varphi^2 + g\cos\varphi)(\dot\varphi - \omega)], \quad \sin\beta = \frac{\rho}{l}, \quad \cos\beta = \frac{\sqrt{l^2 - \rho^2}}{l}. \tag{4.98}$$

Equations (4.97) are valid for all variants of the position of the contact point. Resolving these equations for R and $\ddot\varphi$ results in the equation of motion

$$R = \frac{ml|g\cos\varphi + l\dot\varphi^2|}{\sqrt{(l^2-\rho^2)(1+\mu^2)}}, \quad \ddot\varphi + \frac{\varepsilon\rho}{\sqrt{l^2-\rho^2}}\dot\varphi^2 + \frac{g}{\sqrt{l^2-\rho^2}}\sin(\varphi + \varepsilon\beta) = 0. \tag{4.99}$$

By introducing the new variables

$$\tau = \left(\frac{g}{\sqrt{l^2-\rho^2}}\right)^{1/2} t, \quad \Psi = \frac{d\varphi}{d\tau} \tag{4.100}$$

we can transform the equation of motion to the form

$$\frac{d\Psi^2}{d\varphi} + \frac{2\varepsilon\rho}{\sqrt{l^2-\rho^2}}\Psi^2 = -2\sin(\varphi + \varepsilon\beta). \tag{4.101}$$

It follows from eqs. (4.101) and (4.99) that $\beta \to \pi/2, R \to \infty, \ddot\varphi \to \infty$ when $l \to \rho + 0$. In the case

$$l < \rho, \quad \text{i.e.} \quad \mu > \frac{l}{\sqrt{r^2 - l^2}} \tag{4.102}$$

the values R and $\ddot\varphi$ are no longer real, that is, the solution of the dynamic problem (4.99) no longer exists. It is proved below that the pendulum rotates together with the pin if condition (4.102) is fulfilled.

Up until now, no restriction has been imposed on the coefficient of friction μ. Hence, eqs. (4.97), (4.99), (4.101) and the condition for non-existence of solution (4.102) are valid for both constant μ and for μ depending on the velocity of gliding $(\dot\varphi - \omega)$. The forthcoming analysis is carried out under the assumption that $\mu = \text{const}, l > \rho$. We notice that $\mu = \text{const}$ means $\rho = \text{const}$, and then the general solution of eq. (4.101) can be cast as follows

$$\Psi^2 = c\exp\left[\frac{-2\varepsilon\rho}{\sqrt{l^2-\rho^2}}(\varphi+\varepsilon\beta)\right] + \frac{2(l^2-\rho^2)}{3\rho^2+l^2}\cos(\varphi+\varepsilon\beta)$$
$$- \frac{4\varepsilon\rho\sqrt{l^2-\rho^2}}{3\rho^2+l^2}\sin(\varphi+\varepsilon\beta). \tag{4.103}$$

4.6.2 The equilibrium position and free oscillations

In the equilibrium position $\dot{\varphi} = \ddot{\varphi} = 0$. In accordance with eqs. (4.96) and (4.99), we have $\varepsilon_1 = \text{sign}\cos\varphi, \varepsilon_2 = -1, \sin(\varphi - \beta) = 0$ for $\cos\varphi > 0$ and $\sin(\varphi + \beta) = 0$ for $\cos\varphi < 0$. Hence, we obtain two stationary values of φ

$$\varphi_1 = \beta = \arcsin(\rho/l), \quad \varphi_2 = \pi - \beta = \pi - \arcsin(\rho/l). \qquad (4.104)$$

Here φ_1 and φ_2 describe respectively stable and unstable equilibrium positions.

Let us consider free oscillations about the stable equilibrium position under the following restrictions: $|\varphi| \leq \pi/2$, $|\dot{\varphi}| < \omega$. In this case

$$\varepsilon_1 = 1, \quad \varepsilon_2 = -1, \quad \varepsilon = -1, \qquad (4.105)$$

and eqs. (4.101) and (4.103) respectively take the form

$$\frac{d\Psi^2}{d\Phi} - \frac{2\rho}{\sqrt{l^2 - \rho^2}}\Psi^2 = -2\sin\Phi, \qquad (4.106)$$

$$\Psi^2 = c\exp\frac{2\rho}{\sqrt{l^2 - \rho^2}}\Phi + \frac{2(l^2 - \rho^2)}{3\rho^2 + l^2}\cos\Phi + \frac{4\rho\sqrt{l^2 - \rho^2}}{3\rho^2 + l^2}\sin\Phi, \qquad (4.107)$$

where c denotes an integration constant and $\Phi = \varphi - \beta$, $\Psi = d\Phi/d\tau$. Under the initial condition $\varphi(0) = \beta$, $\dot{\varphi}(0) = 0$ we obtain

$$c_{\min} = \frac{-2(l^2 - \rho^2)}{3\rho^2 + l^2},$$

which corresponds to a singular point $\Psi = \Phi = 0$ in phase space, see Fig. 4.5. For non-zero values of ψ and Φ the phase curves are closed and embedded within each other. They are symmetric about axis Φ and not symmetric about axis Ψ, the extreme left value $-\Phi_{\min}$ being smaller than the extreme right one $\Phi_{\max}$. For this reason, free oscillations of the Zhukovsky-Froude pendulum differ essentially from sinusoidal ones.

Equations (4.106) and (4.107) are valid only under the condition $\varphi_{\max} \leq \pi/2$. The reason for this is that for $\pi/2 < \varphi_{\max} < 3\pi/2$, due to eq. (4.107), we have $\dot{\varphi} = 0$, $\cos\varphi < 0$, $\varepsilon_1 = -1$, $\varepsilon_2 = -1$ in the vicinity of the point $\varphi = \varphi_{\max}$, that is condition (4.105) is not satisfied. The limiting value of c, for which $\varphi_{\max} = \pi/2$, is determined by inserting $\Psi = 0, \Phi = \pi/2 - \beta$ into eq. (4.17). The result is

$$c_{\max} = -\frac{6\rho(l^2 - \rho^2)}{l(3\rho^2 + l^2)}\exp\frac{-\rho(\pi - 2\beta)}{\sqrt{l^2 - \rho^2}}.$$

Therefore, eq. (4.107) and its phase curves, Fig. 4.5, describe free oscillations of the Zhukovsky-Froude pendulum for the following values of c

$$-\frac{2(l^2 - \rho^2)}{3\rho^2 + l^2} \leq c \leq -\frac{6\rho(l^2 - \rho^2)}{l(3\rho^2 + l^2)}\exp\frac{-\rho(\pi - 2\beta)}{\sqrt{l^2 - \rho^2}}.$$

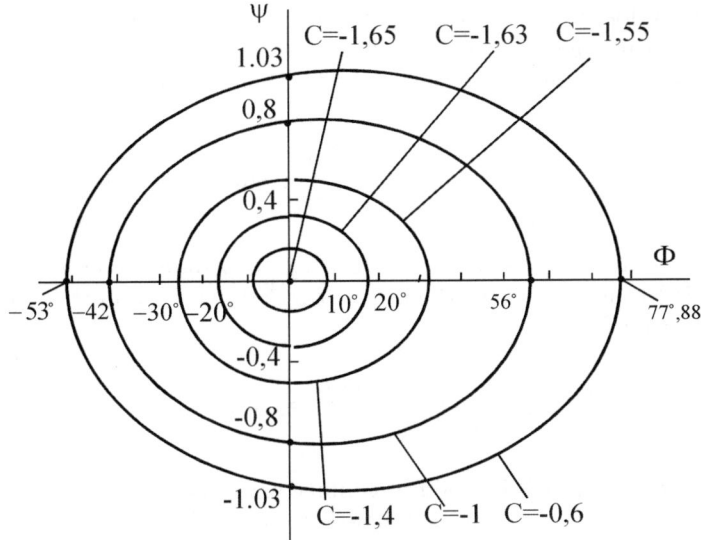

FIGURE 4.5.

Figure 4.5 displays the case of $\mu = 0,5, r = 1, l = 2$. Correspondingly, $\beta = 12,92, c_{\min} = -1,65, c_{\max} = -0,60, \rho = 0,45$.

Let us note that $\sin \Phi \approx \Phi$ for small oscillations. Equations (4.106) and (4.107) reduce to the form considered in [4] as an example of the frictional force which is proportional to the square of velocity.

4.6.3 Regime of joint rotation of the journal and the pin

Let us adopt the notation $\dot{\varphi}_0 = \dot{\varphi}(0)$ and prove two statements for the cases when $l > \rho$ and $l < \rho$.

The first statement. If

$$l > \rho, \quad \dot{\varphi}_0 > \omega, \quad \omega \geq \sqrt{g/\rho}, \tag{4.108}$$

then, as time progresses, the value of $\dot{\varphi}$ decreases to ω, after which the journal rotates together with the pin with angular velocity ω.

Proof. After condition (4.108) we have $g \cos \varphi + l\dot{\varphi}_0^2 > g \cos \varphi + \rho\omega^2 > 0$. Then, due to eq. (4.96) and (4.98) at the initial time instant

$$\varepsilon_1 = \text{sign}(g \cos \varphi + l\dot{\varphi}_0^2) = 1, \quad \varepsilon_2 = \text{sign}(\dot{\varphi}_0 - \omega) = 1, \quad \varepsilon = 1 \tag{4.109}$$

and the equation of motion (4.99) reduces to the following one

$$\ddot{\varphi} = -\frac{\rho}{\sqrt{l^2 - \rho^2}}\dot{\varphi}^2 - \frac{g}{\sqrt{l^2 - \rho^2}}\sin(\varphi + \beta) < 0.$$

Since $\ddot{\varphi} < 0$, velocity $\dot{\varphi}$ decreases until it is equal to ω.

Let the velocity be equal to ω at time instant t_0, i.e. $\dot\varphi(t_0) = \omega$. It is easy to see that acceleration $\ddot\varphi$ is always zero for $t > t_0$. Indeed, when $\ddot\varphi$ is positive at a certain time instant $t_1 > t_0$, then, as follows from eq. (4.99) we obtain

$$\ddot\varphi = -\frac{\rho}{\sqrt{l^2 - \rho^2}}\omega^2 - \frac{g}{l^2 - \rho^2}\sin(\varphi + \beta) > 0$$

which is possible only for $\omega^2 < g/\rho$. But this contradicts condition (4.108).

Provided that $\ddot\varphi(t_1) < 0$, then $\varepsilon_1 = 1, \varepsilon_2 = \varepsilon = -1$ and it follows from eq. (4.99) that

$$\ddot\varphi = \frac{\rho}{\sqrt{l^2 - \rho^2}}\omega^2 - \frac{g}{\sqrt{l^2 - \rho^2}}\sin(\varphi + \beta) < 0.$$

Hence $\omega^2 < g/\rho$, which is again in conflict with (4.33).

Therefore, for $t > t_0$ the acceleration cannot be neither positive nor negative. For this reason it is zero, i.e. joint rotation of the journal and the pin with angular velocity ω is observed.

The second statement. In the case of $l < \rho$ for any initial velocity of rotation of the journal φ_0 the latter rotates together with the pin with the velocity $\dot\varphi(t) = \omega$ for $t > 0$.

Proof. As $l < \rho$, the condition (4.102) for non-existence of the solution is satisfied. By virtue of Theorem 9 and its corollary the relative gliding between the journal and the pin soon comes to a halt, i.e. the value of $\dot\varphi$ becomes instantaneously equal to ω. This means that it is sufficient to consider the case of $\dot\varphi = 0$. If at a certain instant $t > 0$ gliding took place, the vector of the total reaction $\mathbf{R}_\sigma$ would be directed along tangent TI to the circle of radius ρ. Moreover, as follows from Fig. 4.4a, b, c, d $\mu = R_\tau/R = \rho/|\overline{TI}|$.

In the case of $l < \rho$ the particle M is located within the circle of radius ρ, hence the action line of the total reaction force TM intersects this circle, rather than being tangent to it, see Fig. 4.4e. The ratio of the tangential and the normal components of the reaction force is as follows

$$\mu_1 = \frac{R_\tau}{R} = \frac{|\overline{ON}|}{|\overline{TN}|}.$$

As

$$\frac{|\overline{ON}|}{|\overline{TN}|} < \frac{\rho}{|\overline{TI}|},$$

then $\mu_1 < \mu$. This is the condition of relative equilibrium for which $\dot\varphi(t) = \omega$ for $t > 0$.

4.7 A condition of instability for the stationary regime of metal cutting

In the dynamics of metal-cutting machine tools, the instability of motion is usually be explained by the falling dependence of the cutting force on velocity, see [55], [111], [113]. There are also known cases of non-smooth motion of the actuating mechanisms of machine tools, with the reason remaining unexplained.

As the above analysis shows, the greater the frictional coefficient μ, the broader the paradoxical regions in which a tangential impact (a discontinuous change in velocity) occurs. On the other hand, under metal cutting the factor of proportionality between the component of the forces directed along the tangent and the normal to the treated surface is, in general, rather considerable. For this reason, it is expedient to generalise the problem of tangential impacts to the case of metal cutting.

In what follows, an attempt to solve the stated problem leads to a new condition of instability of the regime of stationary cutting resulting in a Painlevé's paradox with increasing hardness of the treated material.

4.7.1 Derivation of the equations of motion

We consider the spindle system of boring, see Fig. 4.6, which consists of a cutting tool 1 and a spindle 2 rotating together about axis $0z$. For the sake of simplicity, the case of planar cutting is taken, the rotation angle being denoted by φ. The effective rigidity is denoted as c, the mass M of the system is assumed to lie in the centre I at a distance r from 0_1 under angle β, $J = Mr_1^2$ denotes the moment of inertia about the axis passing through I parallel to $0z$, r_1 is the radius of gyration, a is the radius of the treated surface of the immovable blank and B denotes a constant torque.

The force of cutting has two components, namely R is the component directed to the centre, i.e. along the normal to the treated surface and R_1 is the component which is perpendicular to the cutting edge T and directed along the tangent to the treated surface. The relationship between R, R_1 and the deepness of cutting τ is represented in the form

$$R = c_1\tau, \quad R_1 = \mu R = \mu c_1 \tau, \tag{4.110}$$

where the positive proportionality coefficients $\tilde{n}_1$ and μ remain nearly unchanged for small changes in τ, [74].

The origin for measuring eccentricity 0_1 is chosen so that elastic force F is balanced by the sum of component R of the cutting force and the projection of the centrifugal force onto 0_1T in the position when 0_1 lies on the central axis $0z$, i.e.

$$\mathbf{F}_0 = (\mathbf{i}_1 \cos\varphi + \mathbf{i}_2 \sin\varphi)(c_1\tau_0 - Mr\dot\varphi_0^2 \cos\beta). \tag{4.111}$$

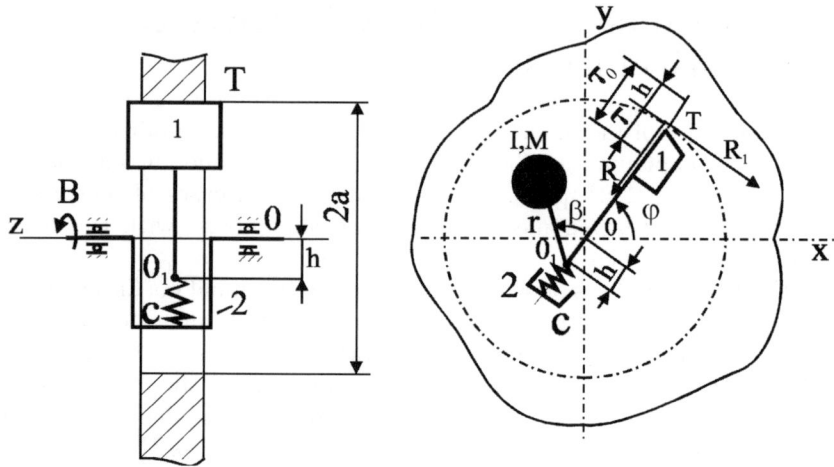

FIGURE 4.6.

Here the subscript 0 corresponds to the case in which the rotational velocity of the spindle $\dot\varphi$ has a stationary value and $\mathbf{F}$ and τ are determined in this equilibrium position.

Under displacement r of point 0_1 the elastic force is

$$\mathbf{F}_0 = (\mathbf{i}_1 \cos\varphi + \mathbf{i}_2 \sin\varphi)(ch + s_1\tau_0 - M r \dot\varphi_0^2 \cos\beta). \tag{4.112}$$

The deepness of cutting is given by

$$\tau = \tau_0 - h. \tag{4.113}$$

Accounting for this in eq. (4.110) we arrive at the following expression for the vector of the cutting force

$$\mathbf{R}_\sigma = [\mu(\mathbf{i}_1 \sin\varphi - \mathbf{i}_2 \cos\varphi) - (\mathbf{i}_1 \cos\varphi + \mathbf{i}_2 \sin\varphi)]c_1(\tau_0 - h). \tag{4.114}$$

The position vectors of the centre of mass I, the origin 0_1 of the eccentricity measurement and the cutting point T can be cast as follows

$$\mathbf{r} = \mathbf{i}_1[r\cos(\varphi+\beta) - h\cos\varphi] + \mathbf{i}_2[r\sin(\varphi+\beta) - h\sin\varphi],$$
$$\mathbf{r}_1 = -h(\mathbf{i}_1 \cos\varphi + \mathbf{i}_2 \sin\varphi), \quad \mathbf{r}_\tau = (a-h)(\mathbf{i}_1 \cos\varphi + \mathbf{i}_2 \sin\varphi). \tag{4.115}$$

Hence,

$$\frac{\partial \mathbf{r}}{\partial \varphi} = \mathbf{i}_1[-r\sin(\varphi+\beta) + h\sin\varphi] + \mathbf{i}_2[r\cos(\varphi+\beta) - h\cos\varphi],$$

$$\frac{\partial \mathbf{r}_\tau}{\partial \varphi} = -(a-h)(\mathbf{i}_1 \sin\varphi - \mathbf{i}_2 \cos\varphi), \quad \frac{\partial \mathbf{r}_1}{\partial \varphi} = h(\mathbf{i}_1 \sin\varphi - \mathbf{i}_2 \cos\varphi),$$

$$\frac{\partial \mathbf{r}}{\partial h} = \frac{\partial \mathbf{r}_1}{\partial h} = \frac{\partial \mathbf{r}_\tau}{\partial h} = -\mathbf{i}_1 \cos\varphi - \mathbf{i}_2 \sin\varphi. \tag{4.116}$$

4.7 A condition of instability for the stationary regime of metal cutting

Thus, we obtain the expressions for the external forces (4.112), (4.114) and the derivatives of vectors (4.116) with respect to the generalised coordinates φ and h. These are used for calculating the coefficients of the kinetic energy and the generalised forces

$$A_{11} = M \left|\frac{\partial \mathbf{r}}{\partial \varphi}\right|^2 + J = M(r^2 + r_1^2 + h^2 - 2rh\cos\beta),$$

$$A_{12} = Mr\sin\beta, \quad A_{22} = M, \quad \frac{\partial A_{11}}{\partial h} = 2M(h - r\cos\beta),$$

$$\frac{\partial A_{ik}}{\partial \varphi} = 0 \quad (i, k = 1, 2), \quad \frac{\partial A_{12}}{\partial h} = \frac{\partial A_{22}}{\partial h} = 0,$$

$$Q_1 = \mathbf{F} \cdot \frac{\partial \mathbf{r}_1}{\partial \varphi} + \mathbf{R}_\sigma \cdot \frac{\partial \mathbf{r}_\tau}{\partial \varphi} + B = \mu a c_1 (h - \tau_0) + B,$$

$$Q_2 = \mathbf{F} \cdot \frac{\partial \mathbf{r}_1}{\partial h} + \mathbf{R}_\sigma \cdot \frac{\partial \mathbf{r}_\tau}{\partial h} = -(c + c_1)h + Mr\dot\varphi_0^2 \cos\beta, \quad (4.117)$$

where subscripts 1 and 2 correspond to coordinates φ and h respectively. Substituting eq. (4.117) into Lagrange's equations and neglecting the small terms yields

$$M(r^2 + r_1^2)\ddot\varphi + Mr\sin\beta \ddot h = B + \mu ac(h - \tau_0),$$

$$Mr\sin\beta \ddot\varphi + M\ddot h = -(c + c_1)h. \quad (4.118)$$

Resolving eq. (4.118) for $\ddot\varphi$, $\ddot h$, replacing variable h by τ and accounting for eq. (4.113) we have

$$\ddot\varphi + (1 + \mu L_1)D_1\tau - E_1 = 0, \quad \ddot\tau + (1 + \mu L_2)D_2\tau - E_2 = 0, \quad (4.119)$$

where the following notation

$$L_1 = \frac{c_1 a}{r(c+c_1)\sin\beta}, \quad D_1 = \frac{(c_1+c)r\sin\beta}{M(r^2\cos^2\beta + r_1^2)},$$

$$E_1 = \frac{B + (c+c_1)\tau_0\sin\beta}{M(r^2\cos^2\beta + r_1^2)}, \quad L_2 = \frac{c_1 ar\sin\beta}{(r^2 + r_1^2)(c+c_1)},$$

$$D_2 = \frac{(r^2 + r_1^2)(c+c_1)}{M(r^2\cos^2\beta + r_1^2)}, \quad E_2 = \frac{Br\sin\beta + (c+c_2)(r^2+r_1^2)\tau_0}{M(r^2\cos^2\beta + r_1^2)} \quad (4.120)$$

has been adopted. We notice that eqs. (4.118) and (4.119) are valid only for positive values of τ and $\dot\varphi$.

4.7.2 Solving the equations

Let us first obtain the solution under the condition

$$(1 + \mu L_2)D_2 > 0. \quad (4.121)$$

158   4. Systems with many degrees of freedom and a single frictional pair

It follows from eq. (4.120) that condition (4.121) is satisfied in two cases

$$\text{a) } 0 < \beta < \pi, \quad \mu \text{ is arbitrary},$$
$$\text{b) } -\pi < \beta < 0, \quad \mu < \frac{(c+c_1)(r^2+r_1^2)}{c_1 ar |\sin \beta|}. \tag{4.122}$$

The solution of eq. (4.119) has the form

$$\tau = \rho \sin(\omega t + \Psi) + \tau_*,$$
$$\dot{\varphi} = (1+\mu L_1)\rho \omega^{-1} \cos(\omega t + \Psi) + \left[E_1 - \frac{(1+\mu L_1)D_1}{(1+\mu L_2)D_2} E_2\right] t + \dot{\varphi}_0, \tag{4.123}$$

where $\dot{\varphi}_0, \rho$ and Ψ are the integration constants, $\omega = [(1+\mu L_2)D_2]^{1/2}$ and

$$\tau_* = \frac{E_2}{(1+\mu L_2)D_2} = \frac{Br \sin \beta + (c+c_1)(r^2+r_1^2)\tau_0}{\mu c_1 ar \sin \beta + (c+c_1)(r^2+r_1^2)}. \tag{4.124}$$

The phase trajectories in the plane $(\tau, \dot{\tau})$ form the family of ellipses

$$(\tau - \tau_*)^2 + \frac{\dot{\tau}^2}{\omega^2} = \rho^2 \tag{4.125}$$

for which the point $(\tau_*, 0)$ is a stable centre.

From a practical perspective, the stationary regime of cutting $\tau = $ const, $\dot{\varphi} = $ const, is of interest. This regime is possible if the coefficient of t in eq. (4.123) and the integration constant ρ vanish, i.e.

$$(1+\mu L_2)D_2 E_1 - (1+\mu L_1)D_1 E_2 = 0, \quad \rho = 0. \tag{4.126}$$

Taking into account eqs. (4.126) and (4.120) we arrive at the condition

$$B = B_0 = \mu a c_1 \tau_0. \tag{4.127}$$

Substituting eq. (4.127) into (4.124) yields

$$\tau = \tau_* = \tau_0. \tag{4.128}$$

Notice that eqs. (4.127) and (4.128) can be obtained by inserting $\ddot{\varphi} = \ddot{\tau} = \dot{\tau} = 0$ into eqs. (4.118) or (4.119). In practice, the oscillatory terms in eq. (4.123) attenuate for any initial condition due to the presence of the structural damping. Hence, under condition $B = B_0$ quantities τ and $\dot{\varphi}$ tend to stationary values τ_0 and $\dot{\varphi}_0$. Such a regime of cutting is referred to as a stable one [74], [111]. In other words, relationships in eq. (4.122) are the conditions of stability of the stationary regime of the considered system.

Let us proceed to seeking a solution of eq. (4.119) under the condition

$$(1+\mu L_2)D_2 \leq 0. \tag{4.129}$$

By virtue of eq. (4.120) it is easy to prove that condition (4.129) is equivalent to the following one

$$-\pi < \beta < 0, \quad \mu \geq \frac{(c+c_1)(r^2+r_1^2)}{c_1 ar|\sin \beta|}. \tag{4.130}$$

For the equality sign in eqs. (4.129) and (4.130), i.e. for $(1+\mu L_2)D_2 = 0$, eq. (4.119) has the solution

$$\tau = \frac{1}{2}E_2 t^2 + \dot{\tau}(0)t + \tau(0),$$

$$\dot{\varphi} = -(1+\mu L_1)D_1 \left[\frac{1}{6}E_2 t^3 + \frac{1}{2}\dot{\tau}(0)t^2 + \tau(0)t\right] + E_1 t + \dot{\varphi}_0 \tag{4.131}$$

which implies that the stationary regime of cutting is unstable.

In the case of the strict inequalities (4.129) and (4.130), that is when $(1+\mu L_2)D_2 < 0$, solution of eq. (4.119) can be cast in the form

$$\tau = \rho_1 e^{\lambda t} + \rho_2 e^{-\lambda t} + \tau^*, \tag{4.132}$$

$$\dot{\varphi} = -\frac{1+\mu L_1}{\lambda} D_1 \left(\rho_1 e^{\lambda t} - \rho_2 e^{-\lambda t}\right) + \left[E_1 - \frac{(1+\mu L_1)D_1}{(1+\mu L_2)D_2}E_2\right]t + \dot{\varphi}_0,$$

where $\dot{\varphi}_0, \rho_1$ and ρ_2 are the integration constants, and $\lambda = [-(1+\mu L_2)D_2]^{1/2}$. The phase trajectories $(\tau, \dot{\tau})$ form a family of hyperbolas

$$(\tau - \tau_*)^2 - \frac{\dot{\tau}^2}{\lambda^2} = 4\rho_1 \rho_2 \tag{4.133}$$

for which point $(\tau_*, 0)$ is an unstable saddle point. Hence, a stationary cutting $\tau = \tau_0$, $\dot{\varphi} = \dot{\varphi}_0$ (for $B = B_0$) is unstable. As a failure of the instability condition leads to the stability condition (4.122), relationship (4.130) is the necessary and sufficient condition of instability of the stationary regime of cutting.

4.7.3 Relationship between instability of cutting and Painlevé's paradox

In order to establish the relationship between the above instability of cutting and Painlevé's paradox with Coulomb friction, we consider the spindle system depicted in Fig. 4.6 in which the blank is assumed to be a rigid body. This replacement is equivalent to letting c_1 tend to infinity which means that τ_0 and h tend to zero. In this case, tool 1 glides along a rigid cylinder. Let μ denote the friction coefficient between them and the contact be taken as being one-sided.

In this case the elastic force $\mathbf{F}$ has a constant absolute value and is given by analogy with eq. (4.111), i.e.

$$\mathbf{F} = \mathbf{F}_0 = F_0(\mathbf{i}_1 \cos \varphi + \mathbf{i} \sin \varphi). \tag{4.134}$$

160 4. Systems with many degrees of freedom and a single frictional pair

The total reaction force acting on the tool 1 from the cylinder is as follows

$$\mathbf{R}_\sigma = R[\varepsilon\mu(\mathbf{i}_1 \sin\varphi - \mathbf{i}_2 \cos\varphi) - (\mathbf{i}_1 \cos\varphi + \mathbf{i}_2 \sin\varphi)], \quad \dot\varphi > 0, \quad (4.135)$$

where R denotes the normal reaction force and $\varepsilon = \operatorname{sign} R$. As before, the driving torque is determined by formula (4.127)

$$B = \mu a F_0. \quad (4.136)$$

The position vectors of points $I, 0_1$ and T under an additional displacement h of tool 1 along the normal and their derivatives are given by relationships (4.115) and (4.116). Besides, in order to compare the forthcoming result with that obtained in the previous subsection, the velocity of gliding is assumed to be positive, i.e. $\operatorname{sign} \dot\varphi = 1$.

Taking into account the above we obtain the following expressions for the coefficients of the kinetic energy

$$A = M(r^2 + r_1^2), \quad A_{12}^0 = Mr\sin\beta, \quad \frac{\partial A_{11}}{\partial h} = \frac{\partial A_{12}^0}{\partial \varphi} = 0, \quad (4.137)$$

where A is the coefficient of the kinetic energy of the system before the contact is removed, $A_{ik}(i,k=1,2)$ are the coefficients after the contact has been removed. Utilising eqs. (4.134)-(4.136) and (4.116) we obtain the following expressions for the generalised forces and the generalised reaction forces

$$Q_1 = B = \mu a F_0, \quad Q_2 = -F_0, \quad S_1 = -\varepsilon_1 \mu a R, \quad S_2 = R. \quad (4.138)$$

Inserting the expressions for the coefficients (4.137) and (4.138) into eq. (2.13) yields the following system of equations

$$\begin{aligned} M(r^2 + r_1^2)\ddot\varphi &= \mu a F_0 - \varepsilon\mu a R, \\ Mr\sin\beta\,\ddot\varphi &= -F_0 + R \end{aligned} \quad (4.139)$$

which in turn results in the equation for the reaction force

$$(1 + \varepsilon\mu L)R = \frac{\mu ar\sin\beta + r^2 + r_1^2}{r^2 + r_1^2} F_0, \quad (4.140)$$

where coefficient L is determined by the formula

$$L = \frac{ar\sin\beta}{r^2 + r_1^2}. \quad (4.141)$$

By Theorem 1, the condition for paradoxes is expressed in the form

$$\mu > \frac{r^2 + r_1^2}{ar|\sin\beta|}. \quad (4.142)$$

It can be proved easily with the help of eqs. (4.140) and (4.141) that

$$\varepsilon_0 = \operatorname{sign} R_0 = \operatorname{sign}(\mu a r \sin \beta + r^2 + r_1^2) = \operatorname{sign} L. \tag{4.143}$$

In this case, symbol ε_1 simultaneously takes two values $\varepsilon_1 = \pm\varepsilon_0 = \pm \operatorname{sign} L$. By Theorem 8 $\varepsilon_1 = \operatorname{sign} L$ corresponds to a stable stationary value of the reaction force whereas $\varepsilon_1 = -\operatorname{sign} L$ corresponds to an unstable value.

On the other hand, only $\varepsilon_1 = +1$ can be realised since the contact is one-sided. Hence, for $0 < \beta < \pi$ we have $\operatorname{sign} L = 1$ and the root of eq. (4.140) under condition (4.142) is a stable stationary value of the reaction force

$$R = \frac{\mu a r \sin \beta + r_1^2 + r^2}{\mu a r \sin \beta + r_1^2 + r} F_0 = F_0, \quad \operatorname{sign} R = \operatorname{sign} R_0 = 1. \tag{4.144}$$

If $-\pi < \beta < 0$, then $\operatorname{sign} L = -1$ and the root of eq. (4.140) under condition (4.142) is an unstable stationary value of the reaction force

$$R = \frac{\mu a r \sin \beta + r_1^2 + r^2}{\mu a r \sin \beta + r_1^2 + r^2} F_0 = F_0, \quad \operatorname{sign} R = -\operatorname{sign} R_0 = 1. \tag{4.145}$$

As one can see, among all of the possible paradoxical cases, the case of

$$-\pi < \beta < 0, \quad \mu \frac{r^2 + r_1^2}{ar|\sin \beta|} \tag{4.146}$$

results in an unstable value of R. Comparing the right hand sides of relationships (4.130) and (4.146) we see that

$$\lim_{c_1 \to \infty} \frac{(c + c_1)(r^2 + r_1^2)}{c_1 a r |\sin \beta|} = \frac{r^2 + r_1^2}{ar|\sin \beta|}. \tag{4.147}$$

Therefore, as $c_1 \to \infty$, i.e. for modelling the blank by a rigid body, the condition of unstable cutting equates to the condition of Painlevé's paradoxes.

4.7.4 Boring with an axial feed

Let us consider the case shown in Fig. 4.7, when the spindle with the cutting tool 1 not only rotates along axis $0z$, but simultaneously moves along axis $0z$ due to the law $z = b\varphi/2\pi$. Here, parameter b is referred to as the feed, a denotes the mid-radius of the conical treated surface $ABCD$, 2α is the opening, R is the component of the cutting force which is normal to surface $ABCD$, R_1 is the component which is tangent to this surface and perpendicular to the cutting edge AB. The values R and R_1 are determined by relationship (4.110). All the remaining symbols are unchanged.

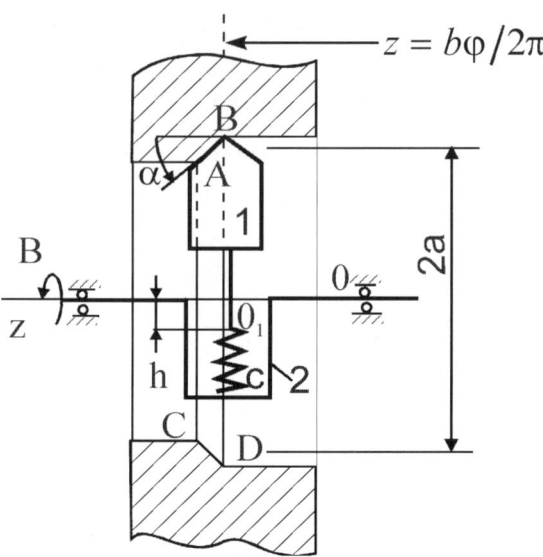

FIGURE 4.7.

Repeating the procedure of applying Lagrange's equations we obtain, instead of eq. (4.119), the following equations

$$\ddot{\varphi} + (1 + \mathsf{L}_1)D_1\tau - E_1 = 0, \quad \ddot{\tau} + (1 + \mathsf{L}_2)D_2\tau - E_2 = 0,$$

where

$$\mathsf{L}_1 = \frac{c_1(\mu a + b/2\pi \sin \alpha)}{r(c + c_1 \cos \alpha) \sin \beta}, \qquad D_1 = \frac{r(c + c_1 \cos \alpha) \sin \beta}{M(r^2 \cos^2 \beta + r_1^2 + b^2/4r^2)},$$

$$E_1 = \frac{B + (c + c_1)r\tau_0 \sin \beta}{M(r^2 \cos^2 + r_1^2 + b^2/4\pi^2)}, \qquad \mathsf{L}_2 = \frac{c_1(\mu a + b/2\pi \sin \alpha)r \sin \beta}{(r^2 + r_1^2)(c + c_1 \cos \alpha)},$$

$$D_2 = \frac{(r^2 + r_1^2)(c + c_1 \cos \alpha)}{M(r^2 \cos^2 \beta + r_1^2 + b^2/4\pi^2)}, \qquad E_2 = \frac{Br \sin \beta + (c + c_1)(r^2 + r_1^2)\tau_0}{M(r^2 \cos^2 \beta + r_1^2 + b^2/4\pi^2)}.$$

Hence, the necessary and sufficient condition for instability of the stationary cutting is expressed in the form

$$-\pi < \beta < 0, \quad \mu \geq \frac{2\pi(c + c_1 \cos \alpha)(r^2 + r_1^2 + b^2/4\pi^2) - c_1 br \sin \alpha |\sin \beta|}{2\pi c_1 ar |\sin \beta|}.$$

5
Systems with several frictional pairs. Painlevé's law of friction. Equations for the perturbed motion taking account of contact compliance

This chapter is concerned with the following problems: i) generalisation of the procedure for deriving equations for systems with Coulomb friction, ii) developing an analytical method of determining Painlevé's force of friction, and iii) detecting an inaccuracy within the equations and theorems suggested by Painlevé for systems with friction.

5.1 Equations for systems with Coulomb friction

5.1.1 System with removed constraints

Let us derive Lagrange's equations for a system with removed constraints. To this end, we consider a system of N particles subjected to $3N - n$ stationary positional constraints. Among them, m two-sided contact constraints with frictional coefficients $\mu_1, \mu_2, ..., \mu_m$ are assumed. The position of each slider is given by the following function of generalised coordinates

$$\mathbf{r}_{T\alpha}^0 = \mathbf{r}_{T\alpha}^0(q_1, ... q_n), \quad (\alpha = 1, ..., m). \tag{5.1}$$

The velocity of the slider glide on the corresponding guide surface is equal to

$$\mathbf{v}_{T\alpha}^0 = \sum_{k=1}^n \frac{\partial \dot{\mathbf{r}}_{T\alpha}}{\partial q_k} \dot{q}_k. \tag{5.2}$$

As above, the superscript 0 corresponds to the case of no additional displacements.

A possible locus of the $\alpha-th$ slider is a part of the $\alpha-th$ guide surface. In the case of an additional displacement, this locus forms a three-dimensional space near this surface. Let the redundant coordinates be $h_1, ..., h_m$, each being determined by analogy with eq. (2.2), i.e.

$$h_\alpha = (\mathbf{r}_{T\alpha} - \mathbf{r}_{T\alpha}^0) \cdot \mathbf{m}_\alpha, \qquad (5.3)$$

where $\mathbf{r}_{T\alpha}$ denotes the position vector of the $\alpha - th$ slider under an additional displacement and $\mathbf{m}$ is the unit vector of the normal to the $\alpha - th$ guide surface at the point of contact. While constructing the equations these coordinates are subjected to the following conditions

$$h_\alpha = \dot{h}_\alpha = \ddot{h}_\alpha = 0. \qquad (5.4)$$

Vectors $\mathbf{r}_{T1}, ..., \mathbf{r}_{Tm}$ are functions not only of the generalised coordinates $q_1, ..., q_n$, but also of the redundant coordinates $h_1, ..., h_m$

$$\mathbf{r}_{T\alpha} = \mathbf{r}_{T\alpha}(q_1, ..., q_n, h_1, ..., h_m). \qquad (5.5)$$

Here, cf. [102], [103],

$$\mathbf{r}_{T\alpha} = (q_1, ..., q_n, 0, ..., 0) = \mathbf{r}_{T\alpha}^0(q_1, ..., q_n). \qquad (5.6)$$

Assuming that this function can be expanded as a Taylor series in the neighbourhood of point $h = 0$ yields the following relationships between the derivatives with respect to the generalised coordinates

$$\left(\frac{\partial \mathbf{r}_{T\alpha}}{\partial q_k}\right)_0 = \frac{\partial \mathbf{r}_{T\alpha}}{\partial q_k} \quad (\alpha = 1, ..., m, \ k = 1, ..., n). \qquad (5.7)$$

It is necessary to mention some conditions for the derivatives with respect to the redundant coordinates. For example, according to eq. (5.3)

$$\mathbf{m}_\alpha \cdot \left(\frac{\partial \mathbf{r}_{T\alpha}}{\partial h_\alpha}\right)_0 = \lim_{T_\alpha \to T_\alpha^0} \frac{(\mathbf{r}_{T\alpha} - \mathbf{r}_{T\alpha}^0) \cdot \mathbf{m}_\alpha}{(\mathbf{r}_{T\alpha} - \mathbf{r}_{T\alpha}^0) \cdot \mathbf{m}_\alpha} = 1 \quad (\alpha = 1, ..., m), \qquad (5.8)$$

where T_α^0 and T_α denote the positions of the $\alpha - th$ slider before and after an additional displacement. An additional displacement of the $\alpha - th$ slider under the condition $h_\beta = 0$ for $\beta \neq \alpha$ can cause only tangential displacements of other sliders on the corresponding surfaces. For this reason,

$$\left(\frac{\partial \mathbf{r}_{T\beta}}{\partial h_\alpha}\right)_0 \cdot \mathbf{m}_\beta = 1 \quad (\beta, \alpha = 1, ..., m, \beta \neq \alpha), \qquad (5.9)$$

i.e. the projection of the $\beta - th$ component of the derivative $(\partial \mathbf{r}_{T\beta}/\partial h_\alpha)_0$ on the normal is equal to zero if $\beta \neq \alpha$.

5.1 Equations for systems with Coulomb friction

The general reaction force $\mathbf{R}^\sigma_\alpha$ acting on the $\alpha-th$ slider from the $\alpha-th$ guidance is expressed as follows

$$\mathbf{R}^\sigma_\alpha = \left(-\varepsilon_\alpha \mu_\alpha \frac{\mathbf{v}^0_{T\alpha}}{|\mathbf{v}^0_{T\alpha}|} + \mathbf{m}_\alpha\right) R_\alpha \quad (\alpha = 1, ..., m), \tag{5.10}$$

where R_α denotes the value of the normal reaction and $\varepsilon_\alpha = \operatorname{sign} R_\alpha$.

The reaction forces corresponding to the generalised and redundant coordinates are respectively given by the formulae

$$S_k = \sum_{\alpha=1}^{m} \mathbf{R}^\sigma_\alpha \cdot \frac{\partial \mathbf{r}^0_{T\alpha}}{\partial q_k}, \quad S'_\beta = \sum_{\alpha=1}^{m} \mathbf{R}^\sigma_\alpha \cdot \left(\frac{\partial \mathbf{r}_{T\alpha}}{\partial h_\beta}\right)_0,$$

$$(k = 1, ..., n, \beta = 1, ..., m).$$

Inserting expressions (5.10) for $\mathbf{R}^\sigma_\alpha$ into these formulae and accounting for relationships (5.7)-(5.9) and the following equalities

$$\frac{\mathbf{v}^0_{T\alpha}}{|\mathbf{v}^0_{T\alpha}|} \cdot \frac{\partial \mathbf{r}^0_{T\alpha}}{\partial \dot{q}_k} = \frac{1}{2|\mathbf{v}^0_{T\alpha}|} \frac{\partial |\mathbf{v}^0_{T\alpha}|^2}{\partial \dot{q}_k} = \frac{\partial v^0_\alpha}{\partial \dot{q}_k},$$

$$\frac{\mathbf{v}^0_{T\alpha}}{|\mathbf{v}^0_{T\alpha}|} \cdot \left(\frac{\partial \mathbf{r}_{T\alpha}}{\partial h_\beta}\right)_0 = \left(\frac{\partial v_{T\alpha}}{\partial \dot{h}_\beta}\right)_0$$

we obtain the expressions

$$S_k = -\sum_{\alpha=1}^{m} \varepsilon_\alpha \mu_\alpha \frac{\partial v^0_{T\alpha}}{\partial \dot{q}_k} R_\alpha, \quad (k = 1, ..., n),$$

$$S'_\beta = -\sum_{\alpha=1}^{m} \varepsilon_\alpha \mu_\alpha \left(\frac{\partial v_{T\alpha}}{\partial \dot{h}_\beta}\right)_0 R_\alpha + R_\beta, \quad (\beta = 1, ..., m), \tag{5.11}$$

where $v^0_{T\alpha} = |\mathbf{v}^0_{T\alpha}|$ and $v_{T\alpha} = |\mathbf{v}_{T\alpha}|$.

A set of positions of the $\beta - th$ slider under a continuous change in h_β, fixed values of $q_k(k = 1, ..., n)$ and zero initial conditions $h_\alpha = 0$ for $\alpha \neq \beta$ is referred to as the trajectory of the $\beta - th$ additional displacement. In the case when these trajectories are orthogonal to the corresponding guide surfaces U_β we have

$$\left(\frac{\partial \mathbf{r}_{T\beta}}{\partial h_\beta}\right)_0 \perp U_\beta, \quad \left(\frac{\partial v_{T\beta}}{\partial \dot{h}_\beta}\right)_0 = \frac{\mathbf{v}^0_{T\beta}}{|\mathbf{v}^0_{T\beta}|} \cdot \left(\frac{\partial \mathbf{r}_{T\beta}}{\partial h_\beta}\right)_0 = 0.$$

In this case, the term with subscript $\alpha = \beta$ does not appear in the formula for S'_β.

Following Lurie [103] and using eq. (5.11) we can represent Lagrange's equations for the system with the removed contact constraint in the fol-

166   5. Systems with several frictional pairs. Painlevé's law of friction

lowing form

$$\sum_{k=1}^{n} A_{ks}^0 \ddot{q}_k + \sum_{k,t=1}^{n} \Gamma_{kt,s}^0 \dot{q}_k \dot{q}_t = Q_s - \sum_{\alpha=1}^{m} \varepsilon_\alpha \mu_\alpha \frac{\partial v_{T\alpha}^0}{\partial \dot{q}_s} R_\alpha, \quad (s=1,...,n),$$

$$\sum_{k=1}^{n} A_{k,n+\beta}^0 \ddot{q}_k + \sum_{k,t=1}^{n} \Gamma_{kt,n+\beta}^0 \dot{q}_k \dot{q}_t = Q'_\beta + R_\beta - \sum_{\alpha=1}^{n} \varepsilon_\alpha \mu_\alpha \left(\frac{\partial v_{T\alpha}}{\partial \dot{h}_\beta}\right)_0 R_\alpha,$$
(5.12)

where Q_s and Q'_β denote respectively the active forces corresponding to the generalised and redundant coordinates, and $\beta = 1, ..., m$.

5.1.2 Solving the main system

Resolving system (5.12) for the reaction forces and accelerations is carried out with the help of block matrices. Let us present system (5.12) as an matrix equation for unknown variables $\ddot{q}$ and R

$$a\ddot{q} + \frac{Dv}{D\dot{q}}\varepsilon\mu R = Q - \Gamma,$$

$$b\ddot{q} + \left(\frac{Dv}{D\dot{q}}\varepsilon\mu - E\right)R = Q' - \Gamma',$$
(5.13)

where

$$a = \begin{bmatrix} A_{11} & ... & A_{n+1} \\ ... & ... & ... \\ A_{1n} & ... & A_{nn} \end{bmatrix}, \quad b = \begin{bmatrix} A_{1n+1} & ... & A_{n,n+1} \\ ... & ... & ... \\ A_{1n+m} & ... & A_{n,n+m} \end{bmatrix},$$

$$\frac{Dv}{D\dot{q}} = \begin{bmatrix} \frac{\partial v_{T1}}{\partial \dot{q}_1} & \frac{\partial v_{T2}}{\partial \dot{q}_1} & ... & \frac{\partial v_{Tm}}{\partial \dot{q}_1} \\ \frac{\partial v_{T1}}{\partial \dot{q}_2} & \frac{\partial v_{T2}}{\partial \dot{q}_2} & ... & \frac{\partial v_{Tm}}{\partial \dot{q}_2} \\ ... & ... & ... & ... \\ \frac{\partial v_{T1}}{\partial \dot{q}_n} & \frac{\partial v_{T2}}{\partial \dot{q}_n} & ... & \frac{\partial v_{Tm}}{\partial \dot{q}_n} \end{bmatrix}, \quad \frac{Dv}{Dh} = \begin{bmatrix} \frac{\partial v_{T1}}{\partial \dot{h}_1} & ... & \frac{\partial v_{Tm}}{\partial \dot{h}_1} \\ \frac{\partial v_{T1}}{\partial \dot{h}_2} & ... & \frac{\partial v_{Tm}}{\partial \dot{h}_2} \\ ... & ... & ... \\ \frac{\partial v_{T1}}{\partial \dot{h}_m} & ... & \frac{\partial v_{Tm}}{\partial \dot{h}_m} \end{bmatrix},$$

$$q = [q_1, ..., q_n], \quad R = [R_1, ..., R_m],$$

$$\Gamma = \left[\sum_{k,t=1}^{n} \Gamma_{kt,1} \dot{q}_k \dot{q}_t, ..., \sum_{k,t=1}^{n} \Gamma_{kt,n} \dot{q}_k \dot{q}_t\right],$$

$$\Gamma' = \left[\sum_{k,t=1}^{n} \Gamma_{kt,n+1} \dot{q}_k \dot{q}_t, ..., \sum_{k,t=1}^{n} \Gamma_{kt,n+m} \dot{q}_k \dot{q}_t\right],$$

$$Q = [Q_1, ..., Q_n], \qquad Q' = [Q'_1, ..., Q'_m],$$
$$\mu = \text{diag}[\mu_1, ..., \mu_m], \qquad \varepsilon = \text{diag}[\varepsilon_1, ..., \varepsilon_m]. \qquad (5.14)$$

In the case in which the trajectories of the additional displacement are orthogonal to the corresponding surfaces, the diagonal elements

$$\partial v_{T\alpha}/\partial \dot{h}_\alpha, \quad (\alpha = 1, ..., m)$$

of matrix $Dv/D\dot{q}$ are equal to zero.

It what follows, we assume that the system of equations (5.13) is linearly independent, i.e. its block matrix

$$\begin{bmatrix} (a) & \left(\dfrac{Dv}{D\dot{q}}\varepsilon\mu\right) \\ (b) & \left(\dfrac{Dv}{D\dot{h}}\varepsilon\mu - E\right) \end{bmatrix} \qquad (5.15)$$

is non-degenerate. Resolving system (5.13) leads to following equations

$$\left[a + \dfrac{Dv}{D\dot{q}}\varepsilon\mu\left(E - \dfrac{Dv}{D\dot{h}}\varepsilon\mu\right)^{-1} b\right]\ddot{q} = Q - \Gamma +$$

$$\dfrac{Dv}{D\dot{q}}\varepsilon\mu\left(E - \dfrac{Dv}{D\dot{h}}\varepsilon\mu\right)^{-1}(Q' - \Gamma'),$$

$$\left[E + \left(ba^{-1}\dfrac{Dv}{D\dot{q}} - \dfrac{Dv}{D\dot{h}}\varepsilon\mu\right)\right] R = ba^{-1}(Q - \Gamma)(Q' - \Gamma'). \qquad (5.16)$$

System (5.13) can be resolved since M and a are not degenerate, and by virtue of the well-known relationship

$$\det\begin{pmatrix}(A) & (B) \\ (C) & (D)\end{pmatrix} = \det A \det(D - cA^{-1}B) = \det D \det(A - BD^{-1}C)$$

matrix $(Dv/D\dot{h})\varepsilon\mu - E$ and the matrix coefficients in front of $\ddot{q}$ and R in eq. (5.16) are also non-degenerate.

The first equation in (5.16) is a system of n differential equations of motion in which the reaction forces are eliminated whereas the second equation in (5.16) is a system of m linear algebraic equations for the unknown reaction forces $R_1, ..., R_m$.

In the forthcoming analysis, the main focus is on the second system. By analogy with eqs. (2.60) and (2.65) the quadratic matrix

$$\mathsf{L} = ba^{-1}\dfrac{Dv}{D\dot{q}} - \dfrac{Dv}{D\dot{h}} \qquad (5.17)$$

is referred to as the influence matrix of the contact constraints, whilst its elements

$$\mathsf{L}_{\alpha\beta} = \sum_{i,k=1}^{n} A_{i,n+a}A_{ki}^{-1}\dfrac{\partial v_{T\beta}}{\partial \dot{q}_k} - \dfrac{\partial v_{T\beta}}{\partial \dot{h}_\alpha}, \quad (\alpha, \beta = 1, ..., m) \qquad (5.18)$$

are termed the indices of influence. Here A_{ki}^{-1} denotes the element of the $i-th$ row and the $k-th$ column of matrix a^{-1}.

The elements

$$^0R_\alpha = \sum_{i,k=1}^n A_{i,n+\alpha} A_{ki}^{-1}\left(Q_k - \sum_{s,t=1}^n \Gamma_{s,t,k}\dot{q}_s\dot{q}_t\right) - \left(Q'_\alpha - \sum_{s,t=1}^n \Gamma_{s,t,n+\alpha}\dot{q}_s\dot{q}_t\right), \quad (\alpha = 1,...,m) \tag{5.19}$$

of the right column in eq. (5.16)

$$^0R = ba^{-1}(Q - \Gamma) - (Q' - \Gamma') \tag{5.20}$$

are the normal reaction forces in the ideal case $\mu = 0$ since the zero frictional coefficients in eq. (5.16) result in equality $R = {}^0R$.

Taking into account eqs. (5.17)-(5.20) the system of equations (5.16) with unknown reaction forces can be set in the form

$$(E + \mathsf{L}\varepsilon\mu)R = {}^0R \tag{5.21}$$

or in the equivalent form

$$\begin{bmatrix} 1+\varepsilon_1\mu_1\mathsf{L}_{11} & \varepsilon_2\mu_2\mathsf{L}_{12} & \cdots & \varepsilon_m\mu_m\mathsf{L}_{1m} \\ \varepsilon_1\mu_1\mathsf{L}_{21} & 1+\varepsilon_2\mu_2\mathsf{L}_{22} & \cdots & \varepsilon_m\mu_m\mathsf{L}_{2m} \\ \cdots & \cdots & \cdots & \cdots \\ \varepsilon_1\mu_1\mathsf{L}_{m1} & \varepsilon_2\mu_2\mathsf{L}_{m2} & \cdots & 1+\varepsilon_m\mu_m\mathsf{L}_{mm} \end{bmatrix} \cdot \begin{bmatrix} R_1 \\ R_2 \\ \cdots \\ R_m \end{bmatrix} = \begin{bmatrix} {}^0R_1 \\ {}^0R_2 \\ \cdots \\ {}^0R_m \end{bmatrix} \tag{5.22}$$

Resolving eq. (5.22) for $R_1, R_2, ..., R_m$ we can write down the equation for the reaction force

$$\left(\lambda_{\beta\beta} + \varepsilon_\beta\mu_\beta \sum_{\alpha=1}^m \mathsf{L}_{\alpha\beta}\lambda_{\alpha\beta}\right) R_\beta = \sum_{\alpha=1}^m {}^0R_\alpha\lambda_{\alpha\beta}, \tag{5.23}$$

where $\lambda_{\alpha\beta}$ denotes the algebraic adjunct of the element of the $\alpha-th$ row and the $\beta-th$ column of the determinant of system (5.22). It is easy to see that in the case of ideal contact constraints

$$\mu = 0, \quad \lambda_{\alpha\beta} = \begin{cases} 1 & \alpha = \beta \\ 0 & \alpha \neq \beta \end{cases}, \quad R = {}^0R, \quad \varepsilon = {}^0\varepsilon.$$

Any of the relationships (5.21)-(5.23) allows us to clarify the question of existence and uniqueness of the dynamical problem (5.16), and thus determine the reaction forces and accelerations. These relationships contain m symbols $\varepsilon_\alpha = \operatorname{sign} R_\alpha (\alpha = 1,...,m)$ each of which has a value of 1 or -1. Therefore, we have altogether 2^m possible combinations of signs of the reactions forces, i.e. 2^m variants of the solution. For example, in the case of $m = 2$ we have four combinations: 1) $\varepsilon_1 = \varepsilon_2 = 1$, 2) $\varepsilon_1 = \varepsilon_2 = -1$, 3) $\varepsilon_1 = -\varepsilon_2 = 1$ and 4) $\varepsilon_1 = -\varepsilon_2 = -1$. Thus it is necessary to make 2^m tests. The combination is considered to be correct if the values of the reaction forces have the same signs when they are substituted into eqs. (5.21)-(5.23). Provided that one and only one combination is correct, the solution of eq. (5.16) exists and is unique. In the case when not a single correct combination exists, or several combinations turn out to be correct, the paradoxical situations of non-existence or non-uniqueness of the dynamical problem (5.16) are observed.

5.1.3 The case of $n = 1, m = 1$

This case if often encountered in practice. For this reason, we demonstrate some relationships obtained from the above results. For instance, the system of equations (5.12) takes the form

$$A_{11}\ddot{q} + \frac{1}{2}\frac{\partial A_{11}}{\partial q}\dot{q}^2 = Q - \sum_{\alpha=1}^{m} \varepsilon_\alpha \mu_\alpha \frac{\partial v_{T\alpha}}{\partial \dot{q}} R_\alpha,$$

$$A_{11+\beta}\ddot{q} + \left(\frac{\partial A_{11+\beta}}{\partial q} - \frac{1}{2}\frac{\partial A_{11}}{\partial h_\beta}\right)\dot{q}^2 = Q'_\beta + R_\beta -$$

$$\sum \varepsilon_\alpha \mu_\alpha \frac{\partial v_{T\alpha}}{\partial h_\beta} R_\alpha, \quad (\beta = 1,...,m). \qquad (5.24)$$

In this particular case

$$a = A_{11}, \quad b = [A_{12}, A_{13},...,A_{1,1+m}],$$
$$\frac{Dv}{D\dot{q}} = \left[\frac{\partial v_{T1}}{\partial \dot{q}},...,\frac{\partial v_{Tm}}{\partial \dot{q}}\right]. \qquad (5.25)$$

Accounting for relationship (5.25) in formula (5.18) we obtain the following values for the indices of influence

$$L_{\alpha\beta} = \frac{\partial A_{11+\alpha}}{\partial A_{11}} \frac{\partial v_{T\beta}}{\partial \dot{q}} - \frac{\partial v_{T\beta}}{\partial h_\beta}, \quad (\alpha, \beta = 1,...,m). \qquad (5.26)$$

Expressions (5.24)-(5.26) are used for the forthcoming analysis of the Painlevé scheme and sliders of the machine tools.

5.2 Mathematical description of the Painlevé law of friction

5.2.1 Accelerations due to two systems of external forces

Let us derive equations for determining the difference between accelerations $^0\mathbf{W}_i$ and $\mathbf{W}_i$ of the ideal mechanical system subjected to two systems of prescribed forces. These equations are applied for an analytical formulation of Painlevè's law of friction. The Painlevè law of friction is understood to be a set of expressions for the force of friction acting on particles, [117], [116].

Let a system of N particles be subjected to $3N - n$ constraints

$$\Phi_j(x_1, y_1, z_1, ..., x_N, y_N, z_N) = 0, \quad (j = 1, ..., 3N - n). \tag{5.27}$$

Let us consider separate actions of two different systems of external forces $^0\mathbf{F}_1, ..., ^0\mathbf{F}_N$ and $\mathbf{F}_1, ..., \mathbf{F}_N$ on the considered mechanical system at a certain time instant and for given $x_i, y_i, z_i, \dot{x}_i, \dot{y}_i, \dot{z}_i$. Applying the D'Alembert-Lagrange principle to these cases we can write the following equations

$$\sum_{i=1}^{N}(M_i\, ^0\mathbf{W}_i - \, ^0\mathbf{F}_i) \cdot \delta\mathbf{r}_i = 0, \quad \sum_{i=1}^{N}(M_i\mathbf{W}_i - \mathbf{F}_i) \cdot \delta\mathbf{r}_i = 0.$$

Subtracting the second equation from the first one and introducing the accelerations $\mathbf{a}_i$ and forces $\mathbf{b}_i$

$$\mathbf{a_i} = \mathbf{W}_i - \, ^0\mathbf{W}_i, \quad \mathbf{b_i} = \mathbf{F}_i - \, ^0\mathbf{F}_i \tag{5.28}$$

we obtain

$$\sum_{i=1}^{N}(M_i\mathbf{a}_i - \mathbf{b}_i) \cdot \delta\mathbf{r}_i = \sum_{k=1}^{n}\delta q_k \sum_{i=1}^{N}(M_i\mathbf{a}_i - \mathbf{b}_i) \cdot \frac{\partial \mathbf{r}_i}{\partial q_k} = 0. \tag{5.29}$$

This yields n equations of the form

$$\sum_{i=1}^{N}(M_i\mathbf{a}_i - \mathbf{b}_i) \cdot \frac{\partial \mathbf{r}_i}{\partial q_k} = 0, \quad (k = 1, ..., n). \tag{5.30}$$

Taking the second time derivative of eq. (5.27) we obtain the following two conditions

$$\left.\frac{d^2\Phi_j}{dt^2}\right|_0 = \sum_{i=1}^{N}\text{grad}_i\Phi_j \cdot \, ^0\mathbf{W}_i + \sum_{i=1}^{N}\mathbf{v}_i \cdot \frac{d}{dt}\text{grad}_i\Phi_j = 0,$$

$$\frac{d^2\Phi_j}{dt^2} = \sum_{i=1}^{N}\text{grad}_i\Phi_j \cdot \mathbf{W}_i + \sum_{i=1}^{N}\mathbf{v}_i \cdot \frac{d}{dt}\text{grad}_i\Phi_j = 0.$$

5.2 Mathematical description of the Painlevé law of friction

Subtracting them and accounting for eq. (5.28) we arrive at the following $3N - n$ equations

$$\sum_{i=1}^{N} \operatorname{grad}_{i} \Phi_{j} \cdot \mathbf{a}_{i} = 0, \quad (j = 1, ..., 3N - n). \quad (5.31)$$

Relationships (5.30) and (5.31) form a system of $3N$ equations with $3N$ unknown variables a_{ix}, a_{iy}, a_{iz}, which are the changes in the acceleration components

$$\sum_{i=1}^{N} M_i \left(\frac{\partial x_i}{\partial q_k} a_{ix} + \frac{\partial y_i}{\partial q_k} a_{iy} + \frac{\partial z_i}{\partial q_k} a_{iz} \right) = S_k,$$

$$\sum_{i=1}^{N} \left(\frac{\partial \Phi_j}{\partial x_i} a_{ix} + \frac{\partial \Phi_j}{\partial y_i} a_{iy} + \frac{\partial \Phi_j}{\partial z_i} a_{iz} \right) = 0,$$

$$(k = 1, ..., n, \; j = 1, ..., 3N - n)$$

$$S_k = \sum_{i=1}^{N} \left(\frac{\partial x_i}{\partial q_k} b_{ix} + \frac{\partial y_i}{\partial q_k} b_{iy} + \frac{\partial z_i}{\partial q_k} b_{iz} \right). \quad (5.32)$$

Here S_k is the change in the generalised forces corresponding to coordinate q_k. System of equations (5.32) is valid for arbitrary values of $^0\mathbf{F}_i, \mathbf{F}_i$ and any prescribed velocities $\dot{q}_i$. In particular this system is homogeneous and valid for $^0\mathbf{F}_i = \mathbf{F}_i = \mathbf{b}_i = 0$. Hence, its determinant must be non-zero otherwise for $\dot{q}_1 = ... = \dot{q}_n = 0$ one can find non-zero values a_{ix}, a_{iy}, a_{iz} such that the mechanical systems moves from the equilibrium position without any external load. Thus, for any strictly positive values $M_1, ..., M_N$ we have that the determinant of the following matrix

$$\begin{bmatrix} M_1 \dfrac{\partial x_1}{\partial q_1} & M_1 \dfrac{\partial y_1}{\partial q_1} & M_1 \dfrac{\partial z_1}{\partial q_1} & \cdots & M_N \dfrac{\partial x_N}{\partial q_1} & M_N \dfrac{\partial y_N}{\partial q_1} & M_1 \dfrac{\partial z_N}{\partial q_1} \\ \cdots & \cdots & \cdots & \cdots & \cdots & \cdots & \cdots \\ M_1 \dfrac{\partial x_1}{\partial q_n} & M_1 \dfrac{\partial y_1}{\partial q_n} & M_1 \dfrac{\partial z_1}{\partial q_n} & \cdots & M_N \dfrac{\partial x_N}{\partial q_n} & M_N \dfrac{\partial y_N}{\partial q_n} & M_N \dfrac{\partial z_N}{\partial q_n} \\ \dfrac{\partial \Phi_1}{\partial x_1} & \dfrac{\partial \Phi_1}{\partial y_1} & \dfrac{\partial \Phi_1}{\partial z_1} & \cdots & \dfrac{\partial \Phi_N}{\partial x_N} & \dfrac{\partial \Phi_N}{\partial y_N} & \dfrac{\partial \Phi_N}{\partial z_N} \\ \cdots & \cdots & \cdots & \cdots & \cdots & \cdots & \cdots \\ \dfrac{\partial \Phi_{3N-n}}{\partial x_1} & \dfrac{\partial \Phi_{3N-n}}{\partial y_1} & \dfrac{\partial \Phi_{3N-n}}{\partial z_1} & \cdots & \dfrac{\partial \Phi_{3N-n}}{\partial x_N} & \dfrac{\partial \Phi_{3N-n}}{\partial y_N} & \dfrac{\partial \Phi_{3N-n}}{\partial z_N} \end{bmatrix} \quad (5.33)$$

does not vanish.

Let us prove the following property of the acceleration difference $\mathbf{a}_i$.

Statement 1. Among all systems of vectors $\mathbf{a}_i$ satisfying condition (5.29) system $\mathbf{a}_i$ is that for which

$$\sum_{i=1}^{N} M_i \mathbf{a}_i^2 = \min \sum_{i=1}^{N} M_i \mathbf{a}'^2_i . \qquad (5.34)$$

Proof. Any system $\mathbf{a}'_i$ can be represented in the form

$$\mathbf{a}'_i = \mathbf{a}_i + \mathbf{c}_i, \quad \sum_{k=1}^{M} M_k \mathbf{c}_k \cdot \delta \mathbf{r}_k = 0 . \qquad (5.35)$$

For condition (5.34) to be satisfied it is necessary and sufficient that, [117], [130],

$$M_i \mathbf{c}_i = \sum_{j=1}^{3N-n} \lambda_j \operatorname{grad}_i \Phi_j , \qquad (5.36)$$

where λ_j are undetermined coefficients which are equal for all particles of the system. It follows from eqs. (5.31), (5.35) and (5.36) that

$$\sum M_i \mathbf{a}'^2_i = \sum M_i (a_i^2 + c_i^2) + 2 \sum_{j=1}^{3N-n} \lambda_j \sum_{i=1}^{N} \operatorname{grad}_i \Phi_j \cdot \mathbf{a}_i$$
$$= \sum M_i (a_i^2 + c_i^2) \qquad (5.37)$$

which yields condition (5.34).

5.2.2 Improved Painlevé's equations

Let us consider the case in which several of constraints (5.27) are non-ideal. Painlevé has shown in [117], [116] that the reaction force $\mathbf{K}_i$ acting on the $i-th$ particle can be uniquely decomposed into two components: i) component ${}^0\mathbf{K}_i$ equal to the reaction force in the case of no friction (all constraints are ideal) and ii) component

$$\boldsymbol{\rho}_i = \mathbf{K}_i - {}^0\mathbf{K}_i = M_i \mathbf{W}_i - M_i {}^0\mathbf{W}_i \qquad (5.38)$$

which is referred to as the force of friction. Here $\mathbf{W}_i$ and ${}^0\mathbf{W}_i$ are the acceleration for the cases when friction is present and absent respectively.

In accordance with [117] and [116] the set of values $\boldsymbol{\rho}$ is referred to as the law of friction. The law of friction is seen to depend on the configuration of the system. The following $3N$ equations with $3N$ unknown variables

5.2 Mathematical description of the Painlevé law of friction

$\boldsymbol{\rho}_{ix}, \boldsymbol{\rho}_{iy}, \boldsymbol{\rho}_{iz}$

$$\sum_{i=1}^{N} \left(\frac{\partial x_i}{\partial q_k} \rho_{ix} + \frac{\partial y_i}{\partial q_k} \rho_{iy} + \frac{\partial z_i}{\partial q_k} \rho_{iz} \right) = \sum \left(\frac{\partial x_i}{\partial q_k} K_{ix} + \frac{\partial y_i}{\partial q_k} K_{iy} + \frac{\partial z_i}{\partial q_k} K_{iz} \right),$$

$$\sum_{i=1}^{N} \left(\frac{\partial \Phi_j}{\partial x_i} \rho_{ix} + \frac{\partial \Phi_j}{\partial y_i} \rho_{iy} + \frac{\partial \Phi_j}{\partial z_i} \rho_{iz} \right) = 0$$

$$(k = 1, ..., n, \quad j = 1, ..., 3N - n) \quad (5.39)$$

were derived in [117], [116] when determining this law. We notice one inaccuracy of the system of equations (5.39). In order to determine the difference in acceleration $\mathbf{a}_i = \mathbf{W}_i -^0 \mathbf{W}_i$ of the non-ideal system compared with the ideal one we use eq. (5.32) with the following values of b_i and S_i

$$\mathbf{b}_i = \boldsymbol{\rho}_i, \quad S_k = \sum \frac{\partial \mathbf{r}_i}{\partial q_k} \cdot \mathbf{b}_i = \sum \frac{\partial \mathbf{r}_i}{\partial q_k} \cdot \mathbf{K}_i \quad (5.40)$$

which are the differences in the prescribed and generalised forces, respectively.

Due to eqs. (5.32), (5.38) and (5.40) we obtain the following system of equations

$$\sum_{i=1}^{N} \left(\frac{\partial x_i}{\partial q_k} \rho_{ix} + \frac{\partial y_i}{\partial q_k} \rho_{iy} + \frac{\partial z_i}{\partial q_k} \rho_{iz} \right) = \sum_{i=1}^{N} \left(\frac{\partial x_i}{\partial q_k} K_{ix} + \frac{\partial y_i}{\partial q_k} K_{iy} + \frac{\partial z_i}{\partial q_k} K_{iz} \right),$$

$$\sum_{i=1}^{N} \frac{1}{M_i} \left(\frac{\partial \Phi_j}{\partial x_i} \rho_{ix} + \frac{\partial \Phi_j}{\partial y_i} \rho_{iy} + \frac{\partial \Phi_j}{\partial z_i} \rho_{iz} \right) = 0$$

$$(k = 1, ..., n, \quad j = 1, ..., 3N - n). \quad (5.41)$$

In the case of Coulomb friction, S_i is due to eq. (5.37) and system (5.41) takes the form

$$\sum_{i=1}^{N} \left(\frac{\partial x_i}{\partial q_k} \rho_{ix} + \frac{\partial y_i}{\partial q_k} \rho_{iy} + \frac{\partial z_i}{\partial q_k} \rho_{iz} \right) = -\sum_{\alpha=1}^{m} \varepsilon_\alpha \mu_\alpha \frac{\partial v_{T\alpha}^0}{\partial q_k} R_\alpha,$$

$$\sum_{i=1}^{N} \frac{1}{M_i} \left(\frac{\partial \Phi_j}{\partial x_i} \rho_{ix} + \frac{\partial \Phi_j}{\partial y_i} \rho_{iy} + \frac{\partial \Phi_j}{\partial z_i} \rho_{iz} \right) = 0. \quad (5.42)$$

As one can see, system (5.41) differs from (5.39) by the presence of multiplier M_i^{-1} in the second equation. The reason for this difference is as follows. While deriving system (5.39) Painlevé considered vectors $\boldsymbol{\rho}_i dt$ as certain virtual displacements which results in the last $3M - n$ equations of the system, i.e. the following conditions

$$\sum_i \text{grad}_i \Phi_j \cdot \boldsymbol{\rho}_i = \sum_i \text{grad}_i \Phi_j \cdot M_i \mathbf{a}_i = 0.$$

This is possible only in the trivial case

$$\text{grad}_i \Phi_j \cdot \mathbf{a}_i = 0, \quad (i = 1, ..., N, \; j = 1, ..., 3N - n)$$

when systems (5.39) and (5.41) are equivalent. In general, vectors $\boldsymbol{\rho}_i dt$ do not always form a virtual displacement. For example, for both particles of the Painlevé-Klein scheme, Fig. 3.3, the value of acceleration difference $\mathbf{a} = (\ddot{x} - ^0\ddot{x})\mathbf{i}_1$ is coincident and proportional to a virtual displacement. At the same time, the friction forces $\boldsymbol{\rho}_s = M_s \mathbf{a}$ $(s = 1, 2)$ given by eq. (5.12) for $M_1 \neq M_2$ do not coincide and can not compose any displacement.

Thus, system of equations (5.39) is not consistent with the definition of the force of friction suggested by Painlevé in the form of eq. (5.38) and can not be utilised to calculate these forces.

From this perspective, we refer to the equations in (5.41) as improved Painlevé's equations for systems with friction.

5.2.3 Improved Painlevé's theorem

Let us consider Painlevé's theorem [117], [116]: among all systems of forces $\mathbf{K'}_i$ whose elementary work is equal to the work of reaction forces $\mathbf{K}_i$, the system $\boldsymbol{\rho}_i$ is such that

$$\sum_{i=1}^{N} \rho_i^2 = \min \sum_{i=1}^{N} K'^2_i. \tag{5.43}$$

Let us prove the inverse statement, namely, that there exists such a system $\mathbf{K'}_i$ that $\sum \mathbf{K'}_i^2 < \sum \rho_i^2$. Indeed, due to the condition of Painlevé's theorem, any system of vectors $\mathbf{K'}_i^2$ admits the following decomposition

$$\mathbf{K'}_i^2 = \boldsymbol{\rho}_i + \boldsymbol{\rho'}_i, \quad \sum_{i=1}^{N} \boldsymbol{\rho'}_i \cdot \delta \mathbf{r}_i = 0, \quad \boldsymbol{\rho'}_i = \sum_{j=1}^{3N-n} \lambda_j \text{grad}_i \Phi_j,$$

where λ_j are indeterminate coefficients. Then

$$\sum_{i=1}^{N} \mathbf{K'}_i^2 = \sum \rho_i^2 + \sum \rho'^2_i + 2 \sum_{j=1}^{3N-t} \lambda_j \sum_{i=1}^{N} M_i \text{grad}_i \Phi_j \cdot \mathbf{a}_i.$$

As mentioned, values $M_i \text{grad}_i \Phi_j \cdot \mathbf{a}_i$ may differ from zero in the general case. For example, let $M_i \text{grad}_i \Phi_j \cdot \mathbf{a}_i \neq 0$, $\lambda_1 \neq 0$, $\lambda_2 = ... = \lambda_{3N-n} = 0$. Then

$$\sum \mathbf{K'}_i^2 = \sum \rho_i^2 + \lambda_1^2 \sum (\text{grad}_i \Phi_1)^2 + 2\lambda_1 \sum \text{grad}_i \Phi_1 \cdot \mathbf{a}_i M_i.$$

Hence, taking value of λ_1 from the interval

$$\left(0, \; \frac{-2 \sum M_i \text{grad}_i \Phi_1 \cdot \mathbf{a}_i}{\sum (\text{grad}_i \Phi_1)^2} \right)$$

we have
$$\lambda_1^2 \sum (\mathrm{grad}_i \Phi_1)^2 + 2\lambda_1 \sum \mathrm{grad}_i \Phi_1 \cdot M_i \mathbf{a}_i < 0$$

and then $\mathbf{K}'^2_i < \rho_i^2$. However, this contradicts condition (5.43), that is, the above theorem is not valid.

Let us prove the following statement which comprises the improved Painlevé's theorem.

Statement 2. Among all systems of forces $\mathbf{K}'_i$, whose elementary work is equal to the work of the reaction forces $\mathbf{K}_i$, the system of Painlevé's forces of friction ρ_i is that for which

$$\sum_{i=1}^{N} \frac{1}{M_i} \rho_i^2 = \min \sum_{i=1}^{N} \frac{1}{M_i} K'^2_i. \tag{5.44}$$

Proof. We represent forces $\mathbf{K}'_i$ in the form $\mathbf{K}'_i = M_i \mathbf{a}'_i$. Due to the condition of the theorem and eqs. (5.38) and (5.40) we have

$$\sum M_i \mathbf{a}'_i \cdot \delta \mathbf{r}_i = \sum \rho_i \cdot \mathbf{r}_i = \sum \mathbf{b}_i \cdot \delta \mathbf{r}_i.$$

Quantities $\mathbf{a}'_i$ satisfy condition (5.29). According to Statement 1, $\sum M_i a_i^2 = \min \sum M_i a'^2_i$, which is equivalent to relationship (5.44).

As one can see, the improved Painlevé's theorem on friction forces (Statement 2) is a corollary of Statement 1 reflecting a general property of acceleration of the system of particles.

Statement 3. The sum of the squares of the constraint forces in the presence of friction divided by M_i is equal to the sum of the squares of the friction forces and squares of the constraint forces in the absence of friction divided by M_i, i.e.

$$\sum_{i=1}^{N} \frac{1}{M_i} K_i^2 = \sum_{i=1}^{N} \frac{1}{M_i} (\rho_i^2 +^0 K_i^2). \tag{5.45}$$

Proof. Since

$$\mathbf{K}_i = \rho_i +^0 \mathbf{K}_i = \rho_i + \sum_{j=1}^{3N-t} \lambda_j \mathrm{grad}_i \Phi_j, \quad \rho_i = M_i \mathbf{a}_i,$$

then

$$\sum_{i=1}^{N} \frac{1}{M_i} K_i^2 = \sum_{i=1}^{N} \frac{1}{M_i} (\rho_i^2 +^0 K_i^2) + 2 \sum_{j=1}^{3N-t} \lambda_j \mathrm{grad}_i \Phi_j \cdot \mathbf{a}_i.$$

Taking into account eq. (5.31) we obtain condition (5.45).

Corollary to Statement 3. The sum of the squares of the constraint forces in the case of no friction divided by mass M_i is less than that in the case of a friction

$$\sum \frac{1}{M_i}{}^0 K_i^2 < \sum \frac{1}{M_i} K_i^2. \tag{5.46}$$

Proof. Condition (5.46) follows immediately from (5.45).

5.3 Forces of friction in the Painlevé-Klein problem

Let us demonstrate the developed method of determining forces $\boldsymbol{\rho}_i$ for the example of the Painlevé-Klein scheme shown in Fig. 3.3. First we calculate forces $\boldsymbol{\rho}_i$ ($i = 1, 2$) by directly using the definition suggested by Painlevé in the form (5.38). Let the condition of absence of paradoxes (3.11) and the conditions

$$\varepsilon_0 = \text{sign}(M_2 P_1 - M_1 P_2) = 1, \quad \varepsilon_2 = -\text{sign}\,\dot{x} = 1$$

be met. The force of Coulomb friction $\mathbf{R}_\tau$ acting on particle M_1 from the guide and the accelerations of the particle are given by the formulae

$$\mathbf{R}_\tau = -\mu \frac{M_2 P_1 - M_1 P_2}{M_1 + M_2 + \mu M_2 + \tan\varphi}\mathbf{i}_1, \quad {}^0\mathbf{W}_1 = {}^0\mathbf{W}_2 = \frac{P_1 + P_2}{M_1 + M_2}\mathbf{i}_1,$$

$$\mathbf{W}_1 = \mathbf{W}_2 = \frac{P_1 + P_2 + \mu P_2 \tan\varphi}{M_1 + M_2 + \mu M_2 \tan\varphi}\mathbf{i}_1.$$

This yields the following expression for the forces of friction corresponding to the particles

$$\boldsymbol{\rho}_s = M_s \mathbf{W}_s - M_s{}^0\mathbf{W}_s = -\mu \frac{M_s(M_2 P_1 - M_1 P_2)\tan\varphi}{(M_1 + M_2)(M_1 + M_2 + \mu M_2 \tan\varphi)}\mathbf{i}_1. \tag{5.47}$$

Let us construct Painlevé's equations in the form (5.39) for the considered system taking account of the constraints

$$\Phi_1 \equiv y_1 = 0, \quad \Phi_2 \equiv z_1 = 0, \quad \Phi_3 \equiv y_2 - l\cos\varphi = 0,$$
$$\Phi_4 \equiv z_2 = 0, \quad \Phi_5 \equiv (x_2 - x_1)^2 + (y_2 - y_1)^2 - l = 0 \tag{5.48}$$

and the expression for the generalised force of friction

$$S = \frac{\mathbf{r}_1}{x_1} \cdot \mathbf{R}_\tau = -\mu \frac{(M_2 P_1 - M_1 P_2)\tan\varphi}{(M_1 + M_2 + \mu M_2 \tan\varphi)}. \tag{5.49}$$

We have a system of six equations

$$\rho_{1x} + \rho_{2x} = -\mu \frac{(M_2 P_1 - M_1 P_2) \tan \varphi}{(M_1 + M_2 + \mu M_2 \tan \varphi)}, \quad \rho_{1y} = 0, \quad \rho_{1z} = 0,$$

$$\rho_{2y} = 0, \quad \rho_{2z} = 0, \quad (x_2 - x_1)(\rho_{2x} - \rho_{1x}) + (y_2 - y_1)(\rho_{2y} - \rho_{1y}) = 0$$

with six unknown variables $\rho_{ix}, \rho_{iy}, \rho_{iz}$ ($i = 1, 2$). The solution of this system

$$\boldsymbol{\rho}_1 = \boldsymbol{\rho}_2 = \frac{(M_2 P_1 - M_1 P_2) \tan \varphi}{2(M_1 + M_2 + \mu M_2 \tan \varphi)} \mathbf{i}_1$$

differs from (5.47), that is an incorrect result is obtained.

If we take equations in the form (5.41) and account for conditions (5.48) and (5.49), we obtain the following system

$$\rho_{1x} + \rho_{2x} = -\frac{(M_2 P_1 - M_1 P_2) \tan \varphi}{M_1 + M_2 + \mu M_2 \tan \varphi},$$

$$\rho_{1y} = 0, \quad \rho_{1z} = 0, \quad \rho_{2y} = 0, \quad \rho_{2z} = 0,$$

$$-\frac{1}{M_2}[(x_2 - x_1)\rho_{1x} + (y_2 - y_1)\rho_{1y}] + \frac{1}{M_2}[(x_2 - x_1)\rho_{2x} + (y_2 - y_1)\rho_{2y}] = 0$$

whose solution coincides with expression (5.47) as expected.

Inserting $\boldsymbol{\rho}_i$ due to (5.47) into the left and right hand sides of eq. (5.45) and taking into account equalities $\mathbf{K}_i = M_i \mathbf{W}_i - \mathbf{P}_i$, we arrive at the identity

$$\sum_{i=1}^{2} \frac{1}{M_i} K_i^2 \equiv \sum_{i=1}^{2} \frac{1}{M_i} (\rho_i^2 + ^0 K_i^2)$$

$$= \frac{(M_1 P_2 - M_2 P_1)^2 [M_1 + M_2(1 + \mu \tan \varphi)^2]}{M_1 M_2 (M_1 + M_2 + \mu M_2 \tan \varphi)^2}.$$

This confirms the validity of Statement 3 and its corollary.

5.4 The contact compliance and equations of perturbed trajectories

5.4.1 Lagrange's equations for systems with elastic contact joints

Let the result of elastic deformation in the contact region be a displacement of the $\alpha-th$ slider from position T_α^0 to position T_α. We denote the projection of the part $T_\alpha^0 T_\alpha$ onto the normal to the $\alpha - th$ guiding surface by h_α

$$h_\alpha = (\mathbf{r}_{T\alpha} - \mathbf{r}_{T\alpha}^0) \cdot \mathbf{m}_\alpha. \tag{5.50}$$

These normal components of the displacements are expressed in terms of the reaction forces by the linear relationship

$$h_\alpha = -c_\alpha^{-1} R_\alpha, \quad (\alpha = 1,...,m), \tag{5.51}$$

where the effective rigidities c_α are infinitely large. It is assumed that values $h_1, ..., h_m$ are independent of each other, i.e. neither can be expressed in terms of the others. The position vectors of the particles and the sliders are then functions of the generalised coordinates $q_1, ..., q_n$ and elastic displacements $h_1, ..., h_m$.

Comparing formulae (5.3) and (5.51) one can see that the redundant coordinates under the removed contacts and the normal components of displacements under elastic joint have the same geometric interpretation. In this case the reaction forces corresponding to $q_1, ..., q_n$ and $h_1, ..., h_m$ are also calculated by means of formulae (5.11) which, by virtue of eq. (5.51), reduce to the following

$$S_k = \sum_{\alpha=1}^m \varepsilon_\alpha \mu_\alpha \frac{\partial v_{T\alpha}}{\partial \dot{q}_k} c_\alpha h_\alpha, \quad S'_\beta = \sum_{\alpha=1}^m \varepsilon_\alpha \mu_\alpha \frac{\partial v_{T\alpha}}{\partial \dot{q}_k} c_\alpha h_\alpha - c_\beta h_\beta$$

$$(k = 1, ..., n, \ \beta = 1, ..., m). \tag{5.52}$$

The kinetic energy is expressed in the form

$$T = \frac{1}{2} \sum_{k=1,s=1}^n A_{ks} \dot{q}_k \dot{q}_s + \sum_{k=1}^n \sum_{\beta=1}^m A_{k,n+\beta} \dot{q}_k \dot{h}_\beta + \frac{1}{2} \sum_{\alpha=1,\beta=1}^m A_{n+\beta,n+\alpha} \dot{h}_\alpha \dot{h}_\beta. \tag{5.53}$$

Using eqs. (5.52) and (5.53) we arrive at the Lagrange differential equations

$$\sum_{k=1}^n A_{ks} \ddot{q}_k + \sum_{\alpha=1}^m A_{n+\alpha,s} \ddot{h}_\alpha + \sum_{k=1,l=1}^n \Gamma_{kl,s} \dot{q}_k \dot{q}_l + 2 \sum_{k=1}^n \sum_{\alpha=1}^m \Gamma_{k,n+\alpha,s} \dot{q}_k \dot{h}_\alpha +$$

$$\sum_{\alpha=1,\beta=1}^m \Gamma_{n+\alpha,n+\beta,s} \dot{h}_\alpha \dot{h}_\beta = Q_s + \sum_{\alpha=1}^m \varepsilon_\alpha \mu_\alpha \frac{\partial v_{T\alpha}}{\partial \dot{q}_s} c_\alpha h_\alpha \quad (s = 1,...,n),$$

$$\sum_{k=1}^n A_{k,n+\gamma} \ddot{q}_k + \sum_{\alpha=1}^m A_{n+\alpha,n+\gamma} \ddot{h} + \sum_{k=1,l=1}^n \Gamma_{kl,n+\gamma} \dot{q}_k \dot{q}_l +$$

$$2 \sum_{\alpha=1,l=1}^{m,n} \Gamma_{n+\alpha,l,n+\gamma} \dot{h}_\alpha \dot{q}_l + \sum_{\alpha,\beta=1}^m \Gamma_{n+\alpha,n+\beta,n+\gamma} \dot{h}_\alpha \dot{h}_\beta =$$

$$Q'_\gamma - c_\gamma h_\gamma + \sum_{\alpha=1}^m \varepsilon_\alpha \mu_\alpha \frac{\partial v_{T\alpha}}{\partial h} c_\alpha h_\alpha \quad (\gamma = 1,...,m). \tag{5.54}$$

5.4 The contact compliance and equations of perturbed trajectories   179

Quantities $h_1, ..., h_m$ oscillate about stationary (or slowly varying) values. As the rigidities are high, these quantities and their time derivatives $\dot{h}_1, ..., \dot{h}_{m-1}, \dot{h}_m$ are small. Under this situation the coefficients in eqs. (5.54) are effectively coincident with those in eq. (5.61)

$$A_{ks} = A_{ks}^0, \quad \frac{\partial v_{T\alpha}}{\partial \dot{q}_s} = \frac{\partial v_{T\alpha}^0}{\partial \dot{q}_s}, \quad \Gamma_{kl,s} = \Gamma_{kl,s}^0, \quad \ldots \tag{5.55}$$

5.4.2 Equations for perturbed reaction forces

We present eq. (5.54) in the form of a system of two matrix equations for unknown variables $\ddot{q}$ and $\ddot{h}$

$$a\ddot{q} + b^T \ddot{h} - \frac{Dv}{D\dot{q}} \varepsilon \mu c h = Q - \Gamma - \Gamma_1,$$

$$b\ddot{q} + a_1 \ddot{h} - \left(\frac{Dv}{Dh}\varepsilon\mu - E\right) ch = Q' - \Gamma' - \Gamma_1'. \tag{5.56}$$

Here we used the old notation for matrices (5.14) and introduced the new notation

$$h = [h_1, h_2, ..., h_m], \quad c = \mathrm{diag}(c_1, ..., c_m). \tag{5.57}$$

$$\Gamma_1 = \left[2\sum_{\alpha=1}^m \sum_{l=1}^n \Gamma_{n+\alpha,l,1}\dot{h}_\alpha \dot{q}_l + \sum_{\alpha,\beta=1}^m \Gamma_{n+\alpha,n+\beta,1}\dot{h}_\alpha \dot{h}_\beta, \ldots, \right.$$

$$\left. \ldots, 2\sum_\alpha \sum_{l=1}^n \Gamma_{n+\alpha,l,n}\dot{h}_\alpha \dot{q}_l + \sum_{\alpha,\beta=1}^m \Gamma_{n+\alpha,n+\beta,n}\dot{h}_\alpha \dot{h}_\beta \right], \tag{5.58}$$

$$\Gamma_1' = \left[2\sum_{\alpha=1}^m \sum_{l=1}^n \Gamma_{n+\alpha,l,n+1}\dot{h}_\alpha \dot{q}_l + \sum_{\alpha,\beta=1}^m \Gamma_{n+\alpha,n+\beta,n+1}\dot{h}_\alpha \dot{h}_\beta, \ldots, \right.$$

$$\left. \ldots, 2\sum_{\alpha=1}^m \sum_{l=1}^n \Gamma_{n+\alpha,l,n+m}\dot{h}_\alpha \dot{q}_l + \sum_{\alpha,\beta=1}^m \Gamma_{n+\alpha,n+\beta,n+m}\dot{h}_\alpha \dot{h}_\beta \right], \tag{5.59}$$

$$a_1 = \begin{bmatrix} A_{n+1,n+1} & A_{n+2,n+1} & \cdots & A_{n+m,n+1} \\ A_{n+1,n+2} & A_{n+2,n+2} & \cdots & A_{n+m,n+1} \\ \cdots & \cdots & \cdots & \cdots \\ A_{n+1,n+m} & A_{n+2,n+m} & \cdots & A_{n+m,n+m} \end{bmatrix}. \tag{5.60}$$

Resolving eq. (5.56) for $\ddot{q}$ and $\ddot{h}$ we obtain

$$(a - b^T a_1^{-1} b)\ddot{q} - \left[b^T a_1^{-1} + \left(\frac{Dv}{D\dot{q}} - b^T a_1^{-1} \frac{Dv}{D\dot{h}} \right) \varepsilon\mu \right] ch$$
$$= Q - \Gamma - \Gamma_1 - b^T a_1^{-1}(Q' - \Gamma' - \Gamma'_1), \qquad (5.61)$$

$$(a_1 - ba^{-1}b^T)\ddot{h} + \left[E + \left(ba^{-1}\frac{Dv}{D\dot{q}} - \frac{Dv}{D\dot{h}} \right) \varepsilon\mu \right] ch$$
$$= Q' - \Gamma' - \Gamma'_1 - ba^{-1}(Q - \Gamma - \Gamma_1), \qquad (5.62)$$

where b^T denotes the transpose to matrix b. Matrix a defined by eq. (5.14) and the matrix

$$D = \begin{bmatrix} (a) & (b^T) \\ (b) & (a_1) \end{bmatrix} \qquad (5.63)$$

are, respectively, the coefficients of the quadratic forms in the velocities for the system with the rigid and elastic contacts. For this reason, both matrices are symmetric and satisfy Sylvester's criterion. Then the coefficients of eqs. (5.61) and (5.62)

$$I_1 = a - b^T a_1^{-1} b, \quad I = a_1 - ba^{-1}b^T \qquad (5.64)$$

are symmetric matrices. Their elements must also satisfy Sylvester's criterion. It is evident that

$$\det D = \det a_1 \cdot \det I_1 = \det a \cdot I. \qquad (5.65)$$

Let us consider expression (5.62). It is a system of equations for accelerations $\ddot{h}_1, ..., \ddot{h}_m$ which are the second time derivatives of the elastic displacements of the sliders. Following the approach of Section 4.3 for the case of a single frictional pair we assume that in the case of high rigidities $c_1, ..., c_m$ system (5.62) can be reduced to the form

$$I\ddot{h} + (E + \mathsf{L}\varepsilon\mu)ch = -{}^0R, \qquad (5.66)$$

where the influence matrix L and the column of the reaction forces ${}^0R$ are given by eqs. (5.17) and (5.20) respectively. By replacing variable h by $-c^{-1}R$ in eq. (5.17) we obtain the following equation for $\ddot{R}$

$$Ic^{-1}\ddot{R} + (E + \mathsf{L}\varepsilon\mu)R = {}^0R. \qquad (5.67)$$

Equation (5.18) differs from the algebraic equation for reaction forces (5.21) in those terms depending upon the second derivatives $\ddot{R}_\alpha (\alpha = 1, ..., m)$. In other words, R_α determined by eq. (5.21) are stationary (or slowly varying) normal reaction forces of the contact constraints. This enables us to draw a conclusion which generalises, to some extent, Theorems 8 and 9 to the case of several frictional pairs.

Conclusion. Paradoxical non-existence and non-uniqueness of problem (5.21) reflect the situation in which the normal reaction forces have either several or no stationary (or slowly varying) values.

As a consequence of this conclusion we can note that in a paradoxical situation of non-existence all the second derivatives $\ddot{h}_\alpha$ and $\ddot{R}_\alpha (\alpha = 1, ..., m)$ are non-trivial and the motion of the mechanical system is no longer smooth for any initial conditions. Equations (5.12), (5.13) and (5.18) confirm this statement.

In the situation of non-uniqueness, in order to analyse the behaviour of the system about the stationary values of the reaction forces, eq. (5.21), we obtain with the help of eqs. (5.17) and (5.18) the following equations for the perturbed elastic displacements

$$I\ddot{\xi} + (E + \mathsf{L}\varepsilon\mu)c\xi = 0 \tag{5.68}$$

and the equations for the perturbed reaction forces

$$Ic^{-1}\ddot{\eta} + (E + \mathsf{L}\varepsilon\mu)\eta = 0 \,. \tag{5.69}$$

It is clear that in the case of no friction ($\mu = 0$) expression (5.19) takes the form

$$I\ddot{\xi} + c\xi = 0 \,. \tag{5.70}$$

Since I and c are positive definite the stationary values of h_α and R_α are stable.

When friction is present then, as follows from eqs. (5.14), (5.17) and (5.18), matrix $(E + \mathsf{L}\varepsilon\mu)c$ is, generally speaking, no longer symmetric. To make it more clear, we decompose matrix $\mathsf{L}\varepsilon\mu$ into symmetric and skew-symmetric parts. Then the equations for perturbed displacements (5.68) take the form

$$I\ddot{\xi} + [E + (\mathsf{L}\varepsilon\mu + \varepsilon\mu\mathsf{L}^T)]c\xi + (\mathsf{L}\varepsilon\mu - \varepsilon\mu\mathsf{L}^T)c\xi = 0 \,. \tag{5.71}$$

As one can see, the symmetric part always contains the friction coefficients $\mu, ..., \mu_m$ for $\mathsf{L} \neq 0$. Referring to the well-known theorem of the theory of stability of motion [2], [75], [107] and judging from the structure of eq. (5.71) we conclude that the force of dry friction can eliminate the stability of the potential system.

5.5 Painlevé's scheme with two frictional pairs

The Painlevé scheme depicted in Fig. 5.1 consists of two particles with masses M_1 and M_2 which are linked to each other by a rigid rod of length

182   5. Systems with several frictional pairs. Painlevé's law of friction

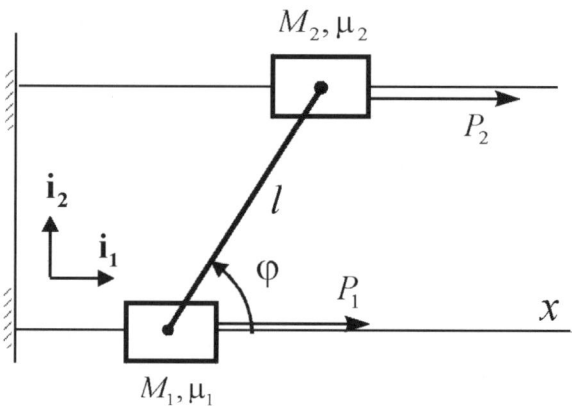

FIGURE 5.1.

l. The particles move along the parallel horizontal straight lines with the friction coefficients μ_1 and μ_2. Forces P_1 and P_2 act on the particles. This scheme differs from the Painlevé-Klein scheme studied above in that a second force of friction exists.

5.5.1 Lagrange's equations, reaction forces and the equations of motion with eliminated reaction forces

In order to obtain expressions for the reaction forces and to derive Lagrange's equations and the differential equations of motion with eliminated reaction forces, we choose the abscissa x_1 of particle $M_1(q = x_1)$ as the generalised coordinate. Then

$$\frac{d\mathbf{r}_1^0}{dq} = \frac{d\mathbf{r}_2^0}{dq} = \mathbf{i}_1. \qquad (5.72)$$

Let us mentally remove the first contact and assume that particle M_1 moves along the normal to the first guide by a value $h_1\mathbf{i}_2$. In this case

$$\left(\frac{\partial \mathbf{r}_1^0}{\partial h_1}\right)_0 = \mathbf{i}_2, \quad \left(\frac{\partial \mathbf{r}_2^0}{\partial h_1}\right)_0 = \tan\varphi\, \mathbf{i}_1. \qquad (5.73)$$

By analogy, under an additional displacement $-h_2\mathbf{i}_2$ of the second particle M_2 we have

$$\left(\frac{\partial \mathbf{r}_2^0}{\partial h_2}\right)_0 = -\mathbf{i}_2, \quad \left(\frac{\partial \mathbf{r}_1^0}{\partial h_2}\right)_0 = -\tan\varphi\, \mathbf{i}_1. \qquad (5.74)$$

By using eqs. (5.1)-(5.3) we obtain the following expressions for the velocities of gliding

$$\mathbf{v}_{T1} = \frac{d\mathbf{r}_1^0}{dq}\dot{q} + \left(\frac{\partial \mathbf{r}_1}{\partial h_1}\right)_0 \dot{h}_1 + \left(\frac{\partial \mathbf{r}_1}{\partial h_2}\right)_0 \dot{h}_2 = (\dot{q} - \tan\varphi \dot{h}_2)\mathbf{i}_1 + \dot{h}_1 \mathbf{i}_2,$$

$$\mathbf{v}_{T2} = \frac{d\mathbf{r}_2^0}{dq}\dot{q} + \left(\frac{\partial \mathbf{r}_2}{\partial h_1}\right)_0 \dot{h}_1 + \left(\frac{\partial \mathbf{r}_2}{\partial h_2}\right)_0 \dot{h}_2 = (\dot{q} + \tan\varphi \dot{h}_1)\mathbf{i}_1 - \dot{h}_2 \mathbf{i}_2,$$

$$v_{T1} = \sqrt{(\dot{q} - \dot{h}_2 \tan\varphi)^2 + \dot{h}_1^2}, \quad v_{T2} = \sqrt{(\dot{q} + \dot{h}_1 \tan\varphi)^2 + \dot{h}_2^2}, \quad (5.75)$$

thus

$$\frac{D v}{D \dot{q}} = [\operatorname{sign}\dot{q} \, \operatorname{sign}\dot{q}], \quad \frac{D v}{D \dot{h}} = \begin{bmatrix} (0) & (\tan\varphi) \\ (-\tan\varphi) & (0) \end{bmatrix} \operatorname{sign}\dot{q}. \quad (5.76)$$

Utilising eqs. (5.1)-(5.3) and formulae (5.11) we can calculate coefficients of the kinetic energy, the generalised forces and the generalised reaction forces. The result is as follows

$$A_{11} = M_1 + M_2, \quad A_{12} = M_2 \tan\varphi, \quad A_{13} = -M_1 \tan\varphi,$$
$$A_{22} = M_1 + M_2 \tan^2\varphi, \quad A_{33} = M_1 \tan^2\varphi + M_2, \quad A_{23} = 0,$$
$$Q = P_1 + P_2, \quad Q_1' = P_2 \tan\varphi, \quad Q_2' = -P_1 \tan\varphi, \quad (5.77)$$
$$S = -\tilde{\varepsilon}_1 \mu_1 R_1 - \tilde{\varepsilon}_2 \mu_2 R_2, \quad S_1' = R_1 - \tilde{\varepsilon}_2 \mu_2 R_2 \tan\varphi, \quad S_2' = \tilde{\varepsilon}_1 \mu_1 R_1 \tan\varphi + R_2$$

where $\tilde{\varepsilon}_\alpha = \varepsilon_\alpha \operatorname{sign}\dot{q} = \operatorname{sign}(R_\alpha \dot{q}), (\alpha = 1, 2)$. The equations in (5.12) take the form

$$(M_1 + M_2)\ddot{q} = P_1 + P_2 - \tilde{\varepsilon}_1 \mu_1 R_1 - \tilde{\varepsilon}_2 \mu_2 R_2,$$
$$M_2 \tan\varphi \ddot{q} = P_2 \tan\varphi + R_1 - \tilde{\varepsilon}_2 \mu_2 R_2 \tan\varphi,$$
$$-M_1 \tan\varphi \ddot{q} = -P_1 \tan\varphi + \tilde{\varepsilon}_1 \mu_1 R_1 \tan\varphi + R_2. \quad (5.78)$$

To obtain expressions for the reaction forces and the differential equation of motion in the form of (5.16) and (5.23) it is necessary to calculate the coefficients with the help of formulae (5.17)-(5.20). The influence matrix is as follows

$$\mathsf{L} = \frac{\tan\varphi \operatorname{sign}\dot{q}}{M_1 + M_2} \begin{bmatrix} (M_2) & (-M_1) \\ (M_2) & (-M_1) \end{bmatrix} \quad (5.79)$$

and the normal reaction force of the ideal system is given by

$$^0 R = \frac{\tan\varphi}{M_1 + M_2} \begin{bmatrix} (M_2 P_1) & (-M_1 P_2) \\ (M_2 P_1) & (-M_1 P_2) \end{bmatrix}. \quad (5.80)$$

For these values of L and ^0R eq. (5.22) reduces to the form

$$\begin{bmatrix} 1+\tilde{\varepsilon}_1\mu_1\dfrac{M_2}{M_1+M_2}\tan\varphi & -\tilde{\varepsilon}_2\mu_2\dfrac{M_1}{M_1+M_2}\tan\varphi \\ \tilde{\varepsilon}_1\mu_1\dfrac{M_2}{M_1+M_2}\tan\varphi & 1-\tilde{\varepsilon}_2\mu_2\dfrac{M_1}{M_1+M_2}\tan\varphi \end{bmatrix}\begin{bmatrix} R_1 \\ R_2 \end{bmatrix}$$

$$=\dfrac{\tan\varphi}{M_1+M_2}\begin{bmatrix} (M_2P_1) & (-M_1P_2) \\ (M_2P_1) & (-M_1P_2) \end{bmatrix}. \qquad (5.81)$$

Resolving eq. (5.81) yields

$$[M_1+M_2+(\tilde{\varepsilon}_1\mu_1 M_2-\tilde{\varepsilon}_2\mu_2 M_1)\tan\varphi]R_1=(M_2P_1-M_1P_2)\tan\varphi,$$
$$[M_1+M_2+(\tilde{\varepsilon}_1\mu_1 M_2-\tilde{\varepsilon}_2\mu_2 M_1)\tan\varphi]R_2=(M_2P_1-M_1P_2)\tan\varphi. \qquad (5.82)$$

As one can see, $R_1=R_2$, hence, $\tilde{\varepsilon}_1=\tilde{\varepsilon}_2=\tilde{\varepsilon}$. Inserting this into eq. (5.82) we obtain the following expression for the reaction force

$$[M_1+M_2+\tilde{\varepsilon}(\mu_1 M_2-\mu_2 M_1)\tan\varphi]R_\alpha=(M_2P_1-M_1P_2)\tan\varphi\ (\alpha=1,2) \qquad (5.83)$$

Applying the first matrix relationship (5.16) under conditions (5.76) and (5.82) we obtain the differential equation of motion

$$[M_1+M_2+\tilde{\varepsilon}(\mu_1 M_2-\mu_2 M_1)\tan\varphi]\ddot{q}=\tilde{\varepsilon}(\mu_1 P_2-\mu_2 P_1)\tan\varphi+P_1+P_2. \qquad (5.84)$$

Of course, this equation can also be obtained by inserting eq. (5.12) into one of the equations in (5.7). Thus we finally obtain expressions for the reaction forces (5.12) and the equations of motion (5.13).

5.5.2 Feasibility of Painlevé's paradoxes

Let us establish the feasibility of Painlevé's paradoxes by means of eq. (5.12). Similar to the example of the Painlevé-Klein scheme considered in Chapter 2 it is sufficient to consider the case of $0<\varphi<\pi/2$. It follows from eq. (5.12) that the paradoxes are absent when

$$\tan\varphi < \dfrac{M_1+M_2}{|\mu_1 M_2-\mu_2 M_1|} \qquad (5.85)$$

and are present when

$$\tan\varphi \geq \dfrac{M_1+M_2}{|\mu_1 M_2-\mu_2 M_1|}. \qquad (5.86)$$

When they are absent, the signs of the normal reaction forces coincide with the sign of the right hand side of eq. (5.12)

$$\varepsilon_1 = \varepsilon_2 = \text{sign}(M_2 P_1 - M_1 P_2). \tag{5.87}$$

In particular this occurs for $\mu_1 M_2 = \mu_2 M_1$ because in this case condition (5.14) holds for any angle φ.

In the paradoxical case, i.e. under condition (5.16) it follows from eq. (5.12) that

$$\varepsilon_1^2 = \varepsilon_2^2 = \text{sign}\left(\frac{M_2 P_1 - M_1 P_2}{\mu_1 M_2 - \mu_2 M_1}\dot{q}\right). \tag{5.88}$$

Hence, the solution of eqs. (5.12) and (5.13) is non-unique ($\varepsilon_1 = \varepsilon_2 = \pm 1$) if

$$\frac{M_2 P_1 - M_1 P_2}{\mu_1 M_2 - \mu_2 M_1}\dot{q} > 0 \tag{5.89}$$

and does not exists ($\varepsilon_1 = \varepsilon_2 = \pm i$) if

$$\frac{M_2 P_1 - M_1 P_2}{\mu_1 M_2 - \mu_2 M_1}\dot{q} < 0. \tag{5.90}$$

In the cases in which the solution exists the law of motion can be determined by integrating eq. (5.13).

5.5.3 Expressions for the frictional force in terms of the friction coefficients

Let us express Painlevé's forces in terms of the coefficients of Coulomb friction μ_1 and μ_2. Let us recall that Painlevé's forces of friction for the particles are given by

$$\boldsymbol{\rho}_i = M_i \mathbf{a}_i = M_i \mathbf{W}_i - M_i{}^0\mathbf{W}_i, \quad (i = 1, 2). \tag{5.91}$$

For the considered system the acceleration vectors are calculated due to eq. (5.13) as follows

$$\begin{aligned}
{}^0\mathbf{W}_1 &= {}^0\mathbf{W}_2 = \mathbf{i}_1 \frac{(P_1 + P_2)}{M_1 + M_2}, \\
\mathbf{W}_1 &= \mathbf{W}_2 = \mathbf{i}_1 \frac{\tilde{\varepsilon}(\mu_1 P_2 - \mu_2 P_1)\tan\varphi + P_1 + P_2}{\tilde{\varepsilon}(\mu_1 M_2 - \mu_2 M_1)\tan\varphi + M_1 + M_2}.
\end{aligned} \tag{5.92}$$

This yields expressions for Painlevé's forces of friction in terms of μ_1 and μ_2

$$\boldsymbol{\rho}_i = M_i \mathbf{a}_i = \frac{M_i}{M_1 + M_2} \frac{(\mu_1 + \mu_2)(M_1 P_2 - M_2 P_1)\tilde{\varepsilon}\tan\varphi}{(\mu_1 M_2 - \mu_2 M_1)\tilde{\varepsilon}\tan\varphi + M_1 + M_2} \mathbf{i}_1 \quad (i = 1, 2). \tag{5.93}$$

For comparison we demonstrate the expressions for the forces of Coulomb friction calculated according to eq. (5.83) for the normal reaction forces

$$\mathbf{R}_{\tau\alpha} = -\tilde{\varepsilon}\mu_\alpha R_\alpha \mathbf{i}_1 = \frac{\tilde{\varepsilon}\mu_\alpha(M_1 P_2 - M_2 P_1)}{\tilde{\varepsilon}(\mu_1 M_2 - \mu_2 M_1)\tan\varphi + M_1 + M_2}\mathbf{i}_1 \quad (\alpha = 1, 2). \tag{5.94}$$

As one can see, though in the considered case Painlevé's forces of friction ρ_i and the forces of Coulomb friction $\mathbf{R}_{\tau\alpha}$ act on the same particle they have different values. Besides, non-uniqueness and non-existence of solution of the dynamical problem result in non-uniqueness and non-existence of solution of not only Coulomb forces but also Painlevé's forces.

5.5.4 Painlevé's scheme for compliant contacts

Let us consider Painlevé's scheme taking account of the contact compliance. Let particles M_1 and M_2 undergo elastic displacements along the normal

$$h_\alpha = -\frac{R_\alpha}{c_\alpha}. \tag{5.95}$$

A minus sign implies that in the case of $R_1, R_2 > 0$ particle M_1 moves downwards whilst particle M_2 moves upwards.

In this case, according to eq. (5.77), we have the following expressions for the matrices

$$a = M_1 + M_2,$$

$$b = \begin{bmatrix} M_2 & \tan\varphi \\ -M_1 & \tan\varphi \end{bmatrix} \begin{bmatrix} M_1 + M_2 \tan^2\varphi & 0 \\ 0 & M_1 \tan^2\varphi + M_2 \end{bmatrix}. \tag{5.96}$$

Hence,

$$I = a_1 - ba^{-1}b^T = \tag{5.97}$$

$$\frac{1}{M_1 + M_2}\begin{bmatrix} M_1[M_1 + M_2(1 + \tan^2\varphi)] & M_1 M_2 \tan^2\varphi \\ M_1 M_2 \tan^2\varphi & M_2[M_2 + M_1(1 + \tan^2\varphi)] \end{bmatrix}.$$

Inserting expressions for matrices I, c, $\mathbf{L}$ and ^0R due to eqs. (5.97), (5.95), (5.79) and (5.77) into eqs. (5.66) and (5.67) we obtain the following systems of differential equations

$$M_1[M_1 + M_2(1 + \tan^2\varphi)]\ddot{h}_1 + M_1 M_2 \tan^2\varphi(M_1 + M_2 + \tilde{\varepsilon}_1\mu_1 M_2 \tan\varphi) \times$$
$$ch_1 \ddot{h}_2 - \tilde{\varepsilon}_2\mu_2 M_1 \tan\varphi ch_2 = (M_1 P_2 - M_2 P_1)\tan\varphi,$$

$$M_1 M_2 \tan^2\varphi \ddot{h}_1 + M_2[M_2 + M_1(1 + \tan^2\varphi)]\ddot{h}_2 + \tilde{\varepsilon}_1\mu_1 M_2 \tan\varphi)ch_1 +$$
$$(M_1 M_2 - \varepsilon_2\mu_2 M_1 \tan\varphi)ch_2 = (M_1 P_2 - M_2 P_1)\tan\varphi, \tag{5.98}$$

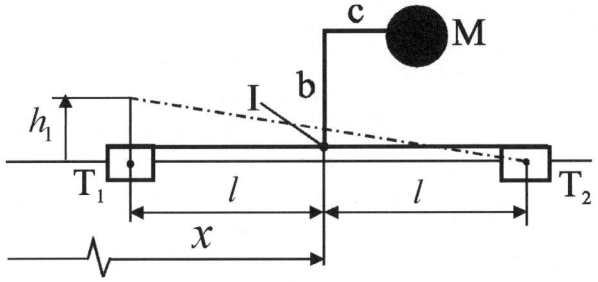

FIGURE 5.2.

$$M_1 c_1^{-1}[M_1 + M_2(1 + \tan^2 \varphi)]\ddot{R}_1 + c_2^{-1} M_1 M_2 \tan^2 \varphi \ddot{R}_2 +$$
$$(M_1 + M_2 + \tilde{\varepsilon}_1 \mu_1 M_2 \tan \varphi) R_1 - \tilde{\varepsilon}_2 \mu_2 M_1 \tan \varphi R_2 = (M_2 P_1 - M_1 P_2) \tan \varphi,$$

$$c_1^{-1} M_1 M_2 \tan^2 \varphi \ddot{R}_1 + M_2[M_2 + M_1(1 + \tan^2 \varphi)]\ddot{R}_2 + \tilde{\varepsilon}_1 \mu_1 M_2 \tan \varphi R_1 +$$
$$(M_1 + M_2 - \tilde{\varepsilon}_2 \mu_2 M_1 \tan \varphi) R_2 = (M_2 P_1 - M_1 P_2) \tan \varphi. \qquad (5.99)$$

5.6 Sliders of metal-cutting machine tools

The model depicted in Fig. 5.2 is widely used in the machine-tool industry. The model consists of two sliders T_1 and T_2, gliding along a straight guide and linked to a particle M by means of the rigid rods. The dimensions are shown in Fig. 5.2. The stability of motion for mechanisms of this class is studied in [37]. According to the approach developed in Sections 5.1 and 5.2 the equations of motion and expressions for the reaction forces will be obtained as well as the feasibility of the situations of non-existence and non-uniqueness is determined.

5.6.1 Derivation of equations of motion and expressions for the reaction forces

Let us imagine that slider T_1 gains an additional displacement h_1 in the vertical direction. Then we can write

$$\mathbf{r}^0 = (x + c)\mathbf{i}_1 + b\mathbf{i}_2, \qquad (5.100)$$
$$\mathbf{r} = \mathbf{r}_{T2}^0 + [(c - l)\mathbf{i}_1 + b\mathbf{i}_2] \cdot 2l(h_1^2 + 4l^2)^{-1/2} + (2l)^{-1}[(c - l)\mathbf{i}_2 + b\mathbf{i}_1]h_1,$$
$$\mathbf{r}_{T1}^0 = (x - l)\mathbf{i}_1, \quad \mathbf{r}_{T1} = (x - l)\mathbf{i}_1 + h_1 \mathbf{i}_2, \quad \mathbf{r}_{T2}^0 = \mathbf{r}_{T2} = (x + l)\mathbf{i}_1,$$

where $\mathbf{r}^0$ and $\mathbf{r}$ denote the position vectors of the particle in the original system and the system with the removed contact of slider T_1 respectively, $\mathbf{r}_{T1}^0, \mathbf{r}_{T1}, \mathbf{r}_{T2}^0$ and $\mathbf{r}_{T2}$ are the position vectors of the sliders in these cases.

It follows from eq. (5.100) that

$$\frac{d\mathbf{r}_{T1}^0}{dx} = \frac{d\mathbf{r}_{T2}^0}{dx} = \frac{d\mathbf{r}^0}{dx} = \mathbf{i}_1, \quad \left(\frac{\partial \mathbf{r}_{T1}}{\partial h_1}\right)_0 = \mathbf{i}_2, \quad \left(\frac{\partial \mathbf{r}_{T2}}{\partial h_1}\right)_0 = 0,$$

$$\left(\frac{\partial \mathbf{r}}{\partial h_1}\right)_0 = \frac{1}{2l}[b\mathbf{i}_1 + (l-c)\mathbf{i}_2], \quad \left(\frac{\partial^2 \mathbf{r}}{\partial x \partial h_1}\right)_0 = 0. \quad (5.101)$$

By analogy, under displacement h_2 of slider T_2 in the vertical direction we have

$$\mathbf{r}_{T1} = \mathbf{r}_{T1}^0, \quad \mathbf{r}_{T2} = (x+l)\mathbf{i}_1 + h_2\mathbf{i}_2, \quad (5.102)$$

$$\mathbf{r} = \mathbf{r}_{T1}^0 + [(c+l)\mathbf{i}_1 + b\mathbf{i}_2] \cdot 2l(h_2^2 + 4l^2)^{-1/2} + (2l)^{-1}[(c+l)\mathbf{i}_2 - b\mathbf{i}_1]h_2,$$

hence,

$$\left(\frac{\partial \mathbf{r}_{T1}}{\partial h_2}\right)_0 = 0, \quad \left(\frac{\partial \mathbf{r}_2}{\partial h_2}\right)_0 = \mathbf{i}_2,$$

$$\left(\frac{\partial \mathbf{r}}{\partial h_2}\right)_0 = -\frac{1}{2l}[b\mathbf{i}_1 - (l+c)\mathbf{i}_2], \quad \left(\frac{\partial^2 \mathbf{r}}{\partial x \partial h_2}\right)_0 = 0. \quad (5.103)$$

Expressions for the derivatives (5.28) and (5.30) are used for calculating coefficients of system (5.12). The result is

$$A_{11} = M, \quad A_{12} = b\frac{M}{2l}, \quad A_{13} = -b\frac{M}{2l},$$

$$Q = X, \quad Q_1' = (2l)^{-1}[bX + (l-c)Y], \quad Q_2' = (2l)^{-1}[-bX + (l+c)Y],$$

$$S = -\mu(\tilde{\varepsilon}_1 R_1 + \tilde{\varepsilon}_2 R_2), \quad S_1' = R_1, \quad S_2' = R_2,$$

$$\frac{A_{1,1+\beta}}{x} - \frac{1}{2}\frac{A_{11}}{h_\beta} = 0 \quad (\beta = 1, 2), \quad (5.104)$$

where subscript 0 is omitted, and X and Y denote the horizontal and vertical components of the external force applied to the particle $\tilde{\varepsilon}_\alpha = \varepsilon_\alpha \operatorname{sign} \dot{x}$ ($\alpha = 1, 2$), $\varepsilon_\alpha = \operatorname{sign} R_\alpha$. The reaction forces are taken as being positive if they are directed upwards. Using eq. (5.104) we arrive at the system of equations

$$\left.\begin{array}{rcl} M\ddot{x} &=& X - \mu(\tilde{\varepsilon}_1 R_1 + \tilde{\varepsilon}_2 R_2) \\ bM\ddot{x} &=& bX + (l-c)Y + 2lR_1 \\ bM\ddot{x} &=& bX - (l+c)Y - 2lR_2 \end{array}\right\}, \quad (5.105)$$

which yields the following expressions for the reaction forces

$$[2l + \mu b(\tilde{\varepsilon}_1 - \tilde{\varepsilon}_2)]R_1 = [\tilde{\varepsilon}_2\mu b - (l-c)]Y,$$
$$[2l + \mu b(\tilde{\varepsilon}_1 - \tilde{\varepsilon}_2)]R_2 = [-\tilde{\varepsilon}_1\mu b - (l+c)]Y \quad (5.106)$$

and the differential equations of motion

$$M[2l + \mu b(\tilde{\varepsilon}_1 - \tilde{\varepsilon}_2)]\ddot{x} = [2l + \mu b(\tilde{\varepsilon}_1 - \tilde{\varepsilon}_2)]X + [\tilde{\varepsilon}_2(l+c) + \tilde{\varepsilon}_1(l-c)]\mu Y. \quad (5.107)$$

5.6.2 Signs of the reaction forces and feasibility of paradoxes

By using eq. (5.106) we intend to establish the values of parameters μ, b, c, l which ensure the validity of each of the following sets of signs

$$
\begin{aligned}
I) & \quad \varepsilon_1 = \varepsilon_2 = 1, & II) & \quad \varepsilon_1 = -\varepsilon_2 = 1, \\
III) & \quad \varepsilon_1 = -\varepsilon_2 = -1, & IV) & \quad \varepsilon_1 = \varepsilon_2 = -1.
\end{aligned}
\qquad (5.108)
$$

Without loss in generality we restrict ourselves to a positive value of the vertical component of the external force, i.e. $Y > 0$. The values of b and c are considered as being positive if particle M resides to the right of and higher than particle I, otherwise it is negative.

A. Let us start with the case $\dot{x} > 0$, i.e.

$$\operatorname{sign} \dot{x} = 1. \qquad (5.109)$$

In this case it follows from equality $\tilde{\varepsilon}_\alpha = \varepsilon_\alpha \operatorname{sign} \dot{x}$ that

$$\tilde{\varepsilon}_\alpha = \varepsilon_\alpha, \quad (\alpha = 1, 2). \qquad (5.110)$$

The first set of signs $\varepsilon_1 = \varepsilon_2 = 1$ can not be realised for any values of μ, b, c, l since the sum of projections on the vertical axis is non-zero for $Y > 0$. This can be easily proved by means of system (5.106). Indeed, accounting for equalities $\varepsilon_1 = \varepsilon_2 = 1$ in eq. (5.106) we obtain inequalities of the type

$$-c + l < \mu b < -c - l.$$

As $l > 0$ this condition is not fulfilled.

Let us consider the set II, i.e. $\varepsilon_1 = -\varepsilon_2 = 1$. For these values of ε_1 and ε_2 it follows from eq. (5.106) that

$$
\begin{aligned}
2(l + \mu b) R_1 &= -(\mu b + l - c) Y, \\
2(l + \mu b) R_2 &= (\mu b - l - c) Y.
\end{aligned}
$$

Hence, conditions $R_1 > 0$ and $R_2 < 0$ are satisfied if

$$\frac{-\mu b - l + c}{2(l + \mu b)} > 0, \quad \frac{-\mu b - l - c}{2(l + \mu b)} < 0. \qquad (5.111)$$

Solving the inequalities in (5.111) we arrive at the following expression for the domain of definition of the set II

$$
\begin{aligned}
-l < \mu b < -l + c, & \quad c > 0, \\
-l + c < \mu b < -l, & \quad c < 0.
\end{aligned}
\qquad (5.112)
$$

In plane $(c, \mu b)$ this domain lies between straight lines $\mu b - c + l = 0$ and $\mu b + l = 0$, see Fig. 5.3.

190   5. Systems with several frictional pairs. Painlevé's law of friction

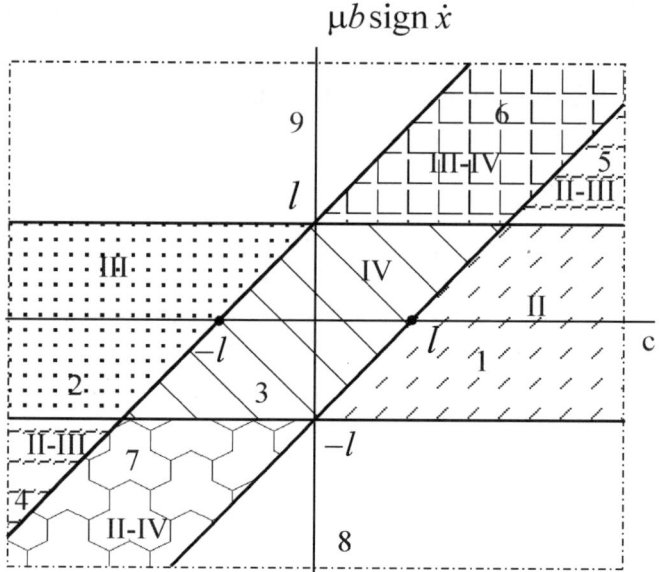

FIGURE 5.3.

In order to construct the domain of definition of set III we insert $\varepsilon_1 = -\varepsilon_2 = -1$ into eq. (5.106). Since

$$2(l - \mu b)R_1 = (\mu b - l + c)Y,$$
$$2(l - \mu b)R_2 = (\mu b - l - c)Y$$

we obtain

$$\frac{\mu b - l + c}{2(l - \mu b)} < 0, \quad \frac{\mu b - l - c}{2(l - \mu b)} > 0. \tag{5.113}$$

Resolving eq. (5.113) we obtain the following expression for domain III

$$l + c < \mu b < l, \quad c < 0,$$
$$l < \mu b < l + c, \quad c > 0. \tag{5.114}$$

In plane $(c, \mu b)$ this domain lies between the straight lines $\mu b - c - l = 0$ and $\mu b - l = 0$.

Let us proceed to set IV. In this case, the system of equations (5.106) takes the form

$$2lR_1 = (-\mu b - l + c)Y,$$
$$2lR_2 = (\mu b - l - c)Y.$$

Hence, condition $R_1 < 0$, $R_2 < 0$ holds true if

$$-l + \tilde{n} < \mu b < l + c. \tag{5.115}$$

Inequalities (5.115) demonstrate that domain IV is determined by the strip between the straight lines $\mu b - c + l = 0$ and $\mu b - c - l = 0$.

B. Let us now investigate the case $\dot{x} < 0$, i.e.

$$\operatorname{sign}\dot{x} = -1, \quad \tilde{\varepsilon}_\alpha = -\varepsilon_\alpha. \tag{5.116}$$

Inserting condition (5.116) into eq. (5.106) and repeating the analysis carried out above for the case of $\operatorname{sign}\dot{x} = 1$ we find the following expressions for the domains of definition of the sets of the signs

$$II: \begin{cases} -c + l < \mu b < l, & c > 0, \\ l < \mu b < l - c, & c < 0, \end{cases} \tag{5.117}$$

$$III: \begin{cases} -l < \mu b < -l - c, & c < 0, \\ -l - c < \mu b < -l, & c > 0, \end{cases} \tag{5.118}$$

$$IV: -l - c < \mu b < l - c. \tag{5.119}$$

C. Let us pool the results of the analysis for the cases $\dot{x} > 0$ and $\dot{x} < 0$. To this end, we plot the domain of definition of the sets according to expressions (5.112), (5.114), (5.115) and (5.117)-(5.119) in plane $(c, \mu b \operatorname{sign}\dot{x})$. As one can see from Fig. 5.3 the domain of uniqueness of solution of the dynamic problem (5.106)-(5.107) is the horizontal strip

$$-l < \mu b \operatorname{sign}\dot{x} < l$$

and consists of subdomains 1, 2, 3, corresponding to the sets of signs II, III, IV.

The domain of non-uniqueness consists of subdomains 4, 5, 6, 7 corresponding to the sets of signs II and III, II and III, III and IV, II and IV.

The domain of non-existence of solution consists of subdomains 8 and 9. Therefore, paradoxes are absent if $\mu b < l$ and occur if $\mu b > l$.

5.6.3 Forces of friction

The forces of Coulomb friction are defined by the formulae

$$\mathbf{R}_{\tau\alpha} = -\mu |R_\alpha| \operatorname{sign}\dot{x} \mathbf{i}_1 = -\tilde{\varepsilon}_\alpha \mu R_\alpha \mathbf{i}_1.$$

Substituting values of R_α, eq. (5.106), into these formulae we obtain the following expressions for the forces of Coulomb friction acting on the sliders from the guide

$$\mathbf{R}_{\tau 1} = \frac{-\tilde{\varepsilon}_1(\tilde{\varepsilon}_2 \mu b - l + c)}{2l + \mu b(\tilde{\varepsilon}_1 - \tilde{\varepsilon}_2)} \mu Y \mathbf{i}_1, \quad \mathbf{R}_{\tau 2} = \frac{-\tilde{\varepsilon}_2(\tilde{\varepsilon}_1 \mu b + l - c)}{2l + \mu b(\tilde{\varepsilon}_1 - \tilde{\varepsilon}_2)} \mu Y \mathbf{i}_1. \tag{5.120}$$

Since the system consists of a single particle, there exists a single Painlevé force of friction given by the formula

$$\rho = M(\ddot{x} - {}^0\ddot{x})\mathbf{i}_1 .$$

Here $\ddot{x}$ is found by eq. (5.107). The value of ${}^0\ddot{x}$ is also obtained from eq. (5.107) under condition $\mu = 0$, i.e. ${}^0\ddot{x} = X/M$. This yields the following expression for the Painlevé force of friction acting on the particle

$$\rho = \frac{\tilde{\varepsilon}_1(l-c) + \tilde{\varepsilon}_2(l+c)}{2l + \mu b(\tilde{\varepsilon}_1 - \tilde{\varepsilon}_2)} \mu Y \mathbf{i}_1 = \mathbf{R}_{\tau 1} + \mathbf{R}_{\tau 2} . \qquad (5.121)$$

As one can see from eqs. (5.120) and (5.121) the Painlevé force of friction is equal to the sum of the Coulomb forces of friction for the studied mechanism.

5.7 Concluding remarks about Painlevé's paradoxes

The previous chapters deal with the specific problems of the theory of mechanical systems with Coulomb friction. The general problems are illustrated by examples. In what follows we summarise the results of the problems considered.

5.7.1 On equations of systems with Coulomb friction

The derivation of equations was carried out by means of Lagrange's equations. The generalised reaction forces and the generalised forces of friction included were explicitly expressed in terms of the frictional coefficients and the normal reaction forces of the contacts. Furthermore, resolving these equations for the accelerations and the reaction forces we obtained equations for the normal reaction forces and the differential equations of motion, the latter containing no reaction forces.

For a single-degree-of-freedom system with a single frictional pair the equation for the reaction force and the differential equation of motion are represented in the form (2.59). For systems with several degrees of freedom and a single frictional pair these equations are given by (4.25) and (4.26) whilst for systems with several degrees of freedom these equations take the form of eqs. (5.16) and (5.21). Clearly, eqs. (4.25), (4.26) and (2.59) follow from eqs. (5.16) and (5.21).

These equations describe the dynamics of the systems outside the paradoxical regions. In the paradoxical regions these equations are not correct since the solution either does not exist or is non-unique.

By considering the transition from a system with elastic contact joints to a system with rigid joints, we obtained differential equations for the reaction forces in the form of (4.43), (5.66) or (5.67) which are valid both in and outside the paradoxical regions.

5.7.2 On conditions of the paradoxes

A rather simple method of determining the paradoxical regions was suggested, see Theorems 1 and 7, and eq. (5.23). For example, in order to determine the paradoxical regions of a single-degree-of-freedom system and a single frictional pair it is sufficient to calculate two coefficients of the kinetic energy A, A_{12}, the derivative $d\mathbf{r}_T^0/dq$ and then determine the sign of the normal reaction force in the absence of friction.

5.7.3 On the reasons for the paradoxes

Up until now, explanations of Painlevé's paradoxes with the help of elastic contacts have been discussed only for particular problems. Hence, the generality of this statement remained open. The validity of this position was proved in Sections 4.3 and 5.4 by means of general equations for systems with friction constructed taking account of elastic contact deformations.

5.7.4 On the laws of motion in the paradoxical situations

As a result of solving the problem of the transition from an elastic to a rigid contact the general laws of motion in the paradoxical regions are obtained. On the one hand, these theorems are a generalisation of the rules established earlier, on the other hand, they essentially supplement and improve these rules. One can convince oneself by means of the following observations.

1. Due to the original Painlevé principle in the situation of non-unique solution only that solution is realised for which the signs of the reaction force with and without friction are coincident. Theorem 8 confirms this, however, it suggests a refinement, namely that such a motion can occur but only for certain initial values of the reaction force and its time derivative. For other initial conditions there occurs a tangential impact and, in particular, a dynamic seizure (an instantaneous stop).

2. According to Klein's suggestion an unstable solution becomes stable under any small perturbation. However Theorem 8 and Fig. 4.1 indicate that this statement is valid only for some perturbations.

3. Analysis of the situation of non-existence of the solution by using particular problems, see [26], [62], [94], [125], allows us to deduce conclusions on the dynamic seizure of some mechanisms. It remains unclear whether this conclusion can be generalised to other mechanisms. If so, what particular closed form expressions for the law of motion should be ascribed to a particular mechanisms. The answer to this question is provided by

Theorem 9 which confirms the appearance of a tangential impact for any initial values of the reaction force and its time-derivative, and expresses a general mathematical formulation of the impact. Moreover, according to Theorems 8 and 9 and their corollary, the dynamic seizure is a particular case of the tangential impact. In the general case, under a tangential impact, the velocity of the particles and the slider experience jumps. However a stop does not necessarily occur.

4. This approach to the problem of tangential impacts differs to some extent from that in [49]. The latter is devoted to a system with a one-sided contact constraint. The motion, in the case of the paradoxical non-uniqueness, is chosen by means of the requirement of continuity of the motion with respect to the impacts due to the rough surface. It is mentioned that such a motion may be accompanied by a tangential impact.

In Section 4.3 the appearance of tangential impacts is considered for a system with a two-sided constraint as $c \to \infty$. It is proved that, in the situation of the paradoxical non-existence of the solution, these impacts are unavoidable regardless of the initial values of the reaction force and its derivative; that is, impacts of the bodies in the frictional contact play no role. In the situation of non-uniqueness the initial values of the reaction forces and its time-derivative affect the feasibility of tangential impacts. For this reason, the influence of impacts must be considered in accordance with the particular case of zero reaction force and its non-zero time-derivative.

In addition to this, as one can see from Fig. 4.1, the condition of continuity of motion with respect to the initial value of the reaction force and its derivative, and thus with respect to the value of the impact, does not hold in general.

Indeed, near the asymptote one can find an infinite number of pairs of adjacent points, one of which moves about the centre of ellipses and the other tends to infinity.

5.7.5 *On the initial motion of an immovable contact*

In Section 2.4 the conditions at the beginning of motion are expressed in terms of the generalised active forces, the coefficients of the kinetic energy and the coefficient of friction, see Theorems 2 and 3. The case in which the dynamic equations admit simultaneously the state of rest and gliding is noted.

5.7.6 *On self-braking*

In Section 2.5 the concept of self-braking is generalised to the case in which the system consists of many particles and is subjected to many forces, and the property of self-braking changes depending on the system configuration. Theorems 4,5,6 allows one to i) partition the system of particles into a set of points of self-braking and a set of points of debraking, ii) determine the and the angle of shift for each particle, and iii) find the paradoxical cases in which a particle can be considered as being simultaneously a point of self-braking and a point of debraking.

5.7.7 On the mathematical description of Painlevé's law

An attempt to develop the problem of determining the frictional forces acting on the particles (Painlevé's forces of friction, [117], [116]) $\boldsymbol{\rho}_i$ is made in Section 5.2. A system of equations (5.32) for calculating the difference in two systems of accelerations corresponding to two systems of external forces is derived. An improved system of Painlevé's equations (5.41) and the resulting system (5.42) are derived and the improved Painlevé's theorem is proved. Equations (5.41) and (5.42) allow one to find the forces of friction $\boldsymbol{\rho}_i$ for prescribed generalised forces of friction and the prescribed coefficients of Coulomb friction respectively. Using eq. (5.42) one understands that forces $\boldsymbol{\rho}_i$ are uniquely determined in terms of the coefficients of Coulomb friction only in the case of no paradoxes. In the paradoxical situations, they are either non-unique or do not exist.

5.7.8 On examples

The general statements formulated above are illustrated by 13 examples. Apart from their explanatory character, nine of the examples are of special interest. These are the extended Painlevé-Klein scheme, the epicyclic mechanism, gear transmission with fixed axes of rotation, the crank mechanism, the link mechanism, the stacker, the Zhukovsky-Froude pendulum, the spindle system of metal cutting and the sliders of machine tools.

The new results obtained by analysis of these mechanism are as follows.

For the extended Painlevé-Klein scheme the conditions for transition from rest to motion for both in and outside the paradoxical regions are obtained. For a plate, the regions of self-braking and debraking are found, and the angles of stagnation and shift for a particle are determined.

For the epicyclic mechanism and the gear transmission the expressions for the reaction forces, the equations of motion, the condition for remaining at rest and the condition for transition to motion and the formula for efficiency are derived. The paradoxes are proved to be absent for these mechanisms.

For the crank mechanism we established the critical relationships between the parameters leading to the paradoxes and self-braking.

For the link mechanism we derived the conditions for the paradoxes stating that the paradoxes are inevitable in the extreme positions, the interval of the paradoxes increasing with the growth of the frictional coefficient and decreasing with the growth of the moment of inertia. The feasibility of self-braking of the mechanism under certain conditions is mentioned.

For the stacker we found a critical height which when exceeded results in the paradoxes. The formulae for the reaction forces, the conditions for transition from the regime of pure rolling to the regime of rolling with sliding and the law of motion under speed-up are obtained.

For the Zhukovsky-Froude pendulum we constructed the equations of motion taking exact account of Coulomb friction and found a non-sinusoidal form of free oscillation. The case of joint rotation of the pendulum and the shaft is also considered.

For the spindle system we found the condition of unstable cutting which results in Painlevé's paradoxes with increasing hardness of the treated metal.

For the sliders of machine tools, the region of existence and the domains of uniqueness and non-uniqueness of the solution are plotted in plane $(c, \mu b \operatorname{sign} \dot{x})$.

In general, these examples indicate that the suggested theory allows one to investigate a rather broad class of mechanisms from a unified position. Due to the formulated general statements for any particular mechanism, one can easily establish the important dynamic characteristics like the value of the normal reaction as a function of the coefficient of Coulomb friction, the condition of transition from rest to motion, the property of self-braking, the feasibility of the appearance of the paradoxes, the laws of motion under the paradoxes, the efficiency and so on. The examples considered indicate that Painlevé's paradoxes are wide-spread in technological problems, and this requires one to prove the feasibility of these paradoxes for all mechanisms with Coulomb friction and to determine the ways to eliminate the negative consequences.

6
Experimental investigations into the force of friction under self-excited oscillations

The mechanical self-excited oscillations in elastic systems are such that the cycle of self-excited oscillations is clearly split into two parts, namely the part of contact and the part of relative sliding. The first part can also be referred to as staying at rest. However we do not use this notation since this part is associated with a small relative displacement, see Chapter 7 for details.

The state of contact is transformed into sliding at the critical value of the force of friction F_+ which is referred to as the maximum friction or the break-down force. The value F_- under which sliding is transformed into the state of contact is referred to as the pick-up force.

As mentioned in Chapter 1 scientists have suggested various characteristics of the force of friction to explain the appearance of self-excited oscillations. Some scientists take the force of break-down as being dependent on the duration of the previous contact [24], [59], [67]-[69] and other scientists are of the opinion that only the force of friction of sliding depends on the velocity of sliding whereas the force of friction of break-down does not depend on the velocity [50], [51], [111]-[114].

The experimental data presented here allows one to assume that the value of F_+, at least for the studied pairs, is determined by the rate of the tangential loading $f = dF/dt$ at the instant of break-down and decreases with the growth of this rate.

It is also established that change in the force of sliding under oscillation does not always correspond to the experimental characteristic obtained in tests with various (but constant) velocities, that is, the value of the force

198   6. Experimental investigations into the force of friction

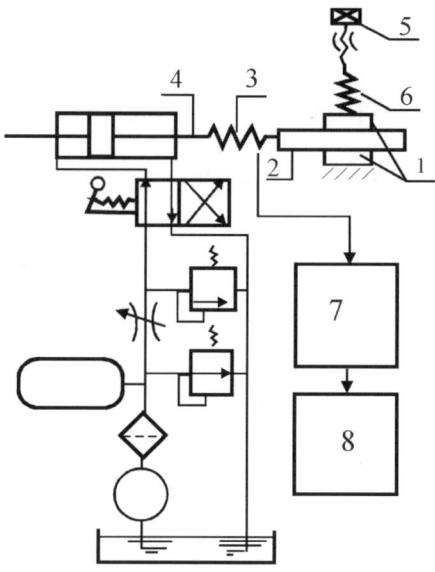

FIGURE 6.1.

of sliding under oscillation differs from the values under stationary motion with corresponding velocities, [76], [79].

6.1 Experimental setups

Three experimental setups were used. We denote these as I, II, III. Setup I is aimed at measuring the force of friction [81]. Setups II are III are utilised for simultaneous measuring the force of friction and the small displacement between the time instants of pick-up and break-down.

6.1.1   The first setup

The scheme used is shown in Fig. 6.1. The actuating mechanisms consists of two guides 1, slider 2 of mass m, spring dynamometer 3 of rigidity c and rod 4 of the drive.

Rod 4 reciprocates due to the hydraulic drive and its velocity is controlled by a throttle with a governor. Motion reversal is performed automatically by means of a special distributor. To damp forced oscillations caused by instability of the motor-pump system, the hydraulic drive is equipped with a damper of the membrane type with an elastic element.

The normal force P and thus the normal pressure p is created by bolt 5 and is measured by the spring dynamometer 6. Displacement y of spring 3

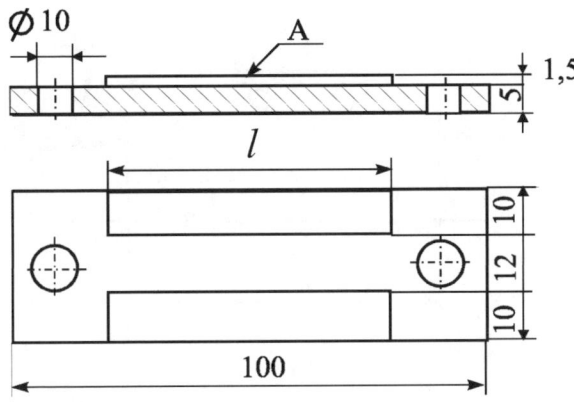

FIGURE 6.2.

is measured by means of the strain gauges. The electric pulses are amplified by amplifier 7 and registered by mirror-galvanometer oscillator 8.

The velocity v of rod 4 can be adjusted continuously from $5 \cdot 10^{-3}$ to 40 mm/s. The nominal area of the contact, i.e. the working surface of the guides take the values $7 \cdot 10^2, 10 \cdot 10^2, 15 \cdot 10^2$ and $20 \cdot 10^2$ mm^2. The nominal force P and the pressure p vary from $P = 200$ N, $p = 0, 1 \text{N/mm}^2$ to $P = 2750$ N, $p = 1, 4$ N/mm^2. Four string dynamometers 3, having rigidities 700, 1380, 5140, 13000 N/mm, were used.

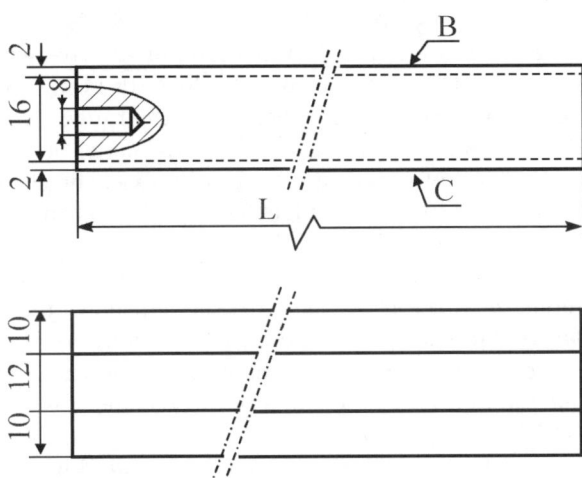

FIGURE 6.3.

The scheme and the dimensions of the slider guide are shown in Figs. 6.2. and 6.3. A change in the nominal area of the contact is achieved by a change of size l of the guide, and a change in mass m of the slider is carried

200   6. Experimental investigations into the force of friction

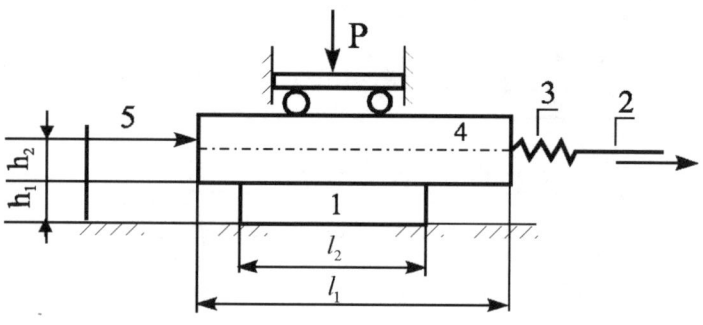

FIGURE 6.4.

out by a change in size L of the slider or by additional masses. The following materials of the frictional pairs "guide-slider" were studied in the experiments: low-alloyed carbon steel with HRC 48 & low-alloyed carbon steel with HB 200, low-alloyed carbon steel with HRC 48 & cast iron, cast iron & cast iron, cast iron & tin-aluminium-copper alloy (a special antifrictional alloy), low-alloyed carbon steel with HRC 48 & tin-aluminium-copper alloy. The working surfaces A, B and C of the guides and sliders are treated with surface finishes of classes 4 to 9.

6.1.2   The second setup

In contrast to setup I, setup II has only one guide 1, see Fig. 6.4. Two variants of drive 2 are used: a hydraulic-mechanical drive with continuous change of velocity v in the range $(1, 16 \cdot 10^{-4}, 2, 5 \cdot 10^2$ mm/s) and a mechanical drive with stepwise change of velocity v in the range $(1, 16 \cdot 10^{-2}, 11, 5$ mm/s). Under a certain condition, motion of the slider is accompanied by self-excited oscillations of various types. In this case, the chosen velocity v of the drive is constant due to the high rigidity of the gear and high power of the motors.

The normal force P is created in the same fashion as for setup I. The spring dynamometer 3 serves to measure the shifting force and the force of friction and dynamometer 5 in the form of a cantilever beam, is aimed at measuring the preliminary displacement. By a proper choice of the galvanometer and the amplifier, the value of displacement is measured to an accuracy of 0,25 μm.

The test specimen (sliders and guides) are fabricated in the same way as those of setup I.

Let us estimate the displacement. It is known [108], that value $x(t)$ measured in tests is the sum of $x = x_c + u$, where x_c is the relative displacement near the contact surface and u is the displacement due to deformation of the slider under tangential load. In setup II $h_1 = 2$ cm, $h_2 = 1, 2$ cm, $l_2 = 30$ cm; $l_2 = 12$ cm. Let us calculate the displacement due to the deformation

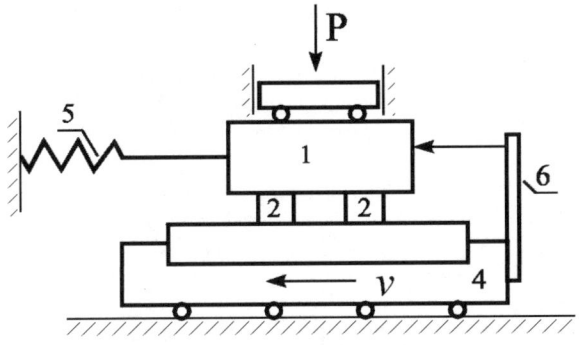

FIGURE 6.5.

for the case in which $l_1 = l_2 = 12$ cm, whilst the tangential force is 1000 N (maximum of the elastic force in the tests) and uniformly distributed over the upper surface of the guide with stress $\tau = 28$ N/cm^2.

Under this assumption, the calculated displacement due to deformation u' is larger than the actual one u, and thus

$$u < u' = \frac{\tau h_1}{G_1} + \frac{\tau h_2}{G_2} = 0,13\,\mu\text{m}\,.$$

Here $G_1 = 4,5 \cdot 10^6$ N/cm^2 and $G_2 = 8,1 \cdot 10^6$ N/cm^2 are the shear moduli. The test results indicated that $x_c \gg u$. Hence, the measured values of x can be understood to be the contact relative displacement $x \simeq x_c$.

The measured displacement due to deformation turned out to be less than the threshold of the amplifier 5, i.e. less than 0,25 μm.

6.1.3 The third setup

This setup is depicted in Fig. 6.5. The normal force P is transmitted to three cylindrical specimens 2 via the rolling bearings. The lower specimen 3 is a plate fixed to slider 4, whose velocity v can be changed in the range $1,17 \cdot 10^{-2} - 11,5$ mm/s by means of a direct current motor. Elastic elements 5 and 6 measure tangential force and small displacement, respectively.

Cylindrical specimens of diameter 5–10 mm are made from mild steel. Their lower face (the working surfaces) are covered by polyamides П.54 and Rilsan of thickness 1 mm. Before the tests these specimens were run-in. Plate 3 is made of low-alloyed carbon steel and has a 9-th class surface finish.

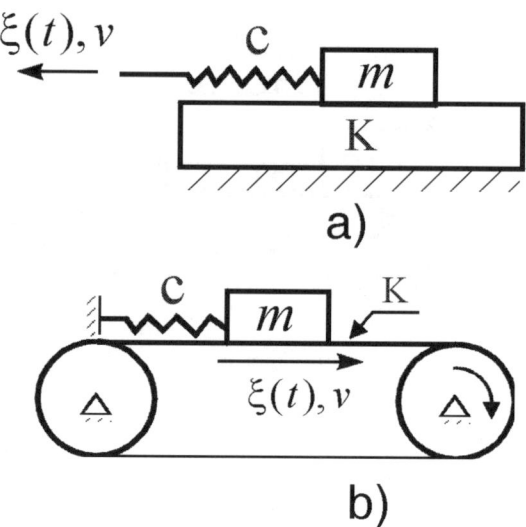

FIGURE 6.6.

6.2 Determining the forces by means of an oscillogram

In order to measure the force of friction F, in particular, value F_+ at breakdown and value F_- at pick-up, we depict in Fig. 6.6 the above setups, [81]. Scheme a corresponds to setups I and II, whilst setup b models setup III. It is worthwhile mentioning that many existing mechanisms with friction can be modelled using these schemes, [23], [67], [76], [81].

In scheme a mass m is attached to a spring of rigidity c, the motion of its left end being prescribed by $\xi(t)$. The counterbody K does not move. Scheme b differs from the scheme a in that body K executes the motion obeying law $\xi(t)$, whereas the left end of the spring is immovable. In other words, according to scheme b one of the contacting bodies, namely body K is rigidly related to the drive.

Let x denote the relative position of the counterbody K with respect to mass m and y denote the position of mass m with respect to the left end of the spring. Under this notation, the absolute motion of mass m is described by $x(t)$ for scheme a and $y(t)$ for scheme b. The coordinates are related to each other as follows

$$y + x - \xi = 0. \tag{6.1}$$

The equation for the dynamics for scheme a takes the form

$$m\ddot{x} = cy - F, \tag{6.2}$$

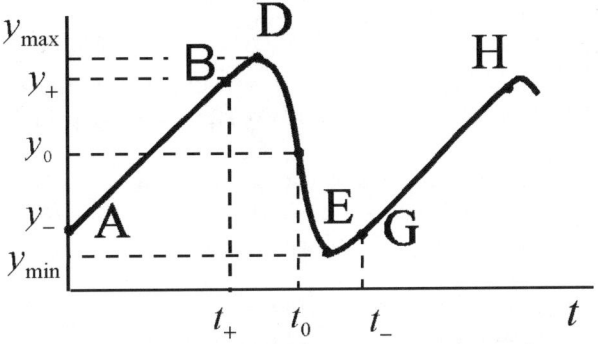

FIGURE 6.7.

where c denotes the spring rigidity and F is the force of external friction. As mentioned in Chapter 1 some displacement takes place on the contacting parts. Acceleration $\ddot{x}$ of this displacement is small which allows us to neglect it in the dynamical equation. The force of its resistance is small and is not taken into account in this equation, hence

$$F = cy. \tag{6.3}$$

In particular, the resistance forces at the break-down F_+ and at the pick-up F_-, see Fig. 6.7, are determined as follows

$$F_+ = cy_+, \quad F_- = cy_-. \tag{6.4}$$

Equalities (6.3) and (6.4) show that for scheme a the force of friction F, the limiting values F_+ and F_- included, is equal to the elastic shifting force cy. Let us notice that for setup I the value F equals the double force of friction ($F = 2\mu P$) since the slider contacts simultaneously two guides.

In the case of the scheme b the equation of dynamics is given by

$$m\ddot{y} = -cy + F.$$

Acceleration $\ddot{y}$ at the contact is coincident with the acceleration of the prescribed motion $\ddot{\xi}$, thus

$$F = cy + m\ddot{\xi}, \quad F_\pm = cy_\pm + m\ddot{\xi}_\pm. \tag{6.5}$$

Here, in contrast to eqs. (6.3) and (6.4), value $F(F_+, F_-)$ is not equal to the elastic force $cy(cy_+, cy_-)$, but the sum of this force and the force of inertia $m\ddot{\xi}$. Hence, for experimental determination of the force of friction by means of scheme b it is necessary to measure not only the elastic displacement y but also the acceleration $\ddot{\xi}$.

Let us demonstrate how to determine the values of the force of break-down F_+ and the force of pick-up F_- with the help of the oscillogram of the relaxation self-excited oscillations provided that acceleration $\ddot{\xi}$ is given. By means of substitution $F' = F - m\ddot{\xi}$, we can reduce equality (6.5) to the form of eqs. (6.3) and (6.4). For this reason, the further analysis of this section is performed on eqs. (6.3) and (6.4), or on the principle scheme of Fig. 6.6a.

Figure 6.7 displays a record of the self-excited oscillation $y(t)$. Points B and G correspond to the break-down and pick-up. When these points are found, F_+ and F_- are determined by means of eq. (6.4). However the oscillogram $y(t)$ does not allow one to indicate the points of break-down and pick-up. Usually, the extrema D, E are understood to be these points and estimates of values F_+, F_- are obtained using

$$\hat{F}_+ = cy_{\max}, \quad \hat{F}_- = cy_{\min}. \tag{6.6}$$

As one can see from Fig. 6.7 this estimation has a systematic error as $\hat{F}_+ - F_+ = c(y_{\max} - y_+) > 0$, $\hat{F}_- - F_- = c(y_{\min} - y_-) < 0$.

Now we derive a formula for calculating F_+, F_- in terms of $y_{\max}, y_{\min}$ without the mentioned systematic error. Within the parts of the immovable contact AB, GH the velocity of the elastic displacement is equal to the velocity of the prescribed motion

$$\dot{y} = \dot{\xi} = v(t), \quad 0 \le t \le t_+, \quad t > t_-. \tag{6.7}$$

On the sliding parts $BDEG$, function $y(t)$ is known to be approximated by a sine-function, i.e.

$$y = y_0 - a\sin(\omega t - t_0). \tag{6.8}$$

Here $y_0 = (y_{\min} + y_{\min})/2$, $\omega^2 = c/m$ and a denotes the amplitude. It follows from eqs. (6.7) and (6.8) that the force of break-down is equal to

$$\dot{\xi}(t_+) = v_+ = -\omega a \cos\omega(t_+ - t_0),$$
$$y(t_+) = y_+ = y_0 - a\sin\omega(t_+ - t_0),$$

hence

$$\frac{y_{\max} - y_{\min}}{2} = a = \left[(y_+ - y_0)^2 + \frac{v_+^2}{\omega^2}\right]^{1/2}.$$

Resolving these equations for y_+ we obtain

$$\frac{F_+}{c} = y_+ = y_0 + \left[\left(\frac{y_{\max} - y_{\min}}{2}\right)^2 - \frac{v_+^2}{\omega^2}\right]^{1/2}. \tag{6.9}$$

Using the condition for pick-up

$$\dot{\xi}(t_-) = v_- = -\omega a \cos\omega(t_+ - t_0),$$
$$y(t_-) = y_- = y_0 - a\sin\omega(t_+ - t_0)$$

we obtain by analogy

$$\frac{F_-}{c} = y_- = y_0 - \left[\left(\frac{y_{\max} - y_{\min}}{2}\right)^2 - \frac{v_-^2}{\omega^2}\right]^{1/2}. \quad (6.10)$$

Formulae (6.9) and (6.10) suggest exact relationships between the forces F_+, F_- and the displacement of the spring $y_{\max}, y_{\min}$ under the assumption of sinusoidal oscillation (6.8) of the sliding part. When terms v_+^2/ω^2, v_-^2/ω^2 are sufficiently small in comparison with $(y_{\max} - y_{\min})^2/2$, the estimated $\hat{F}_+, \hat{F}_-$ obtained from eq. (6.6) approach the values of F_+, F_- from eqs. (6.9) and (6.10). In other words, expression (6.6) is acceptable only under the condition

$$\frac{v_\pm^2}{\omega^2} \ll \frac{(y_{\max} - y_{\min})^2}{2} \quad (6.11)$$

which implies that points D and E coincide respectively with points B and G, see Fig. 6.7. Taking into account eq. (6.6) we can set condition (6.11) as follows

$$cmv_\pm^2 \ll \frac{(F_+ + F_-)^2}{4}. \quad (6.12)$$

Velocity v and rigidity c are often the specified parameters, [76], [81], [91], [92]. Inequality (6.11) can be satisfied by means of decreasing mass m. If the setup contains a mechanism for the control of the normal force P, the latter can be varied by means of changing the mass m. In this case the force is $P = mg$, where g denotes the acceleration due to gravity. In this case m can not be taken arbitrarily and using formula (6.6) can lead to considerable error, see [81] for details.

Equations (6.6), (6.9) and (6.10) result in expressions for systematic errors in the estimation of forces Δ_+, Δ_- in the form

$$\frac{\Delta_+}{c} = \frac{\hat{F}_+ - F_+}{c} = \frac{y_{\max} - y_{\min}}{2} - \left[\left(\frac{y_{\max} - y_{\min}}{2}\right)^2 - \frac{v_+^2}{\omega^2}\right]^{1/2},$$

$$\frac{\Delta_-}{c} = \frac{\hat{F}_- - F_-}{c} = \frac{y_{\max} - y_{\min}}{2} + \left[\left(\frac{y_{\max} - y_{\min}}{2}\right)^2 - \frac{v_-^2}{\omega^2}\right]^{1/2}. \quad (6.13)$$

If values v_+^2/ω^2, v_-^2/ω^2 are sufficiently small, so that Δ_+ and Δ_- can be expanded in power series in terms of the above values, then the first approximation yields

$$\Delta_+ = \frac{mv_+^2}{y_{\max} - y_{\min}} = \frac{cmv_+^2}{\hat{F}_- - \hat{F}_-},$$

$$\Delta_- = \frac{-mv_-^2}{y_{\max} - y_{\min}} = \frac{-cmv_-^2}{\hat{F}_- - \hat{F}_-}. \quad (6.14)$$

Let us consider a numerical example. Let the mass and the rigidity of the system be $m = 16 \cdot 10^3$ kg, $c = 16 \cdot 10^5$ N/m. In the regime of self-excited oscillations for the velocities $v_+ = v_- = v_1 = 10^{-3}$ m/s and $v_+ = v_- = v_2 = 2 \cdot 10^{-3}$ m/s we observed respectively $\hat{F}_+ = 3950$ N, $\hat{F}_- = 2700$ N, and $\hat{F}_+ = 3870$ N, $\hat{F}_- = 2650$ N. Inserting these values into eq. (6.14) we find

$$\Delta_{\pm 1} = \pm 21 \text{ N}, \quad \Delta_{\pm 2} = \pm 85 \text{ N}.$$

Led by the above reasoning in the test for measuring the force of break-down F_+ we took the minimum feasible mass m of the slider. In the case of considerable difference $y_{\max} - y_{\min}$ the recorded oscillograms are seen to be jagged, see Fig. 6.8, and thus the estimation for F_+ by eq. (6.6) is acceptable. Equations (6.2)-(6.4) are used for processing the oscillograms by means of the following relationships

$$\begin{aligned} F &= c\xi(t) + cy_{\min}, & \text{at} \quad 0 \le t \le t_+, \\ F_+ &= cy_{\max}, & \text{at} \quad t = t_+. \end{aligned} \quad (6.15)$$

Moreover, the duration of the phase of sliding $t_- - t_+$ is negligibly small in comparison with the duration of the contact $t_- - t_+ \ll t_+$. Thus, the oscillation period is actually equal to t_+.

When the difference $y_{\max} - y_{\min}$ is small, then instead of eq. (6.15) we can use the following expressions

$$F = c\xi(t) + cy, \quad 0 \le t \le t_+,$$

$$F_\pm = cy_0 \pm \left[\frac{(y_{\max} - y_{\min})^2}{4} - \frac{v_\pm^2}{\omega^2} \right]^{1/2}. \quad (6.16)$$

In tests, the normal load P was taken from the range 200 N - 1500 N and mass m was 1,5 kg. For such a small mass inequality (6.12) holds for all values of P, c, v and this allows one to process the test results by means of eq. (6.15) without considerable error.

6.3 Change in the force of friction under break-down of the maximum friction in the case of a change in the velocity of motion

The material of this and the following sections are based on [76]. In order to determine changes in the character of self-excited oscillations due to a change in velocity v one often carries out experiments with the constant values of this velocity, that is the velocity takes discrete values. The results obtained in such tests for each frictional pair sum up and give a general

6.3 Change in the force of friction

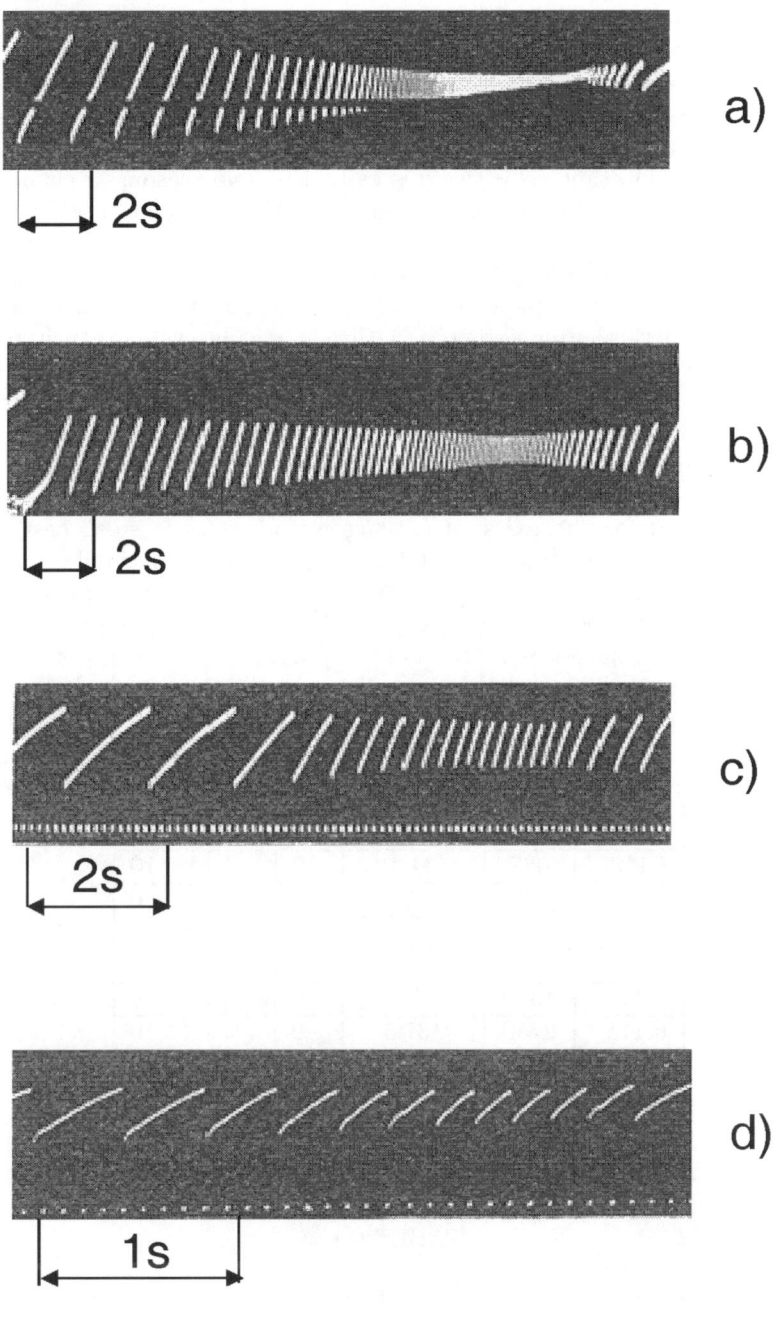

FIGURE 6.8.

impression about the character of the oscillations (amplitude, frequency etc.) depending upon the chosen velocity v.

It is evident that for each frictional pair the physical and mechanical properties of the contacting surfaces and thus, the frictional conditions are individual for each test since the surface layers are destroyed due to friction. Hence, the difference in the results obtained in different tests with corresponding constant velocities v is caused by two reasons: a change in the frictional conditions and the dependence of the character of self-excited oscillations on velocity v. If the dependence of the character of self-excited oscillations on v is established without account being taken for the change in the frictional condition, then the dependence obtained will be inaccurate. In addition to this, after each test with a particular value of v it takes time to prepare next test with another value of v. In this interval of time the functioning of the measuring system can change due to an inherent instability and this can affect the accuracy of the test.

For these reasons, tests to determine relationship $F_+(v)$ were carried out in setup I under continuous (rather than stepwise) changes in v. The tests were performed for the pairs: steel & cast iron, steel & steel, cast iron & cast iron, cast iron & tin-aluminium-copper alloy and tin-aluminium-copper alloy & steel. Figure 6.8 displays some parts of the oscillograms of oscillations due to the frictional pair steel & cast iron. The corresponding values of the normal force P, the normal pressure p, the class of surface finish of the guides ∇_1 and the slider ∇_2, and the frequency of the time marker η are given in the upper part of Table 6.1.

Figure	P, N	p, N/cm^2	∇_1	∇_2	η, Hz
6.9a	2750	137,5	9	6	10
6.9b	687	34,3	9	7	10
6.9c	340	17	9	6	10
6.9d	250	12,7	9	7	10
6.11a	2750	137,5	9	5	10
6.11b	1370	68,3	9	6	10
6.11c	2750	137,5	9	7	10
6.11d	340	17,0	9	7	10
6.11e	687	34,3	9	7	10
6.11f	687	34,3	9	8	10
6.11g	687	34,3	9	5	10

TABLE 6.1

One can see that velocity v whose value is proportional to the steepness of the bold parts of the oscillograms first increases and then decreases. As v grows, force F_+ and amplitude a decrease unless the oscillations become unrecordable. The values of F_+ and a increase as v increases.

Thus, the maximum friction F_+ depends on velocity v. This dependence is of a descending character with the growth of v and results in a descending amplitude a of the relaxation self-excited oscillations.

The dependence $F_+(v)$ obtained by means of the oscillograms of Fig. 6.8 is shown in Fig. 6.9. Curves 1 and 2 correspond to F_+ and F_- for $P = 2750$ N, whilst curves 3 and 4 correspond to F_+ and F_- for $P = 687$ N. The obtained results are qualitatively coincident with the experimental data from [67]. The only difference is that the present curves $F_+(v)$ and $F_-(v)$ are approximately symmetric about a straight line parallel to the axis of abscissas. For instance, such a symmetry takes place for $F_c = \text{const}$.

6.4 Dependence of the friction force on the rate of tangential loading

The following expressions for the rate of the tangential loading f and the duration of contact T

$$f = \frac{dF}{dt} = c\frac{d\xi}{dt} = cv, \quad t \le t_+ = T,$$

$$T = \frac{1}{c}\int_{F_-}^{F_+} \frac{dF}{v} = \int_{F_-}^{F_+} \frac{dF}{f} \tag{6.17}$$

follow immediately from eqs. (6.1)-(6.3). In the case of a slowly varying v we can write

$$T = \frac{F_+ - F_-}{cv}. \tag{6.18}$$

Clearly, increasing and decreasing values of v lead respectively to an increase and decrease of the loading rate f. On the other side, according to Fig. 6.9 the integration limits F_+ and F_- in eq. (6.18) approach each other as v increases, and hence, the duration T of the phase of contact decreases and tends to zero.

This means that a decrease in F_+ with growth of v indicates the existence of at least one of two dependences: increasing dependence $F_+(T)$ and decreasing dependence $F_+(f)$. At first sight one gains the impression that these two dependences are identical since the time interval T of loading from pick-up until break-down is determined by the loading rate f. However, this is not the case. For the same value of the loading rate at the instant of the break-down $f = cv_+$ we can get any value of T by changing the loading scheme, i.e. by varying velocity $v(t)$ for $t < T$. The inverse is valid as well, that is, for any value of T we can get any value of f.

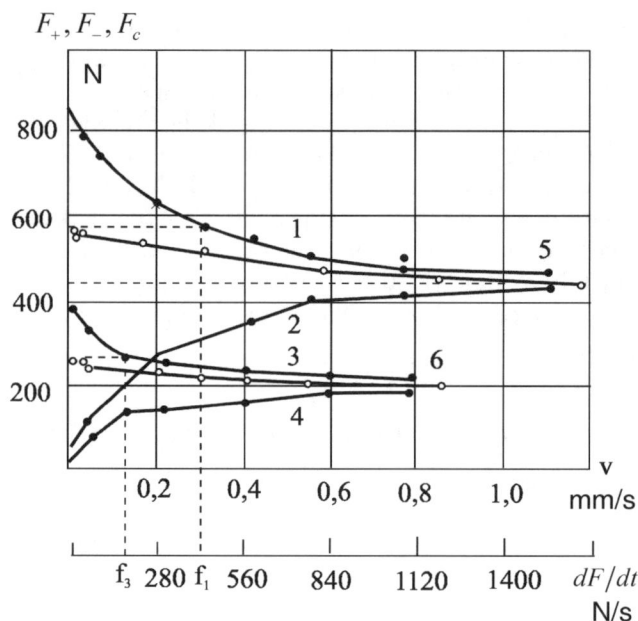

FIGURE 6.9.

In [67] experimental curves $F_+(v)$ is considered to have the following dependence on T

$$F_+ = F_\infty - (F_\infty - F_0)\exp(-kT), \qquad (6.19)$$

where constants F_0, F_∞, k can be determined with the help of Fig. 6.9 by using formulae (6.18) and (6.19). This approach is valid only in the case when the force of break-down F_+ does not depend on the rate of loading f.

To clarify the question of the dependence of F_+ on T or f the curves $F_+(v)$ are not sufficient. In addition to these, one needs to perform tests in which one of the variables T and f varies whereas the other remains constant. It is desirable to carry out the tests using the same setups and specimens which were utilised for curve $F_+(v)$. Such tests were carried out and some results are shown in Fig. 6.10 for the steel & cast iron pair. The numerical values obtained from these tests are displayed in the lower part of Table 6.1.

Let us consider the case in which the rate $f = dF/dt$, at the instant of break-down, remains constant and the duration T varies. As Fig. 6.10b shows, between the third and sixth oscillations the values of f and T are as follows

$$\begin{aligned}f_3 &= f_4 = f_5 = f_6 = 215\,\mathrm{Ns}^{-1},\\ T_3 &= 20\,\mathrm{s},\quad T_4 = T_5 = T_6 = 2\,\mathrm{s}.\end{aligned}$$

6.4 Dependence of the friction force on the rate of tangential loading 211

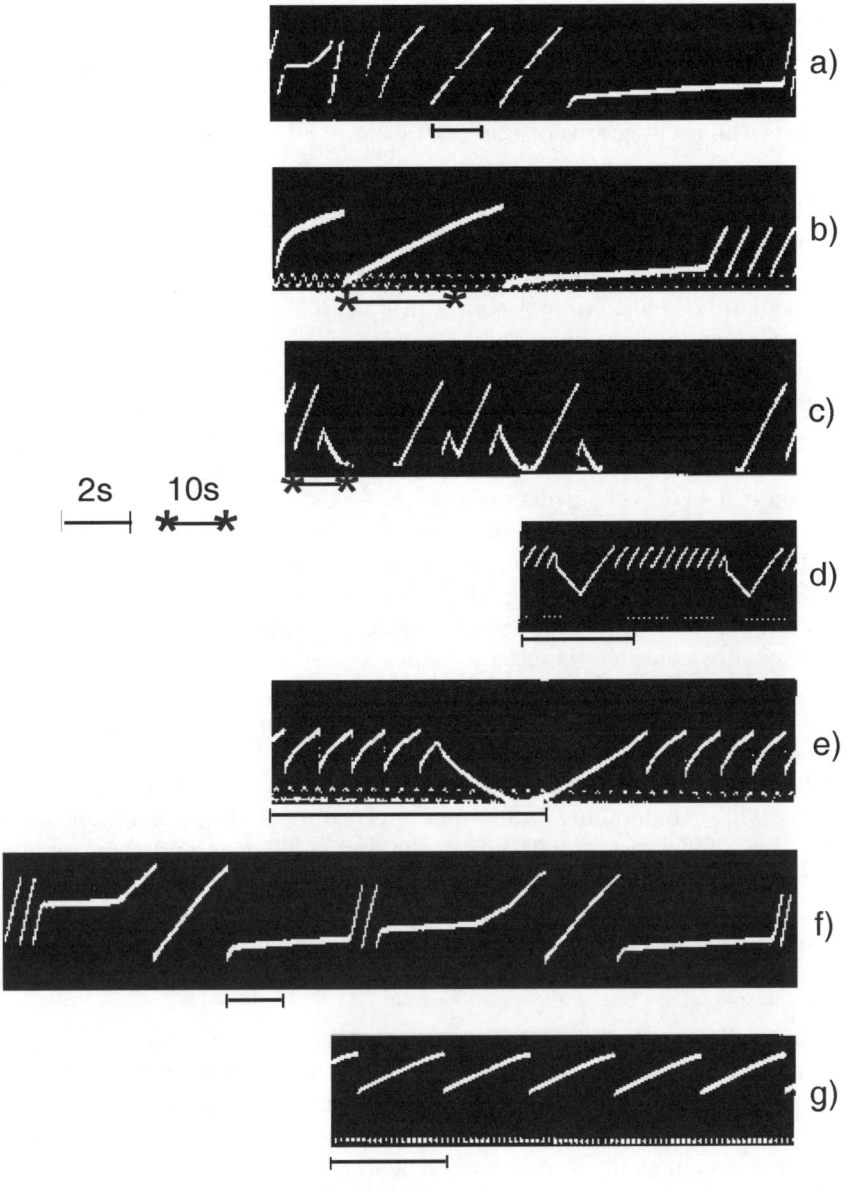

FIGURE 6.10.

Here subscripts 3-6 indicate the number of oscillation. According to [67] when the duration of the contact T_i ($i = 3, ..., 6$) is known, then the corresponding values of force F_+ could be determined from Fig. 6.9. This would lead to the following results

$$F_{+3} = 840\,\text{N} > F_{+4} = F_{+5} = F_{+6} = 660\,\text{N}.$$

However the oscillograms of Fig. 6.10 yield

$$F_{+3} = F_{+4} = F_{+5} = F_{+6} = 660\,\text{N}$$

which contradict the above assumption.

The above and the analogous analysis of other oscillations (among them the oscillations which are not shown in Figs. 6.8 and 6.10) show that force F_+ is constant for various T and constant f. This is valid for T over the interval $0,08\,\text{s}-300\,\text{s}$. In other words, for tested frictional pairs the force of break-down F_+ does not depend on the duration of the contact T.

Earlier we established the dependence of F_+ on v which proved the existence of at least one of the dependences $F_+(T)$ and $F_+(f)$. But F_+ is independent of T, thus it depends on f.

The existence of dependence $F_+(f)$ can be detected by varying f for constant T. For instance, it follows from Fig. 6.10f that for $T_7 = T_9$ and $f_7 < f_9$ we have $F_{+7} > F_{+9}$ and, thus, for different values of f and a constant value of T, the values of F will be different.

Let us also consider the case in which both T and f vary. One can see from Fig. 6.10a that for $T_6 = 0,5\,\text{s}$, $T_{10} = 10\,\text{s}$ and $f_6 < f_{10}$ one obtains $F_{+6} > F_{+10}$. Clearly, this result contradicts formula (6.19) and confirms the above conclusion.

Let us draw our attention to a specific feature which is the process of loading whilst undergoing oscillations. According to the oscillogram of Fig. 6.10a, load F increases first during the first oscillation and then remains constant, then it increases again and finally increases with loading rate f_1 which is equal to the rate f_2 of the second oscillation. Though the first and second oscillations have different loading processes, their loading rates are coincident at the instant of break-down. Thus, the forces of break-down are also coincident, $F_{+1} = F_{+2}$.

The oscillogram in Fig. 6.10c shows other loading processes between the second and sixth oscillations. The values of f at the instant of break-down coincide. Under different loading schemes, but for the same values of f at the instant of break-down, we have $F_{+1} = F_{+2} = F_{+3} = F_{+4} = F_{+5} = F_{+6}$. Phenomena analogous to these can be observed for other oscillations in Fig. 6.10.

Therefore, the maximum force of friction does not depend on the process of tangential loading. However it depends only on the loading rate at the instant of break-down $(dF/dt)_{F=F_+}$. This effect has been confirmed in a series of other tests with different loading processes.

We can conclude that for frictional self-excited oscillations, the maximum force of friction F_+ depends on the rate of tangential loading f at the instant of the break-down. This dependence has a descending character, that is, force F_+ decreases and tends to a certain limit as f increases. The growth of force F_+ with decreasing f leads to the appearance of the phase of immovable contact within a separate cycle of oscillation and this is the reason for self-excited oscillations of the relaxation type.

Curve 1 in Fig. 6.9 displays the dependence of $F_+(f)$ for the frictional pair steel & cast iron with the normal force $P = 2750$ N whilst curve 3 presents this dependence for $P = 687$ N. This dependence can be approximated by the formula

$$F_+ = A\exp(-\beta f) + F_0, \qquad (6.20)$$

where F_0 is the value of F_+ as $f \to \infty$, $A + F_0$ is the value of F_+ for $f = 0$ and β is constant. For example, for the pair steel & cast iron under the normal force $P = 2750$ N we have, as a result of statistical processing, that

$$A = 420 \text{ N}, \quad F_0 = 450 \text{ N}, \quad \beta = 0,003 \text{ s/N},$$

hence

$$F_+ = 420\exp(-0,003f) + 450 \text{ (N)}. \qquad (6.21)$$

6.5 Plausibility of the dependence $F_+(f)$

6.5.1 Control tests

No matter how accurate the working surfaces of the contacting bodies are treated, the physical and mechanical properties of the contact layers of the sliders are inhomogeneous along their length. One can assume that this inhomogeneity can considerably affect the accuracy of the tests. In order to clarify this question, control tests were carried out for each frictional pair, the velocity v of rod 4 of setup I being kept constant. The oscillograms are shown in Fig. 6.10g. As one can see, for a constant velocity, the amplitude a and forces F_+, F_- remain practically constant. For example, in the case of sliding between steel and cast iron at a distance of 100 mm force F_+ lies in the interval 560–570 N, whilst the amplitude a is in the interval $0, 17-0, 18$ mm. Thus, a change in the quality of the working surface of the slider is so small that it does not influence the validity of the test results.

6.5.2 Estimating the numerical characteristics

For the low-alloyed carbon steel & cast iron pair, one hundred tests were carried out for the velocities $v = 0,006$ mm/s and $v = 1,105$ mm/s and the normal force $P = 2750$ N. The recorded values of F_+ are represented by the following statistical series

i) for $v = 0,006$ mm/s

force, N	F_+	760-770	770-780	780-790	790-800
frequency	p	0,01	0,08	0,08	0,155

force, N	F_+	800-810	810-820	820-830	830-840	840-850
frequency	p	0,27	0,185	0,15	0,06	0,01

i) for $v = 1,105$ mm/s

force, N	F_+	415-425	425-455	455-465	465-475
frequency	p	0,02	0,04	0,10	0,15

force, N	F_+	475-485	485-495	495-505	505-515	515-525
frequency	p	0,29	0,17	0,10	0,05	0,01

The result of the processing yields respectively: i) $\tilde{F}_+ = 806,2$ N; $\tilde{D}_F = 2.78$ N^2; $I_{0,95} = (803; 809, 5)$; ii) $\tilde{F}_+ = 462,8$ N; $\tilde{D}_F = 3.40$ N^2; $I_{0,95} = (459; 466, 5)$, where $\tilde{F}_+$ and $\tilde{D}_+$ denote respectively the mean value and the variance, and $I_{0,95}$ is the 0,95% confidence interval.

6.5.3 Statistical properties of the dependences

The above dependence $F_+(f)$ is obtained by means of two regularities, first, force F_+ decreases (increases) as velocity v increases (decreases) and second, the greater the loading rate at the instant of the break-down $f = cv$, the smaller the value of F_+ for the same duration of the contact T_+. As an example, we consider the statistical properties for the steel & cast iron pair. Six groups of observations corresponding to five values of the normal force P were performed. Each group consisted of 100 tests, the value of velocity v being continuously changed in the interval $5 \cdot 10^{-3}$mm/s–10 mm/s (first increased and then decreased). The result consisted of 500 oscillograms, each of which distinctively indicates a decrease (increase) in value F_+ with increase (decrease) in v, see Fig. 6.8. Thus, a decrease in F_+ with a growth in v manifests itself with statistical probability 1.

Two hundred measurements of F_+ with constant T and various f were performed. The above regularity manifested itself 198 times, i.e. the statistical probability is 0,99.

6.5.4 Test data of other authors

The above results were published in 1972 in [76]. In 1976 Dorfman [39] carried out tests with the help of a routing machine and confirmed the conclusion on the dependence $F_+(f)$ for an the industrial application. Moreover, having observed the properties of the relaxational self-excited oscillations after a long-term stop the author concluded that the value of F_+ depends on both the rate f and the duration of the contact T.

The dependence $F_+(f)$ was proved once again by Shmakov in 1979, [132].

6.6 Characteristic of the force of sliding friction

The experimental curve $F_c(\dot{x})$ obtained in tests with various constant values of the sliding velocity $\dot{x} = v = \text{const}$ is referred to as the characteristic of the force of sliding friction. Curves 5 and 6 in Fig. 6.9 represent the characteristics F_c for the steel & cast iron pair. Curves 5 and 6 correspond to $P = 2750$ N and $P = 687$ N. In both cases the guide and the slider have the surface finishes of the ninth and sixth classes, respectively.

The solution of the problem of frictional self-excited oscillations is often based upon the existence of the falling dependence $F_c(v)$. However, Kudinov [74] pointed out that in the case of rapid change of the velocity (for example under oscillations) the force of friction changes moderately, at least much less than under a stationary motion with the corresponding velocities. This is due to the inertia of the thermal processes which lead to an essential decrease in the influence of the velocity on the force of friction F_c.

Let F_u denote the true force of sliding friction under oscillations. Let us clarify whether the instant values of F_u coincide with the characteristic $F_c(v)$. According to Fig. 6.9 we have

$$F_+(f > f_i) < F_c(0), \quad (i = 1,3). \tag{6.22}$$

Here f_i denotes the value of f_i for which $F_+ = F_c = (0)$. Inequality (6.22) is obtained by comparing the experimental dependences $F_c(v)$ and $F_+(f)$. In order to prove the above inequality we carried out the following experiments. Dynamometer 3 had a rather high rigidity ($c = 13 \cdot 10$ N/mm) which allowed the setup to function without oscillations at low velocities. The hydraulic drive was adjusted such that

$$\frac{dF}{dt} = cv > f_i.$$

After switching on the drive the values $F_+(cv) = cy_{\max}$ and $F_c(v)$ were determined. It was observed that $F_c(v) < F_+(cv)$ and the sliding decelerated to a stop ($v \approx 0$). Force F_c increased gradually until the maximum value and took the value $F_+(cv)$. Hence, force F_+ at high rate f is less than force F_c at low rate v which corresponds to relationship (6.22).

Let us prove that by virtue of inequality (6.22) the values of F_u and $F_c(v)$ do not coincide. As follows from Fig. 6.6 after the break-down, the acceleration must be negative and thus,

$$m\ddot{y} = F_u - F_+(cv) < 0. \qquad (6.23)$$

Equations (6.22) and (6.23) yield that $F_u < F_+(cv) < F_c(0)$ which is required. Therefore, immediately after break-down, the true force of the sliding friction F_u was observed to be smaller than force $F_{\tilde{n}}$ due to characteristic $F_c(v)$. This enables us to assume that the sliding friction force under the relaxation self-excited oscillation does not obey (at least immediately after the break-down) the experimental characteristic $F_c(v)$ obtained for various constant velocities.

This observation, cf. [76], confirms the validity of the prediction by Kudinov [74]. However this observation only indicates the existing difference between F_u and F_c. However it does not render the true values of force F_u in the oscillatory regime. For this reason, this conclusion was considered to be incomplete, see [83], [77] for details.

Later Dorfman [39] and Shmakov [132] succeeded in obtaining the true values of the force $F_c(\dot{x})$ under self-excited oscillations by analysing the oscillograms of the displacement and acceleration. By comparing $F_c(\dot{x})$ with the characteristic $F_c(v)$ obtained for constant values of v the authors detected considerable difference.

We can conclude the following:

1. For all tested frictional pairs the maximum friction F_+ depends on the rate of the tangential loading $f = dF/dt$ at the instant of break-down. This dependence is of the descending character, i.e. force F_+ decreases and tends to a certain limit with growth of f. The growth of force F_+ with decrease in f results in the appearance of a phase of immovable contact in a separate oscillatory cycle which is the reason for self-excited oscillations of the relaxation type.

2. The force of sliding friction under the relaxation self-excited oscillation does not obey (at least immediately after the break-down) the experimental characteristic $F_c(v)$ obtained for various constant velocities $\dot{x} = v$.

7
Force and small displacement in the contact

Under an increasing tangential load a so-called preliminary displacement occurs before the relative sliding. This phenomenon has been studied extensively both theoretically and experimentally, see [152], [31], [66], [71], [106], [108], [109]. Thus, a small relative motion with small velocity during contact is typical for oscillatory systems with friction. In what follows we will use the terminology "small displacement" to describe this motion. We do not use the term "preliminary" because the displacement under consideration does not rigorously correspond to the theoretic definition of the preliminary displacement suggested for example in [152], [72], [106].

As mentioned in Section 6.2 the analysis is performed by means of the scheme shown in Fig. 6.6a. The measurements for the metal pairs were carried out by the author together with Shmakov [92], whilst those for polymer & metal pairs with Bashkarev [90], [91].

7.1 Components of the small displacement

7.1.1 Definition of break-down and initial break-down

The transition from contact to sliding is referred to as break-down. Depending upon the parameters of the system, the normal force P and velocity v, see Fig. 6.8, there may exist different types of break-down. These are i) break-down accompanied by a jump and ii) break-down without jump. The jumps are observed in the case when both the rigidity c and the velocity v are small. For small mass m of the slider, as product cv increases the value

218   7. Force and small displacement in the contact

of the jump decreases and tends to zero (see oscillograms of Fig. 6.8) and the first type of break-down transforms into the second one.

Let us introduce the concept of initial break-down. Let the contacting bodies be free of the normal force P and the tangential force F

$$P = 0 \quad \text{for} \quad t < 0, \quad F = 0 \quad \text{for} \quad t \leq 0. \tag{7.1}$$

At time instant $t = 0$ the normal force P is applied and then the tangential force F is gradually increased due to the motion of the drive

$$P = \text{const} > 0 \quad \text{for} \quad t \geq 0, \quad F \neq 0, \quad \frac{dF}{dt} \neq \infty \quad \text{for} \quad t > 0. \tag{7.2}$$

Under such a loading a break-down occurs for a critical relationship between $F(t)$ and P. This break-down is referred to as the initial one since it is the first break-down after the normal force P has been applied. This definition differs from the concept of the first break-down given for example in [67], [68]. In these books, the first break-down is that occurring first after the drive has been switched one, i.e. measured from the beginning of the increase or decrease in force F. This definition does not take into account, or at least does not point out, whether the system after the previous motion is completely unloaded or remains loaded. In the first case, when the drive is stopped ($v = 0$) contact is lost and the first break-down is the initial one for further motion. In the second case, the first break-down is clearly not an initial one.

An example of the initial break-down accompanied with a jump appears in the oscillogram of Fig. 7.1a, recorded under a continuous increase in force $F(t)$. Force $F(t)$ and displacement $x(t)$ marked respectively by 1 and 2 are shown and the points of break-down D, d and the point of set-up E are marked in this figure. The point of set-up e on curve $x(t)$ is beyond the oscillogram. The point of break-down is defined as the point at which functions $F(t)$ and $x(t)$ experience discontinuities. Such discontinuities occur at rather high eigenfrequencies and not very high velocity of the drive v, see [76], [81], [92].

7.1.2 Reversible and irreversible components

We consider the displacement from the moment of applying the normal force until the moment of the initial break-down for the steel & cast iron pair on setup II with the nominal contact area 20 cm^2. The oscillograms of Fig. 7.1 are recorded for $P = 840$ N and four various forces $F(t)$ corresponding to a change in direction of motion of the drive with an intermediate stop ($v = 0$). Curves 1 and 2 denote $F(t)$ and $x(t)$ respectively. The ascending parts of $F(t)$ correspond to $v > 0$ (where $v = 2,33 \cdot 10^{-2}$ mm/s), the horizontal parts correspond to $v = 0$ and the descending parts correspond to a smooth decrease in v ($v < 0$).

7.1 Components of the small displacement 219

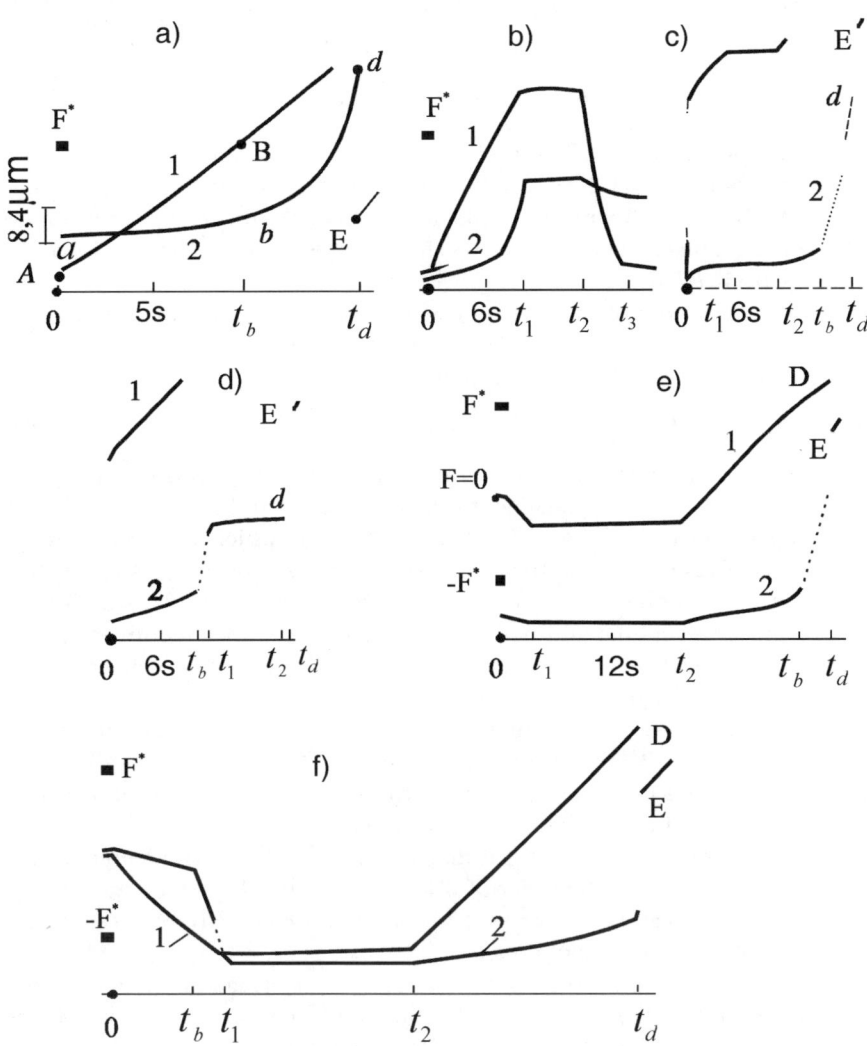

FIGURE 7.1.

In the case of continuous loading until break-down, see Fig. 7.1a, curve $x(t)$ in the time interval $(0, t_\alpha)$ is divided into two parts: part ab with small slope, which can be viewed as being linear, and nonlinear part bd with progressively increasing slope. Curve $F(t)$ is practically linear on AB and nonlinear on BD. At point B force F is approximately equal to the arithmetical average of the forces of break-down F_+ and set-up F_-

$$F(t_b) = \frac{F_+ + F_-}{2} = F^*. \qquad (7.3)$$

The displacements on ab and bd are denoted by x_α and x_β respectively. The total displacement is $x_+ = x_\alpha + x_\beta$. On the considered oscillogram $F^* = 170$ N, $x_\alpha = 3$ μm, $x_\beta = 35$ μm, $x_+ = 38$ μm. Hence, the main part of the displacement belongs to the nonlinear part bd where force F increases from F^* to F_+.

Figure 7.1b represents the case in which the system is loaded in $(0, t_1)$, stays at rest $(v = 0)$ in (t_1, t_2), and then is unloaded in $(t_2, t_3]$. During the stop the elastic force F is greater than F^*:

$$F^* < F(t) < F_+ \qquad \text{for} \qquad t_1 < t < t_2. \qquad (7.4)$$

Observation indicates that $x_\alpha = 3$ μm, $x(t_1) = x(t_2) = 11,5$ μm, $x(t_3) = 8$ μm. Thus, there is no displacement during the stop and under the unloading the slider moves back by $x(t_1) - x(t_3) = 3,5$ μm, which is approximately equal to x_α. This allows one to assume that the component x_α is reversible whereas the component x_β is irreversible. More precisely, two simultaneous processes (reversible and irreversible) take place in the nonlinear part. However the first process, if it exists, is so small in comparison with the second one that it can be neglected.

Experiments were carried out in which the system was loaded until the value of F was less than F^* and then unloaded to $F = 0$. The value of x first increased and then decreased to zero. Thus, if $F < F^*$, then contact in the tangential direction behaves like an elastic joint.

As shown in Chapter 5, elastic displacement u of the slider under the tangential force F is of the order of $0,1$ μm, i.e. much less than x_α, x_β and x_+. For this reason, the above said is completely applicable for the small displacements in the contact zone.

The presence of the two components x_α, x_β is mentioned in many papers on the preliminary displacement, e.g. [66], [106]. However, the values of the displacement in these papers is one order of magnitude smaller than those in the described tests.

7.1.3 Influence of the intermediate stop and reverse on the irreversible displacement

Oscillograms in Fig. 7.1c and d are obtained in the processes in which the system experiences a break-down after a stop in (t_1, t_2). For $F(t_1) < F^*$, see Fig. 7.1c, the stop has no influence on the values of $x(t_b)$ and $x(t_d)$, i.e. the values of x_α, x_β and x_+ coincide with those in the case of continuous loading before break-down. Here t_d denotes the duration of the time interval between the time instant of applying force P until the instant of the initial break-down. This duration differs from the duration T_+ of the contact between the set-up and the break-down in the regime of relaxation self-excited oscillations.

Provided that $F(t_1) < F^*$, Fig. 7.1d, the small displacement under further loading is practically absent. As a result, the irreversible component $x(t_d) - x(t_b)$ and thus the total small displacement $x(t_d)$ become smaller than the corresponding values of x_β and x_+, obtained under the continuously increasing loading.

Figures 7.1e and f illustrate the cases of reverse in which velocity v is negative in $(0, t_1]$, vanishes in $(t_1, t_2]$ and is positive in $(t_2, t_d]$, that is the drive moves back and forth. Oscillograms d and e differ from each other in the value of force $F(t_1)$ at the stop.

For $0 > F(t_1) > -F^*$, Fig. 7.1e, curve $x(t)$ in $(t_2, t_d]$ has two different parts, namely the part in $(t_2, t_b]$ with a lower velocity $\dot{x}$ and the part in $(t_b, t_d]$ with a higher velocity $\dot{x}$. In the oscillogram $x(t_b) - x(t_2) = 5$ μm, and $x(t_d) - x(t_b) = 36$ μm. The first value is the reversible component which is the sum $-x(t_1) + x_\alpha$. The second value is of irreversible character and is equal to x_β. Here, as above, x_α and x_β denote respectively the reversible and irreversible displacements due to the continuous loading.

Another situation is observed for $F(t_1) < -F^*$. The value of the displacement in $(t_2, t_\alpha]$ is only 7 μm, i.e. it is approximately equal to $2x_\alpha$. This means that the reversible displacement is doubled whereas the irreversible displacement actually vanishes.

Comparing the results of processing oscillograms of Fig. 7.1c, d, e and f we can see that in the case of loading with both an intermediate stop and a reverse the irreversible displacement after the stop depends on force $F(t_1)$ at the stop. The value of the irreversible displacement is equal to x_β if $|F(t_1)| < F^*$ and is equal to zero if $|F(t_1)| > F^*$.

The reversible displacement must clearly increase under reverse and must double if $|F(t_1)| > F^*$. Thus, the doubled displacement mentioned in [71] is related to the irreversible component rather than to the total displacement.

The influence of the intermediate stop and the reverse on the small displacement before the initial break-down is determined above. Let us consider the case of the first break-down when, after the previous motion, the drive stops and then moves again until the first break-down, the normal load being permanently applied. According to the above definition, such a first break-down is not an initial one. Tests also show the influence of the stop and the reverse on the displacement. Thus, the values of the displacement before the first break-down can vary from zero up to values close to x_+, depending upon the tangential force acting on the contact during the previous stop.

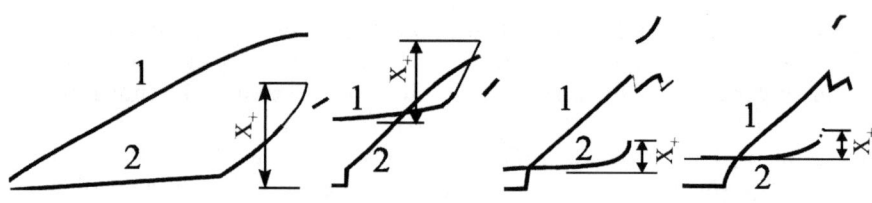

FIGURE 7.2.

7.1.4 Dependence of the total small displacement on the rate of tangential loading

Tests were carried out in which velocity v changed from 0,012 to 0,467 mm/s for each value of P. Fig. 7.2 shows four oscillograms recorded for $P = 840$ N, $c = 900$ N/mm, $v = 0,012, 0,047, 0,233, 0,467$ mm/s (from left to right). As one can see, with growth of v the total small displacement x_+ decreases which is basically the result of a decrease in the reversible component x_β.

It is clear that the tangential force F_y rather than the velocity v directly affects the frictional contact of the bodies. Hence, we can assume that x_+ does not depend on v but it depends on the rate of tangential loading f which is related to v as follows

$$f = \frac{dF}{dt} = cv - c\frac{dx}{dt} = cu, \qquad (7.5)$$

where u denotes the rate of displacement of the spring, Fig. 6.4.

In order to prove this assumption, tests were carried out for various values of c : $c_1 = 360$ N/mm, $c_2 = 900$ N/mm, $c_3 = 180$ N/mm. It was observed that the value of x_+ decreases with growth of c. In the case of $c_1 u_1 = c_2 u_2 = c_3 u_3$ which can be provided by a proper choice of velocity v the equality $x_{+1} = x_{+2} = x_{+3}$ holds.

Summarising the result obtained we can write $x_+(\alpha c, u) = x_+(c, \alpha u) = x_+(\alpha, cu)$, where α denotes an arbitrary positive constant. Assuming $\alpha = 1$ and taking into account eq. (7.1) we have

$$x_+(c, u) = x_+(1, cu) = \varphi(cu) = \varphi(f).$$

Thus, the above dependence of value x_+ on v and c reflects its dependence on the rate of loading f.

Figure 7.3 shows dependences $x_+(f)$ marked by 1, 2, 3 for $P = 560, 840$ and 1120 N respectively. The value of f is calculated by formula (7.5) where instead of dx/dt the mean value of this derivative in the irreversible part of the displacement is taken. The dependence of x_+ on f is considerable. For example, with growth of f from $3, 67$ N/s to 385 N/s ($P = 840$ N) the total small displacement x_+ decreases monotonically from 70 to 10 μm, i.e. it becomes seven times smaller.

7.1 Components of the small displacement 223

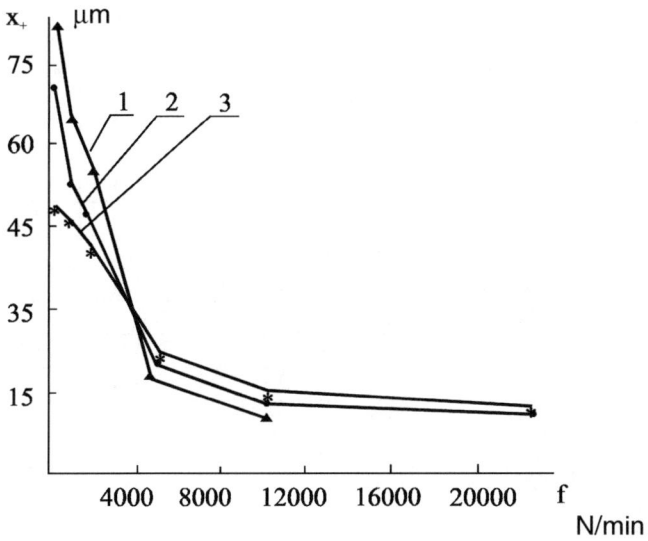

FIGURE 7.3.

7.1.5 Small displacement of parts of the contact

Six groups of observations corresponding to six values of force P were performed for each frictional pair. Each group consisted of 6-12 independent tests in which value of v varied from 0,005 to 0,233 mm/s.

Table 7.1 contains the results of observation of the total displacement $\tilde{x}_+$ in μm from the instant of set-up to the instant of break-down for $P =420$ N for the steel & steel pair. Figure 7.4 displays the oscillograms of the third test of the above group for the progressively increased velocity of the drive $v =0,005, 0,012, 0,023, 0,047, 0,113, 0,047, 0,113, 0,233$ mm/s. The time scale (interval s) is respectively equal to 4, 4, 4, 2, 2, 2, 0,2 s.

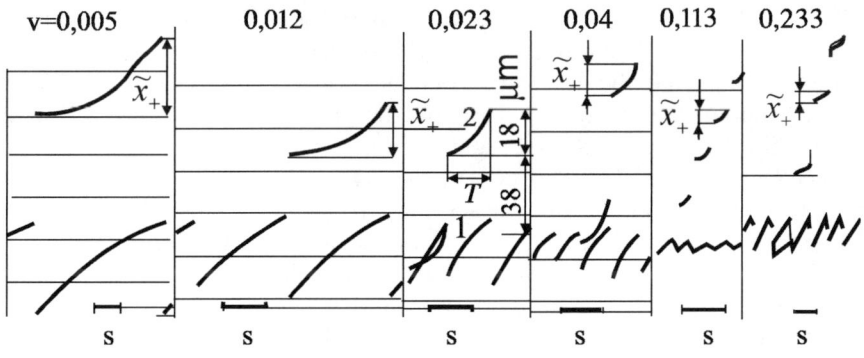

FIGURE 7.4.

v=0,005	v=0,012	v=0,023	v=0,047	v=0,113	v=0,233
23	20,20	16,16	12,12	5,4,3	3,2,3
22	20,18	16,14	10,8	4,4,5	3,3,3
36	22,22	16,20	16,16,14	6,4,6	3,2,3
32	20,20	20,18	16,12	5,5,5	3,2,3
20	16,10	10,10	8,6,10	5,4,4	3,2,3
22	18,20	16,18	12,12,12	4,4,6	2,2,3
25	24,24	18,18	12,20	5,6,5	4,4,2
20	10,24	18,10	12,10	4,4,5	3,3,3
21	20,20	14,14	12,8	5,5,5	3,3,3
18	10,6	18,8	12,12	4,5,5	2,3,3
22	20,23	20,18	12,12	5,5,4	3,3,3
20	20,20	18,16	12,10,18	5,5,5	3,3,3

TABLE 7.1

In order to find the contribution of the displacement $\tilde{x}_+$ to the total displacement, let us consider for example the third oscillogram in Fig. 7.4. Within period T the slider moves 56 μm from point 1 to point 2 relative to the guide. In this case $\tilde{x}_+ = 18$ μm, i.e. the contact displacement is about one third of the total relative motion. Hence, the very existence of this displacement is an important property of the frictional self-excited oscillations.

The oscillograms in Fig. 7.4 show that the total displacement is 36 μm at $v = 0,005$ mm/s and that it monotonically decreases with the growth of v and is equal to 2 μm at $v = 0,233$ mm/s (the figure with $v = 0,233$ has been scaled by a factor of 2,5 times compared to the others). Such a decrease in $\tilde{x}_+$ is observed in all tests, see Table 7.1. Hence, by analogy with the above, we conclude that the value of $\tilde{x}_+$ depends on the rate of tangential loading f and decreases with growth of this rate.

Let us recall that the descending dependence $F_+(f)$ was proved in Chapter 6. Comparing the latter with dependence $x_+(f)$ one can see a similar descending character. At first sight, one gets the impression that dependences $F_+(f)$ and $x_+(f)$ (or $\tilde{x}_+(f)$) reflect the same influence of the loading rate f on the process in the contact zone. Apparently, this was the reason for unifying curves $F_+(f)$ and $x_+(f)$ in the so-called frictional characteristic in [132]. However, this approach is not absolutely convincing. Indeed, for all tested metal frictional pairs the value of $F_+(f)$ depends only on rate f at the instant of break-down, but it does not change due to intermediate stops or reverse for which f becomes zero. At the same time value x_+ can essentially decrease as a result of such intermediate stops or reverse.

7.1.6 Comparing the values of small displacement with existing data

Summarising the results of investigations of the preliminary displacement carried out by other authors in [152], [31], [66], [71], [106] it is necessary to mention certain discrepancies. For instance, according to the data of [106] this displacement is about one micrometer. On the other hand, in [71] this displacement is reported to reach tens of micrometers.

The discrepancy becomes even more considerable if one compares the published results with the values of small displacement from Fig. 7.3 and Table 7.1. Indeed, for $f \simeq 0$ the total small displacement reaches 70 μm.

In order to clarify the reasons for the above discrepancy we compare the definitions of the preliminary displacement adopted in [152], [72], [106] with the definition of the small displacement utilised in the present tests. For example, the preliminary displacement is defined in [152] as the displacement before "visual sliding". It remains unknown what instant of time is considered as the moment when visual motion begins and if the breakdown occurs at this time instant. Judging from the oscillogram of Fig. 7.1a one can take both t_b and t_α. At these time instants the value of the preliminary displacement takes different values $x(t_b) = x_\alpha = 3$ μm and $x(t_b) = x_+ = 38$ μm. Thus, in order to have a rigorous definition of the preliminary displacement the term of "visual sliding" needs to be refined.

An exact definition is given in [72], [106]. In these papers the preliminary displacement is defined as the relative motion of the bodies for which sliding does not take place within the total area of the contact, that is, there exist zones of adhesion within the total area of the contact. This displacement results in the sliding (visual motion) at the instant when all adhesion zones vanish.

In the theory of the relaxation self-excited oscillations [48], [67], [76], [78]-[81], [121] it is accepted that the transition to sliding occurs at the instant of break-down whereas the transition to contact occurs at the instant of set-up. For this reason we consider the small displacement as a relative motion from the instant of set-up until the instant of break-down. The displacement in the oscillatory system defined in such a fashion is identified with the preliminary displacement studied in [72] and [106] only in the case when the break-down (accompanied with a jump) occurs at the instant when all adhesion zones vanish.

Two critical values of the tangential force F^* and F_+ were observed in the tests. When force F increases from zero up to F^* the displacement is only $2 \div 4$ μm, whilst in the case of a low-rate loading up to F_+ it can reach tens of micrometers. This suggests that probably the first part $(F < F_*)$ is characterised by sliding not within the total contact area (the preliminary displacement) whereas the second part $(F^* < F < F_+)$ corresponds to total sliding. The break-down for $F = F_+$ should be understood as a transition from one phase of the total sliding characterised by a low velocity to the second phase with a higher velocity of sliding. At the same time, as the loading rate f increases, the second critical value F_+ (the maximum friction) decreases and tends to the first critical value F^* whereas the total value of the small displacement x_+ tends to the total value of the preliminary displacement.

7.2 Remarks on friction between steel and polyamide

Test on steel & polyamide pairs were performed using setup III. Polyamide П54 and Rilsan were tested. The normal force P took the values 62−500 N, see Sections 6.1 and 6.2.

7.2.1 On critical values of the force of friction

Based upon experimental data we notice the following.

Firstly, the maximum friction F_+ decreases with the growth of v for the same duration of the contact. This confirms the existence of the dependence $F_+(f)$.

Secondly, for the same rate f the value of F_+ and thus the amplitude a of the relaxation self-excited oscillations is larger at the first break-down than at the others. Their values decrease progressively which is explained by the presence of dependence $F_+(T)$ [48]. Thus, if the time-dependence is insignificant for the metal pairs this dependence is considerable for steel & polyamide pairs.

Thirdly, similar to the case of the metal pairs, the small displacement before the initial break-down is also split into linear and nonlinear parts. However, the transition from the first part to the second one takes place under the tangential force F^* which is smaller that the arithmetic average of the values F_- and F_+, that is $F^* < (F_- + F_+)/2$.

Fourthly, a time lag of the small displacement with respect to the tangential force F is observed which causes so-called frictional creep, see [90] and [91].

Among these facts only the second and fourth differ from the results of tests on the metal pairs. Moreover, the phenomenon of time lag has not yet been reported elsewhere. To this end, an analysis of this phenomenon is given in what follows.

7.2.2 Time lag of small displacement

The question is stated as follows: What is the time-dependence of the elastic tangential force $F(t)$ and small displacement $x(t)$ if the drive is switched off when $F(t) < F_+$. It is clear from the data of Chapter 6 that for metal

7.2 Remarks on friction between steel and polyamide

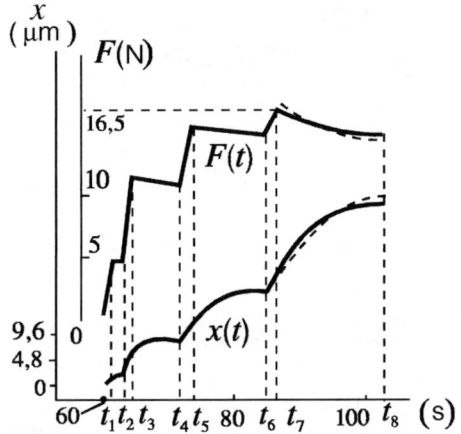

FIGURE 7.5.

pairs the small displacement $x(t)$ is practically absent, i.e. values of x and F remain practically constant.

Let us consider now the case of the steel & polyamide pair. The experiments were carried out for two regimes of loading: under initial loading (see eqs. (7.1) and (7.2)) with intermediate stop of the drive and for the relaxation self-excited oscillations with intermediate stops of the drive within the phase of the contact.

The first regime of loading was utilised as follows

$$P = \begin{cases} 0 & t < 0 \\ \text{const} & t \geq 0 \end{cases}, \quad \dot{\xi} = \begin{cases} 0 & t_{2k-1} < t \leq t_{2k} \\ v & t_{2k} < t \leq t_{2k+1} \end{cases},$$
$$k = 0, 1, 2, ..., n, \quad t_{-1} = 0, \quad F(0) = 0, \quad F(t_{2n+1}) = F_+. \quad (7.6)$$

This condition indicates that the drive is switched on at instant t_0 and switched off at t_1, then it is switched on again at t_2 and so on. The process continues until the value of F reaches the maximum F_+ for which breakdown occurs. According to this regime we carried out five series of tests for five nominal values of the normal force P =19,6, 49, 98, 147, 196 N with nominal pressure p =0,25, 0,624, 1,25, 1,87, 2,50 kPa for the low-alloyed carbon steel & Pilsan and low-alloyed carbon steel & П54 pairs. Each series consisted of six tests with various values of the velocity. Ten measurements were performed for each test. It is worthwhile mentioning that the regularities listed below were observed in all measurements.

As an example let us consider oscillogram of Fig. 7.5 for the steel & Rilsan pair for $P = 98$ N, v =0,114 mm/s. For $i = 0 \div 8$ t_i = we have 60, 60,4, 62, 62,5, 71,4, 71,7, 84, 84,2, 101,4 N, F^* =7 N and F_+ =18 N. As one can see from the oscillogram the first stop $t_1 < t \leq t_2$ corresponds to $F(t_1) = 5$ N$< F^*$. In this case coordinate x and force F do not change, i.e. $x(t) = x(t_1) = 1,5$ μm and $F(t) = F(t_1) = 5$ N for $t_1 < t < t_2$.

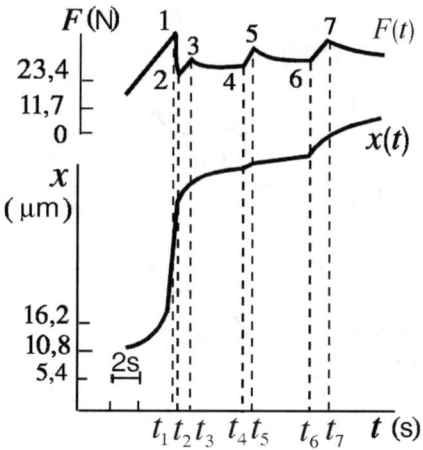

FIGURE 7.6.

The second, third and fourth stops $t_3 < t \leq t_4$, $t_5 < t \leq t_6$, $t_7 < t \leq t_8$ took place under condition $F^* < F(t_3) = 11,5 < F(t_5) = 15,5 < F(t_6) = 17,5 < F_+$. In this case, the value of x at each stop increases and tends to the corresponding limit and this results in a decrease in force F to the limiting value F_∞. Moreover, the increase in x and decrease in F at the $k-th$ stop are proportional to the difference $F(t_{2k-1}) - F^*$. For example, for the second stop we have $F(t_3) - F^* = 4,5$ N and changes in x and F are $x(t_4) - x(t_3) = 2$ μm, $F(t_3) - F(t_4) = 0,2$N. For the third stop we have $F(t_5) - F^* = 8,5$ N, $x(t_6) - x(t_5) = 10$ μm, $F(t_5) - F(t_6) = 1$N.

Let us notice that velocity $\dot{x}$ decreases monotonically to zero during the stop of the drive. For instance, in $t_7 < t \leq t_8$ it changes according to the following table

t, s	84,2÷85	85÷87	87÷89	89÷91	91÷93	93÷95	95÷97
$v, \dfrac{\mu m}{s}$	8,7	5,5	2,5	1,5	0,6	0,4	0,25

where v denotes the mean value of $\dot{x}$ in the given time interval.

Let us now proceed to consider the lag in the displacement under relaxation self-excited oscillations. The oscillogram in Fig. 7.6 was recorded in this regime of motion for $P = 796$ N, $v = 0,114$ mm/s. The numbers on curve $F(t)$ denote the following: 1 is the point of break-down, 2 is the point of set-up, points 3, 5 and 7 are the points of switch-off of motion of the slider and 4 and 6 are points of switch-on of motion of the slider. One can see from plots $F(t)$ and $x(t)$ that the small displacement continues at each stop of the drive, i.e. a lag takes place. The greater force F, the more intensive the small displacement $x(t)$ at the stop.

7.2.3 Immovable and viscous components of the force of friction

Let us turn our attention to the limiting values F_∞, which the shifting elastic force F tends to in the case of stop of the drive. The resisting force of one body on the other, i.e. the force of friction, has the value of F_∞ when the rate of displacement $\dot{x}$ (the rate of creep) is zero. For this reason, value F_∞ can be referred to as the force of static friction, This definition however differs from the conventional concept of the incomplete force of static friction. The incomplete force of static friction is often considered as a force of resistance to the preliminary displacement and its value is taken to be equal to the value of the shifting force F, the force of inertia $m\ddot{x}$ being neglected. However for the cases in which the phenomenon of creep is not observable, for example for the steel & steel pair, values F_∞ and F are practically coincident.

Following the hypothesis on the presence of viscous friction on the contact (Kragelsky's hypothesis [71]) we assume that the small displacement and in particular the frictional creep is caused by three forces: the shifting force F, the force of static friction F_∞ and the force of viscous friction Φ. Neglecting the force of inertia we can present force F as a sum of immovable F_∞ and viscous Φ components of friction.

Let us assume that the equalities $F_{\infty k} = \text{const}$, $\Phi = h_k \dot{x}$, $h_k = \text{const}$ hold for the $k - th$ stop, i.e. for $t_{2k-1} < t < t_{2k}$. Then we can set the differential equation of creep in the form

$$h_k \dot{x} + c(x - x_{2k-1}) + F_{\infty k} - F_{2k-1} = 0.$$

Here subscript $2k-1$ indicates that the corresponding values are determined at time instant t_{2k-1}. The particular solution of this equation is given by

$$x = x_{2k-1} + \frac{F_{2k-1} - F_{\infty k}}{c} \left\{ 1 - \exp\left[-\frac{c}{h_k}(t - t_{2k-1})\right] \right\}, \quad t_{2k-1} < t < t_{2k}. \tag{7.7}$$

Taking into account equality $F(t) = F_{2k-1} - c(x - x_{2k-1})$ we can write for this time interval that

$$F(t) = F_k + (F_{2k-1} - F_k) \exp\left[-\frac{c}{h_k}(t - t_{2k-1})\right], \quad t_{2k-1} < t < t_{2k}. \tag{7.8}$$

The coefficients for the fourth stop ($k = 4$) are $F_{\infty 4} = 14{,}6$ N, $h_4 = 397$ kNs/m, $x_{2k-1} = x_7 = 18 \cdot 10^{-3}$ m, $F_{2k-1} = F_{\infty 7} = 16{,}5$ N/m and were determined with the help of the oscillogram of Fig. 7.5. By means of eqs. (7.7) and (7.8) and the obtained coefficients we calculated $x(t)$ and $F(t)$ which are presented by the dashed lines in Fig. 7.5. As one can see, these lines practically coincide with the experimental curves $x(t)$ and $F(t)$. Similar coincidences take place for other stops, for example, in $(t_3, t_4]$ and $(t_5, t_6]$.

Thus, the phenomenon of creep in the phase of small displacement can be explained by means of the hypothesis of the presence of viscous friction in the contact.

7.3 Conclusions

1. Small displacement in the self-excited system is considered as a relative motion of the bodies unless the shifting force reaches maximum friction which causes break-down.

2. The value of the total small displacement $x_+(\tilde{x}_\alpha)$ is split into a reversible component $x_+(\tilde{x}_\alpha)$ and an irreversible component x_β. The reversible displacement transforms into the irreversible one at the instant when the tangential force is approximately equal to the arithmetical average $F^* = (F_+ + F_-)/2$.

3. When the rate of tangential loading f is low, the value of $x_+(\tilde{x}_\alpha)$ is much less than x_β. As f increases, x_β and in turn x_+ decrease.

4. In the case of reverse, the reversible displacement increases and if $|F(t_1)| > F^*$ (t_1 denotes the instant of the reverse) it is doubled, i.e. it becomes equal to $2x_\alpha$.

5. Provided that the loading stops during the irreversible displacement, then the displacement stops as well and under further loading to the break-down the bodies remain practically immovable relative to each other.

6. It is possible to make an assumption that for $F < F^*$ the sliding on the contact area is not global, i.e. the sliding is local. If $F^* < F < F_+$ then the sliding is spread over the total area of the contact.

The above conclusions are valid for the tested metal pairs of friction.

7. For the steel & polyamide pairs $F^* < (F_+ + F_-)/2$.

8. For these pairs when the drive stops at position where $F > F^*$ the small displacement $x(t)$ continues and its rate $\dot{x}$ tends to zero. This can be explained by a hypothesis on viscous friction at the contact [71].

8
Frictional self-excited oscillations

As explained in Chapter 6 the maximum friction F_+ depends on the rate of tangential loading f at the break-down and decreases as this rate increases. The existence of this dependence causes a jump of the force of friction $F_+ - F_c$ at the instant of break-down and thus results in a hard excitation of the frictional self-excited oscillations. Besides, a "traditional" falling characteristic of the force of sliding friction F_c as a function of the sliding velocity $\dot{x}$ leads to a soft excitation of the system. The present Chapter is concerned with the analysis of frictional self-excited oscillations by means of the two dependences $F_+(f)$ and $F_c(\dot{x})$.

8.1 Self-excited oscillations due to hard excitation

Under certain conditions, for example for good lubrication, the falling part of curve $F_c(\dot{x})$ does not exist, see [71], [73], [132]. In this case self-excited oscillations appear only due to the difference between F_+ and F_c at the instant of break-down. An analysis of the oscillations under this assumption is in what follows.

8.1.1 The case of no structural damping

Let the oscillatory system consist of rod 1 of the drive, spring 2 of rigidity c, slider 3 of mass v and guide 4, see Fig. 8.1. The rod is assumed to move

8. Frictional self-excited oscillations

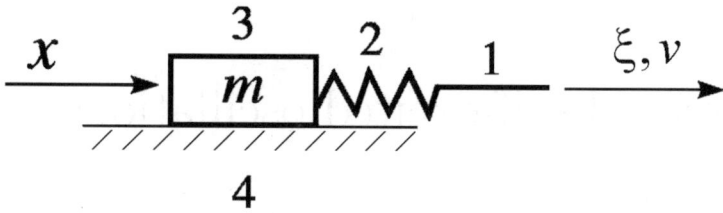

FIGURE 8.1.

according to the law

$$\xi = vt + \xi_0, \qquad (8.1)$$

where velocity v is constant. The value of the frictional force at the break-down is taken according to eq. (6.20), i.e.

$$F_+(f) = Ae^{-\beta f} + F_0 = Ae^{-\beta cv} + F_0, \qquad (8.2)$$

where $f = dF/dt$ at the break-down. The force of sliding friction is assumed to be constant for any $\dot{x}$ and equal to F_0

$$F_c = F_0. \qquad (8.3)$$

The elastic force is as follows

$$F = c(\xi - x) = c(vt + \xi_0 - x). \qquad (8.4)$$

Let us notice that account for the preliminary displacement causes no principle difficulties but it complicates the description. For this reason, the slider is assumed to be immovable from the instant of set-up until the instant of break-down, that is, it is assumed to be at rest. The dynamic equation for the system has the form

$$\left. \begin{array}{ll} m\ddot{x} = F - F_c & \text{when sliding}, \\ x = \text{const} & \text{at rest}. \end{array} \right\} \qquad (8.5)$$

Let the break-down occur at $t = 0$, then inserting eqs. (8.3) and (8.4) into the first equation in (8.5) yields

$$\ddot{x} + \omega^2(x - vt + F_0/c - \xi_0) = 0, \quad 0 < t \leq T_-, \qquad (8.6)$$

where $\omega = \sqrt{c/m}$, T_- denotes the duration of the phase of sliding. The solution of the differential equation (8.6) can be set in the form

$$x = \xi_0 + vt - F_0/c + a\sin(\omega t + \varphi), \quad 0 < t \leq T_-, \qquad (8.7)$$

where a is a constant amplitude which is equal to the amplitude at the instant of break-down. Condition $0 < t \leq T_-$ shows that expressions (8.6) and (8.7) describe the motion only during the sliding phase.

8.1 Self-excited oscillations due to hard excitation   233

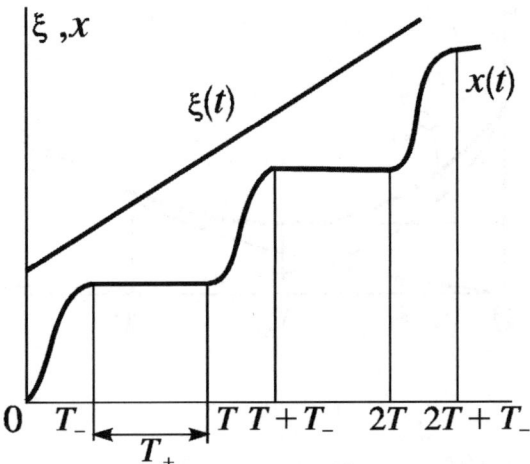

FIGURE 8.2.

Equations (8.5) and (8.7) yield the following equation of motion of the system within the whole period T of the relaxation oscillation

$$x = \begin{cases} \xi_0 + vt - F_0/c + a\sin(\omega t + \varphi) & 0 < t \leq T_-, \\ \xi_0 + vT_- - F_0/c + a\sin(\omega T_- + \varphi) & T_- < t \leq T = T_- + T_+. \end{cases} \quad (8.8)$$

Curves $\xi(t)$ and $x(t)$ are plotted in Fig. 8.2.

Let us determine the amplitude a, phase φ and period T of the relaxation self-excited oscillation. At the break-down the elastic force F is equal to the force of the break-down F_+. At the initial instant of time (immediately after the break-down) velocity $\dot{x}$ is equal to zero, whereas acceleration $\ddot{x}$ must compensate for the forces $(F_+ - F_0)$ acting on the slider. Taking into account eqs. (8.2), (8.4) and (8.8) we obtain

$$\dot{x}(t=0) = v + a\omega\cos\varphi = 0,$$

$$\ddot{x}(t=0) = -a\omega^2\sin\varphi = F_+ - F_0 = \frac{A\exp(\beta cv)}{m}.$$

This yields the formulae

$$a^2 = \frac{v^2}{\omega^2} + \frac{A\exp(\beta cv)}{c^2},$$

$$\varphi = -\pi + \arctan\frac{\omega A\exp(\beta cv)}{cv} = -\pi + \arccos v/\omega a. \quad (8.9)$$

Dependence $a^2(v)$ is plotted in Fig. 8.3.

Taking the derivative of $a^2(v)$ with respect to v we obtain

$$\frac{da^2}{dv} = \frac{2v}{\omega^2} - \frac{2\beta^2 A\exp(-2\beta cv)}{c}.$$

234   8. Frictional self-excited oscillations

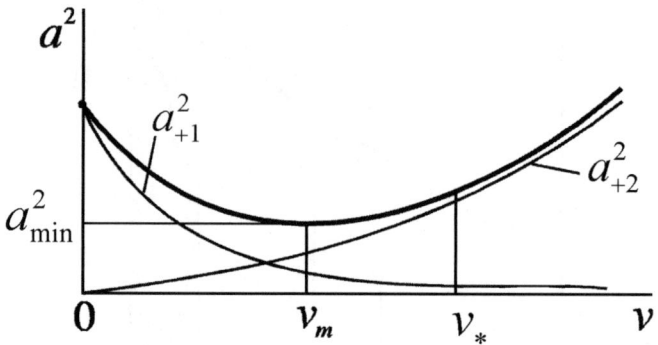

FIGURE 8.3.

Denoting the value of v for which $a = a_{\min}$ by v_m we have

$$v = v_m : \quad \frac{da^2}{dv} = 0$$

or

$$mv_m - \beta A^2 \exp(-2\beta c v_m) = 0. \tag{8.10}$$

Solving eq. (8.10) we can find v_m, substituting the result into the first expression in eq. (8.9) we obtain the minimum amplitude $a_{\min}$.

Using the notation

$$a_{+1} = A\exp(-\beta cv)/c, \quad a_{+2} = v/\omega, \tag{8.11}$$

we have

$$a^2 = a_{+1}^2 + a_{+2}^2.$$

It is easy to see from eqs. (8.9) and (8.11) that

$$a = \begin{cases} a_{+1} = A\exp(-\beta cv)/c, & \text{for low velocity } v, \\ a_{+2} = v/\omega, & \text{for high velocity } v. \end{cases} \tag{8.12}$$

When velocity v is small, the self-excited oscillations have an explicit relaxation character. With growth of v the contact part decreases whilst the sliding part increases. For sufficiently large v the shape of the oscillation is close to sinusoidal. An appropriate name for such a relaxation oscillation is sinusoidal self-excited oscillation with a short stop.

The period of self-excited oscillation is the sum of time intervals T_- and T_+ of the sliding part and rest, i.e. $T = T_- + T_+$. At the moment of set-up

$$\dot{x}(T_-) = v + a\omega\cos(\omega T_- + \varphi) = 0,$$
$$\ddot{x}(T_-) = -a\omega^2 \sin(\omega T_- + \varphi) < 0.$$

8.1 Self-excited oscillations due to hard excitation

Taking into account eq. (8.9) we obtain the following expression for T_-

$$T_- = \frac{2}{\omega}\left(\pi - \arccos\frac{v}{a\omega}\right) = \frac{2}{\omega}\left(\pi - \arctan\frac{\omega A\exp(-\beta cv)}{cv}\right). \quad (8.13)$$

At the time instant of the second break-down (at $t = T_- + T_+$) coordinate x of the slider is equal to the value $x(T_-)$ at the set-up

$$x(T_- + T_+) = x(T_-) = \xi_0 - \frac{F_0}{c} + vT_- + a\sin(\omega T_- + \varphi). \quad (8.14)$$

The elastic force is equal to the force of break-down

$$F = c[(T_- + T_+)v + \xi_0 - x(T_-)] = F_+ = F_0 + A\exp(-\beta cv). \quad (8.15)$$

Taking into account eqs. (8.9) and (8.13) we obtain from eqs. (8.14) and (8.15) the following expression for T_+

$$T_+ = \frac{2A\exp(-\beta cv)}{cv}. \quad (8.16)$$

Using eqs. (8.13) and (8.16) we obtain the following formula for T

$$\begin{aligned}T = T_- + T_+ &= \frac{2}{\omega}\left(\pi - \arccos\frac{v}{a\omega}\right) + \frac{2A}{cv}\exp(-\beta cv) \\ &= \frac{2}{\omega}\left(\pi - \arctan\frac{\omega A}{cv}\exp(-\beta cv)\right) + \frac{2A}{cv}\exp(-\beta cv).\end{aligned} \quad (8.17)$$

A graph of the dependence $T(v)$ is presented in Fig. 8.4. Taking into account condition (8.12) we can use expression (8.17) to determine the following one

$$T = \begin{cases} \dfrac{2A}{cv}\exp(-\beta cv) + \dfrac{\pi}{\omega}, & \text{for low velocity } v, \\ \dfrac{2\pi}{\omega}, & \text{for high velocity } v. \end{cases} \quad (8.18)$$

The graph in Fig. 8.4 and the formulae in eq. (8.18) show that period T decreases and tends to $2\pi/\omega$ as v increases.

Let us plot the relaxation self-excited oscillation in the phase plane. To this end, we introduce the notation

$$X = \frac{\dot{x}}{\omega}, \quad Y = \frac{\ddot{x}}{\omega^2}. \quad (8.19)$$

It follows from (8.8) that

$$X = \frac{x}{\omega} + a\cos(\omega t + \varphi), \quad (8.20)$$

236   8. Frictional self-excited oscillations

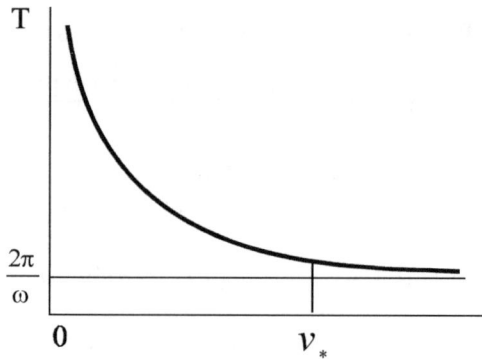

FIGURE 8.4.

$$Y = -a\sin(\omega t + \varphi). \tag{8.21}$$

In phase space shown in Fig. 8.5, the characteristic of the force of breakdown

$$Y = \frac{F_+ - F_0}{m\omega^2} = \frac{A}{m\omega^2}\exp(-\beta c\dot{x}) = \frac{A}{m\omega^2}\exp(-\beta c\omega X) \tag{8.22}$$

is presented by curve 1. The sliding part described by eqs. (8.20) and (8.21) is depicted by arch CDE of the circle with the centre at point $G(v/\omega, 0)$ whereas the contact part is represented by line EC. Let us notice that when the small displacement is neglected, the terms "contact" and "rest" are identical.

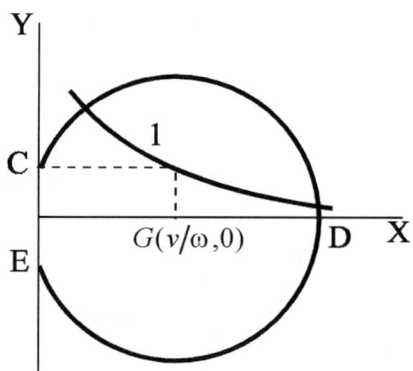

FIGURE 8.5.

Qualitative coincidence of the theoretical results with the experimental data is seen from Figs. 8.3 and 8.4 and the oscillogram of Fig. 8.6, the latter being recorded for a steel & cast iron pair in the case of decreasing

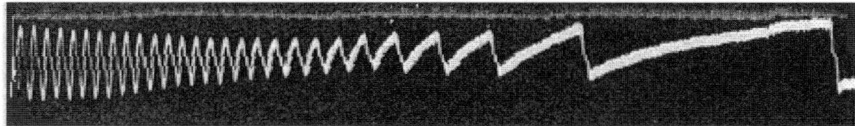

FIGURE 8.6.

v. Judging from the oscillogram, the amplitude of oscillation decreases to the minimum value $a_{\min}$ and then increases as v decreases. The period T increases and tends to infinity when v tends to zero. One can also see that the contact parts are considerable for low velocities v, that is, the oscillations have a clear relaxation character. At high velocities v the shape of the oscillation is very close to being sinusoidal, i.e. the relaxation oscillation transforms into sinusoidal oscillations with a short stop.

The stationary motion $\dot{x} = v, \ddot{x} = 0$ is stable. According to eq. (8.8) under the initial condition

$$a|_{t=0} < \frac{v}{\omega}$$

velocity $\dot{x}$ never vanishes. The motion has no stops and is not accompanied by self-excited oscillations. In the case when

$$a|_{t=0} \geq \frac{v}{\omega}$$

the condition for a stop $\dot{x} = 0$ is fulfilled at a certain time instant which converts the system into the regime of relaxation self-excited oscillations. In other words, the systems has a property of the hard excitation of oscillations.

8.1.2 Including damping

The results of the analysis of the system without structural damping show that if $v > v_m$ then the amplitude of the self-excited oscillation with a short stop continuously increases with the growth of v. However, due to structural damping the relaxation self-excited oscillations can not be observed for an arbitrary high velocity v. There exists a critical velocity v_*, see Fig. 8.3, which causes transformation of the discontinuous motion into sliding without stops. Hence, as damping increases, one can expect that for $v > v_m$ the amplitude of oscillation increases with the growth of v. However this does not increase to infinity, but is bounded by the value

$$a = \left(\frac{v_*^2}{\omega^2} + \frac{A^2}{c^2} \exp(-2\beta c v_*) \right)^{1/2}. \tag{8.23}$$

Thus, taking account for damping is reduced to determining the critical value v_*, Fig. 8.4. The damping force can be set in the form

$$F_d = \delta(\dot{x} - v), \tag{8.24}$$

where $\dot{x} - v$ denotes the rate of change of the length of the spring and δ is a constant factor. Instead of eq. (8.5) we have

$$\left.\begin{array}{l}\ddot{x} + \lambda(\dot{x} - v) + \omega^2(x - vt + F_0 c^{-1} - \xi_0) = 0, \quad \dot{x} > 0, \\ x = \text{const}, \hfill \dot{x} = 0,\end{array}\right\} \quad (8.25)$$

where $\lambda = \delta/m$. The solution of system (8.27) can be found in the form

$$x = \begin{cases} \xi - F_0 c^{-1} + vt + a(t)\sin(\omega t + \varphi), & t \leq T_-, \\ \xi - F_0 c^{-1} + vT_- + a(T_-)\sin(\omega T_- + \varphi), & T_- \leq t \leq T_- + T_+, \end{cases} \quad (8.26)$$

where

$$a(t) = a_0 \exp(-\lambda t/2)$$

and a_0 denotes the initial value of the amplitude.

In the regime of relaxation self-excited oscillations, a change in amplitude a is negligibly small within each separate phase of sliding. Hence, this amplitude can be taken as being constant and equal to amplitude $a(0)$ at the break-down which is given by formula (8.9). Equation (8.26) then reduces to eq. (8.8). In other words, in the case of structural damping motion with periodic stops can be described by the same expression (8.8) which is used in the case of no damping.

Let us consider now the regime of sliding without periodic stops. In this case $T_+ = 0$, $T_- = \infty$ and the motion of the slider obeys the following law

$$x = \xi - F_0 c^{-1} + vt + a_0 \exp(-\lambda t/2)\sin(\omega t + \varphi). \quad (8.27)$$

Let the slider be at rest at the initial time instant, i.e. $\dot{x}(0) = 0$, and the velocity of rod 1 be $v = \text{const}$. Then

$$a_0 = \left(\frac{v^2}{\omega^2} + \frac{A^2}{c^2}\exp(-2\beta cv)\right)^{1/2}. \quad (8.28)$$

Hence,

$$a_0 = a_0 \exp(-\lambda t/2) = \left(\frac{v^2}{\omega^2} + \frac{A^2}{c^2}\exp(-2\beta cv)\right)^{1/2} \exp(-\lambda t/2). \quad (8.29)$$

By using eq. (8.29) we can see that the difference

$$a_0^2 - v^2\omega^{-2} = A^2 c^{-2} \exp(-2\beta cv) \quad (8.30)$$

decreases and tends to zero as v increases. Hence, for a sufficiently high velocity v the amplitude $a(t)$ is smaller than $v^2\omega^{-2}$ after the break-down due to the damping. But then

$$\dot{x} = v + a\omega\cos(\omega t + \varphi) > 0 \quad \text{for} \quad t > 0, \quad (8.31)$$

8.1 Self-excited oscillations due to hard excitation

i.e. the velocity of the slider after the break-down is always positive and thus the system executes sliding without stops.

When the velocity of driving v decreases, difference (8.30) increases. In the limit, when v reaches the critical value v_*, the velocity $\dot{x}$ and acceleration $\ddot{x}$ may simultaneously vanish at a certain time instant

$$\dot{x}(t_*) = v_* + a(t_*)\omega \cos(\omega t_* + \varphi) = 0,$$
$$\ddot{x}(t_*) = -a(t_*)\omega^2 \sin(\omega t_* + \varphi) = 0 \qquad (8.32)$$

which causes the slider to stop and transition from the regime of decaying free oscillations to the regime of relaxation self-excited oscillations. Condition (8.32) holds only for $\sin(\omega t_* + \varphi) = 0$. Then

$$a_* = a(t_*) = v_*\omega^{-1}, \quad t_* = [\pi - \varphi(v_*)]\omega^{-1}. \qquad (8.33)$$

Since t_* and da/dt are small, the increment of the amplitude $\Delta a = a_0 - a_*$ is also small. Equations (8.30) and (8.33) yield

$$A^2 c^{-2} \exp(-2\beta c v_*) \approx 2a(a_0 - a_*) = -2a(\widetilde{\frac{da}{dt}})t_*, \qquad (8.34)$$

where $(\widetilde{da/dt})$ denotes the mean value of da/dt in $(0, t_*)$. Now one can see that value $c^{-1}A\exp(-\beta c v_*)$ is also small. This enables us to obtain φ_* and t_* from eqs. (8.9) and (8.33)

$$\varphi_* = -\pi, \quad t_* = 2\pi/\omega. \qquad (8.35)$$

Substituting expressions (8.33) and (8.35) for a and t into eq. (8.29) yields

$$\left(\frac{v_*^2}{\omega^2} + \frac{A^2}{c^2}\exp(-2\beta c v_*)\right)\exp\left(-\frac{2\pi\lambda}{\omega}\right) = \frac{v_*^2}{\omega^2}, \qquad (8.36)$$

thus

$$v_* = \frac{\omega A \exp(-2\beta c v_* - \lambda\pi/\omega)}{c\left(1 - \exp(-2\lambda\pi/\omega)\right)^{1/2}}. \qquad (8.37)$$

Solving eq. (8.36) or eq. (8.37) we can obtain the critical velocity v_*. In the particular case $\lambda = 0$ value v_* approaches infinity.

The mechanical system considered here is a system with a hard excitation of oscillations. For example, sliding without stops can be transformed into sliding with relaxation self-excited oscillations by stopping the slider, i.e. by putting $\dot{x} = 0$. Such an excitation is, in particular, an initial motion from the state of the rest, i.e. $\dot{x}(0) = 0$, $x(0) = \xi_0 - F_+ c^{-1}$. In the case of structural damping a hard excitation is possible only for $v < v_*$.

As shown in Chapter 6 the decrease in force F_+ with a growth in velocity v leads to two different assumptions: dependence $F_+(T_+)$ and dependence

240   8. Frictional self-excited oscillations

$F_+(f)$. This gives the impression that both dependences lead to the same result. But comparing the above said with the calculations made on $F_+(T_+)$, [47], [48], one can see that the force of break-down F_+ and the properties of self-excited oscillations differ from each other. The difference is as follows.

Firstly, by virtue of the ascending dependence of F_+ on the duration of the contact T_+, the amplitude a at the first break-down should increase monotonically with the growth of velocity v and should be larger than the amplitude at successive break-downs. The oscillations after the first break-down attenuate and tend to a limiting cycle. On the other hand, calculation using dependence $F_+(f)$, eq. (8.2), shows that the amplitude of self-excitation and, in particular, the amplitude at the first break-down first decreases and then increases with growth of v.

Secondly, taking into account an ascending dependence $F_+(T_+)$ without structural damping indicates that the stationary self-excited oscillations are observed only at velocities v under the critical value v_k. However, due to $F_+(f)$ such a critical velocity exists only due to the structural damping.

8.2 Self-excited oscillations under both hard and soft excitations

In this section the frictional self-excited oscillations are analysed with account for dependences $F_+(f)$ and $F_c(\dot{x})$.

8.2.1 Equations of motion

The system under consideration is depicted in Fig. 8.1. The motion of rod 1 of this system is given by eq. (8.1). The force of break-down F_+ is presented in the form of eq. (8.2) and the dependence of the force of sliding F_c on velocity $\dot{x}$ is approximated by the following third order polynomial

$$F_c = F_0 = -\alpha \dot{x} + \frac{\alpha}{3\gamma^2}\dot{x}^3, \qquad (8.38)$$

where α denotes a positive constant, γ is the length of the falling part of the curve $F_c(\dot{x})$, i.e. that value of $\dot{x}$ for which $F_c(\gamma) = F_{c\min}$. The force of structural damping F_d is assumed to obey eq. (8.24).

First we consider motion of the slider on the guide without periodic stops. In this case, substituting expressions (8.4), (8.38) and (8.24) for forces F, F_c, F_d into the equation of dynamics

$$m\ddot{x} = F - F_c - F_d$$

we arrive at the following differential equation of motion

$$\ddot{x} + \lambda(\dot{x} - v) - \varepsilon \dot{x} + \frac{\varepsilon}{3\gamma^2}\dot{x}^3 + \omega^2\left(x - vt - \xi_0 + \frac{F_0}{c}\right) = 0, \qquad (8.39)$$

8.2 Self-excited oscillations under both hard and soft excitations

where $\lambda = \delta/m$, $\varepsilon = \alpha/m$, $\omega = \sqrt{c/m}$.

Denoting

$$y = x - vt - \xi_0 + F_0 c^{-1} \tag{8.40}$$

we can write eq. (8.39) in the form

$$\ddot{y} + \omega^2 y = \varepsilon v - (\lambda - \varepsilon)\dot{y} - \frac{\varepsilon}{3\gamma^2}(\dot{y} + v)^3. \tag{8.41}$$

Applying asymptotic methods [16] we obtain the first approximation to the solution of eq. (8.4) in the form

$$y = a \sin(\omega t + \Psi), \tag{8.42}$$

$$\frac{da}{dt} = \left(\frac{\varepsilon - \lambda}{2} - \frac{\varepsilon v^2}{2\gamma^2}\right) a - \frac{\varepsilon \omega^2 a^3}{8\gamma^2}, \tag{8.43}$$

where the small aperiodic component is neglected. Integrating eq. (8.6) and taking into account eq. (8.3) yields the following equation for the motion of the slider

$$x = \xi_0 - F_0 c^{-1} + vt + a \sin(\omega t + \Psi), \tag{8.44}$$

where

$$a^2 = \frac{4a_0^2[(\varepsilon - \lambda)\gamma^2 - \varepsilon v^2]\exp\left(\frac{\gamma^2 - v^2}{\gamma^2}\varepsilon t - \lambda t\right)}{4(\varepsilon - \lambda)\gamma^2 - 4\varepsilon v^2 + \varepsilon a_0^2 \omega^2 \left[\exp\left(\frac{\gamma^2 - v^2}{\gamma^2}\varepsilon t - \lambda t\right) - 1\right]} \tag{8.45}$$

and a_0 is the value of a at $t = 0$. In the latter equation the small aperiodic component is also neglected. Equation (8.6) has two stationary solutions

$$a = 0, \qquad \text{for any } v,$$

$$a = a_1 = \frac{2}{\omega}\left(\gamma^2 - \lambda\frac{\gamma^2}{\varepsilon} - v^2\right)^{1/2} = \frac{2}{\omega}(v_H^2 - v^2)^{1/2}, \quad \text{for } v < v_H, \tag{8.46}$$

where $v_H = (1 - \lambda/\varepsilon)^{1/2}\gamma$. The first stationary solution ($a = 0$) describes the equilibrium $\dot{y} = \ddot{y} = 0$ which is stable if $v \geq v_H$. If $0 < v < v_H$ the solution is unstable, i.e. for any initial perturbation $a_0 \neq 0$ the oscillation amplitude increases according to law (8.8). The second solution ($a = a_1$) represents a stable self-excited oscillation

$$y_1 = \frac{2}{\omega}(v_H^2 - v^2)^{1/2}\sin(\omega t + \Psi). \tag{8.47}$$

8. Frictional self-excited oscillations

Thus, motion of the slider without periodic stops can be accompanied by a stable self-excited oscillation which, in the limit, transforms to stationary quasi-harmonic auto-oscillation with amplitude a_1.

Let us proceed to the case in which the slider moves with periodic stops. The auto-oscillations are of the relaxation type, the auto-oscillatory cycle consists of two parts, the rest part and the sliding part. If $t = 0$ corresponds to the instant of break-down, then we have the following equation of motion for the slider

$$x(t) = \begin{cases} \xi_0 - F_0 c^{-1} + vt + a(t)\sin(\omega t + \Psi), & 0 \leq t \leq T_-, \\ \xi_0 - F_0 c^{-1} + vT_- + a(T_-)\sin(\omega T_- + \Psi), & T_- \leq t \leq T_- + T_+, \end{cases} \tag{8.48}$$

where T_- and T_+ denote the duration of sliding and rest respectively. Within the sliding part the change in amplitude a is negligibly small. Hence the amplitude can be taken as being constant and equal to the amplitude a_+ at the break-down, i.e.

$$a(t) = a_0 = a_+. \tag{8.49}$$

At the moment of break-down, velocity $\dot{x}$ is equal to zero whereas acceleration $\ddot{x}$ must compensate the jump in the force of friction $(F_+ - F_-)$. Then it follows from eqs. (8.11) and (8.12) that

$$\dot{x}(0) = v + a_0\cos\Psi = 0, \quad m\ddot{x}(0) = -ma_0\omega^2\sin\Psi = A\exp(-\beta cv).$$

Resolving these relationships we obtain the following expressions for amplitude a_0 and phase Ψ

$$a_0 = \left(\frac{v^2}{\omega^2} + \frac{A^2}{c^2}\exp(-2\beta cv)\right)^{1/2} = a_+,$$

$$\Psi = -\pi + \arctan\frac{\omega A\exp(-\beta cv)}{cv} = -\pi + \arccos\frac{v}{a_0\omega}. \tag{8.50}$$

The above analysis shows that two forms of stationary auto-oscillation are possible in the elastic system with friction, namely, quasi-harmonic auto-oscillations with amplitude a_1 and relaxation quasi-harmonic auto-oscillations with amplitude a_+. The motion of the slider is described by eqs. (8.7) and (8.11) respectively. The region for the existence of these oscillations and the conditions governing their form are determined in the next subsection.

8.2.2 Critical velocities

As mentioned above, there are three stationary regimes: i) the equilibrium state (the state of uniform motion $\dot{x} = v, \ddot{x} = 0$), ii) the stationary quasi-harmonic auto-oscillation and iii) the relaxation auto-oscillation which is

8.2 Self-excited oscillations under both hard and soft excitations

the sinusoidal one with short stops included. It was also shown that the state of equilibrium is stable if $v > v_H$. Let us define the intervals of velocity v for each regime. The values of v at which one form transforms into another are referred to as the critical velocities. Without loss of generality we restrict ourselves only to two typical initial conditions

$$\dot{x}(0) = 0, \quad x(0) = \xi_0 - F_+ c^{-1} = \xi_0 (F_0 + A \exp(-\beta cv)) c^{-1}, \qquad (8.51)$$

$$\dot{x}(0) = v, \quad \ddot{x} = 0. \qquad (8.52)$$

Any other initial condition can be finally reduced to one of conditions (8.14) and (8.15).

Let us determine first the critical velocity under condition (8.14). This condition holds for example if the slider is initially at rest though rod 1 already moves with constant velocity and the values of the elastic force F and the force of break-down F_+ are coincident. Then $a_0 = a(0) = a_+$. According to eq. (8.13) the difference

$$a_0^2 - v^2 \omega^{-2} = A^2 c^{-2} \exp(-2\beta cv) \qquad (8.53)$$

is small for high velocities v. In this case, due to attenuation, the amplitude a, eq. (8.8), becomes smaller than $v\omega^{-1}$ in a very short time after the first break-down. For this reason, $\dot{x} = v + a\omega \cos(\omega t + \Psi) > 0$ for any $t > 0$ and the stationary oscillations are not of the relaxation type. As v decreases difference $(a_0^2 - v^2/\omega^2)$ increases. In the limit when v reaches the critical value v_* velocity $\dot{x}$ and acceleration $\ddot{x}$ simultaneously vanish, i.e.

$$\begin{aligned}\dot{x} &= v_* + a(t_*)\omega \cos(\omega t_* + \Psi) = 0, \\ \ddot{x} &= -a(t_*)\omega^2 \sin(\omega t_* + \Psi) = 0 \end{aligned} \qquad (8.54)$$

which causes periodic stops in the motion of the slider and thus the transition to the relaxation auto-oscillation. Condition (8.54) is similar to condition (8.32). Therefore, the analysis resulting in eq. (8.35) is valid as well and this allows us to write

$$\Psi(v_*) = -\pi, \quad t_* = \frac{2\pi}{\omega}, \quad a_* = a(t_*) = \frac{v_*}{\omega}. \qquad (8.55)$$

Here an asterisk denotes that the values are determined at the instant when velocity $\dot{x}$ and acceleration $\ddot{x}$ simultaneously vanish.

Because $da/dt < 0$ for $t = t_*$, hence $a_* > a_1$. Using eqs. (8.46) and (8.55) we arrive at the following relationship

$$v_* > \frac{2}{\sqrt{5}} \left(1 - \frac{\lambda}{\varepsilon}\right)^{1/2} \quad \gamma = \frac{2}{\sqrt{5}} v_H. \qquad (8.56)$$

8. Frictional self-excited oscillations

Substituting $a = a_*$, a_0 and t_*, eqs. (8.50) and (8.55), into eq. (8.45) yields the following equation for the unknown parameter v_*

$$-\frac{5\varepsilon v_*^4}{\omega^2} + \frac{4\eta\gamma^2 v_*^2}{\omega^2} - \frac{\varepsilon A^2 v_*^2 r^{v_*}}{c^2} =$$
$$\left(-\frac{5\varepsilon v_*^4}{\omega^2} + \frac{4\eta\gamma^2 v_*^2}{\omega^2} + \frac{4\eta A^2 \gamma^2 r^{v_*}}{c^2} - \frac{5\varepsilon A^2 v_*^2}{c^2}\right) \exp\left[\left(\eta - \varepsilon\frac{v_*^2}{\gamma^2}\right)\frac{2\pi}{\omega}\right], \quad (8.57)$$

where $\eta = \varepsilon - \lambda$, $r = \exp(-2\beta c)$.

In the particular case of $\lambda \gg \varepsilon$ we can neglect ε in comparison with λ. Then we approximately have $a_* = a_0 \exp(-\lambda/2t_*)$ which, by virtue of eq. (8.50), yields

$$v_* = \frac{\omega A \exp(\beta c v_* - \lambda\pi/\omega)}{c\left(1 - \exp(-2\pi\lambda/\omega)\right)^{1/2}}. \quad (8.58)$$

Solving eq. (8.57) or eq. (8.58) we can find v_* which is the critical velocity under initial condition (8.51).

Let us determine now the critical velocity under initial condition (8.52) which corresponds to the state of equilibrium $\dot{y} = \ddot{y} = 0$. According to eq. (8.45), it is unstable for $v < v_H$. Hence, the oscillations can develop and transform into stationary auto-oscillations. If v is sufficiently large (under condition $v < v_H$), then $\dot{x} = v + a_1 \cos(\omega t + \Psi) > 0$ for any $t > 0$, i.e. $\dot{x}_{\min} = v - a_1\omega > 0$, and the stationary auto-oscillations are quasi-harmonic. As v decreases amplitude a_1, eq. (8.46), increases. In the limit, at a critical v_{**} we have $\dot{x}_{\min} = v_{**} - a_1\omega = 0$ which causes periodic stops in the motion of the slider and in turn transition to relaxation auto-oscillations. Taking into account a_1, eq. (8.46), we obtain

$$v_{**} = \frac{2}{\sqrt{5}} v_H = \frac{2}{\sqrt{5}}\left(1 - \frac{\lambda}{\varepsilon}\right)^{1/2} \gamma. \quad (8.59)$$

Comparing eqs. (8.56) and (8.59) we see that $v_* > v_{**}$.

Thus, according to two initial conditions, there exist two critical velocities v_* and v_{**} which cause transformation of one form of the auto-oscillation into the other, see Fig. 8.7.

8.2.3 Amplitude of auto-oscillation

According to eqs. (8.46) and (8.50) the amplitude of stationary auto-oscillation can be presented as follows

$$a^2 = \begin{cases} a_+^2 = \dfrac{v^2}{\omega^2} + \dfrac{A^2 \exp(-2cv)}{c^2} & (v < v_k = v_* \text{ or } v_{**}) \\ a_1^2 = \dfrac{4}{\omega^2}\left(\gamma^2 - \dfrac{\lambda\gamma^2}{\varepsilon} - v^2\right) & (v > v_k) \end{cases} \quad (8.60)$$

8.2 Self-excited oscillations under both hard and soft excitations 245

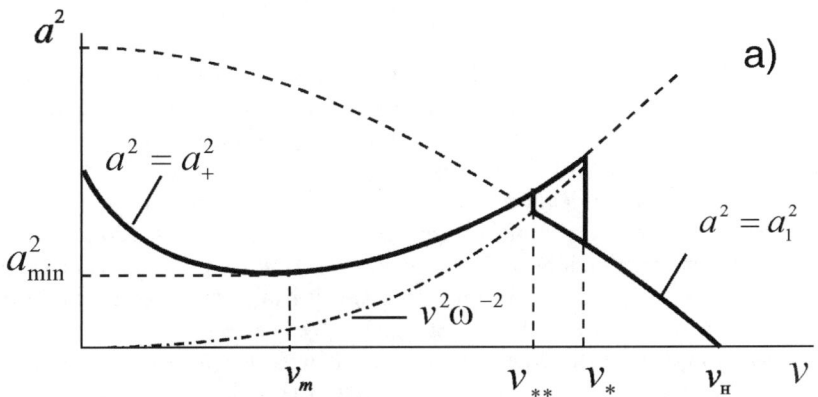

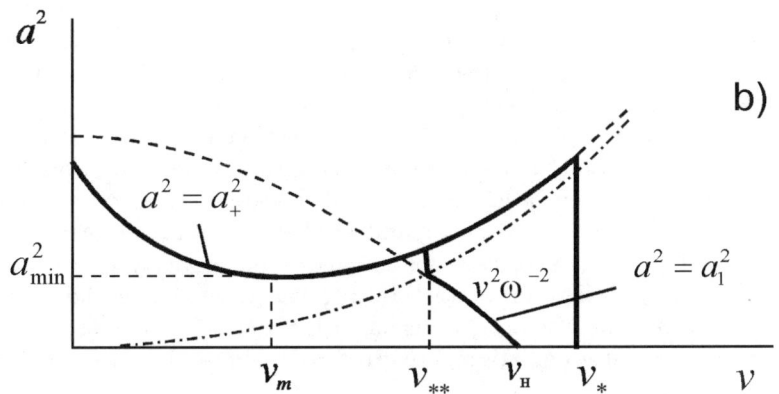

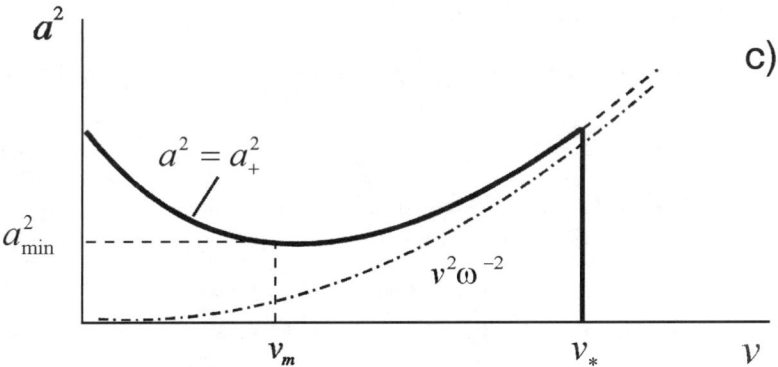

FIGURE 8.7.

246   8. Frictional self-excited oscillations

$$v_k = \begin{cases} v_* & \text{under condition (8.51)}, \\ v_{**} & \text{under condition (8.52)}. \end{cases} \qquad (8.61)$$

Curves $a^2(v)$ have hysteresis loop, cf. Fig. 8.7, whose upper and lower branches are described by conditions (8.51) and (8.52) respectively. There are three typical cases: i) $0 < v_{**} < v_* < v_H$ (Fig. 8.7a), ii) $0 < v_{**} < v_H < v_*$ (Fig. 8.7b), and iii) $v_{**} = v_H = 0 < v_*$ (Fig. 8.7c).

In case i), under initial condition (8.51), the motion of the system is accompanied by relaxation auto-oscillations if $v < v_*$ and stationary quasi-harmonic auto-oscillations if $v_* < v < v_H$. If $v > v_H$ then the system executes free decaying oscillation after the initial break-down. The oscillation amplitude at instant $t = 0$ is given by eq. (8.50) and decreases in time due to eq. (8.45). On the other hand, according to initial condition (8.52) the auto-oscillation is of the relaxation type if $v < v_{**}$ and is quasi-harmonic if $v_{**} < v < v_{H*}$. Thus, a hysteretic character is observed within (v_{**}, v_*).

In case ii), under initial condition (8.52), there exist the same two types of auto-oscillation as in case i). However under condition (8.51) only relaxation auto-oscillation occurs.

In case iii), under condition (8.51), the relaxation auto-oscillation is possible if $0 < v \leq v_*$, and no auto-oscillation appears under condition (8.52).

The properties of relaxation auto-oscillations are discussed in the previous subsection. These properties manifest themselves in the present case too provided that the system is self-excited. One property is worthwhile mentioning among these properties. In interval $v_m < v < v_k$ the phase of rest is negligible and the oscillation has a form close to a sinusoidal one. Such relaxation auto-oscillation is referred to as sinusoidal auto-oscillation with a short stop.

8.2.4   Period of auto-oscillation

If $v_k < v < v_H$ then the period T_1 of quasi-harmonic auto-oscillation is equal to

$$T_1 = 2\pi/\omega. \qquad (8.62)$$

The period of relaxation auto-oscillation T_2 is determined as the sum $T_+ + T_-$. At set-up (transition from sliding to rest) we have

$$\dot{x}(T_-) = v + a_+\omega \cos(\omega T_- + \Psi) = 0,$$
$$\ddot{x}(T_-) = -a_+\omega^2 \sin(\omega T_- + \Psi) < 0. \qquad (8.63)$$

Taking into account eqs. (8.50) and (8.63) we obtain

$$T_- = \frac{2}{\omega}\left(\pi - \arccos\frac{v}{a_+\omega}\right) = \frac{2}{\omega}\left(\pi - \arctan\frac{\omega A}{cv}\exp(-\beta cv)\right), \qquad (8.64)$$

8.2 Self-excited oscillations under both hard and soft excitations 247

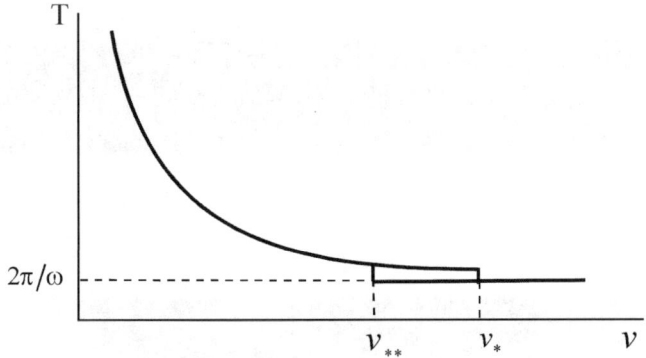

FIGURE 8.8.

$$\ddot{x}(T_-) = a_+\omega^2 \sin \Psi = -A\exp(-\beta cv)/m. \tag{8.65}$$

According to eqs. (8.48) and (8.49) at the instant of the second break-down

$$\ddot{x}(T_- + T_+) = \ddot{x}(0) = -a_+\omega^2 \sin \Psi = A\exp(-\beta cv)/m. \tag{8.66}$$

It follows from eqs. (8.65) and (8.66) that

$$\ddot{x}(T_- + T_+) - \ddot{x}(T_-) = 2A\exp(-\beta cv)/m. \tag{8.67}$$

This change in acceleration within the time interval of the rest T_+ must be compensated by the change in the elasticity force. Hence, $2A\exp(-\beta cv) = cvT_+$, that is

$$T_+ = \frac{2A\exp(-\beta cv)}{cv}. \tag{8.68}$$

Finally, using eqs. (8.62), (8.64) and (8.68), we can write the general formula for the period of auto-oscillation

$$T = \begin{cases} T_1 = 2\pi/\omega, & v > v_k, \\ T_2 = \dfrac{2}{\omega}(\pi - \arctan\dfrac{\omega A\exp(-\beta cv)}{cv}) + \dfrac{2A\exp(-\beta cv)}{cv}, & v < v_k. \end{cases} \tag{8.69}$$

Curve $T(v)$ is plotted in Fig. 8.8.

8.2.5 Self-excitation of systems

As one can see from eq. (8.59) if $\lambda < \varepsilon$ then $v_{**} > 0$, see Fig. 8.7a,b. The relaxational (including sinusoidal with short stop) and quasi-harmonic auto-oscillations are observed in the system. These oscillations are excited

248   8. Frictional self-excited oscillations

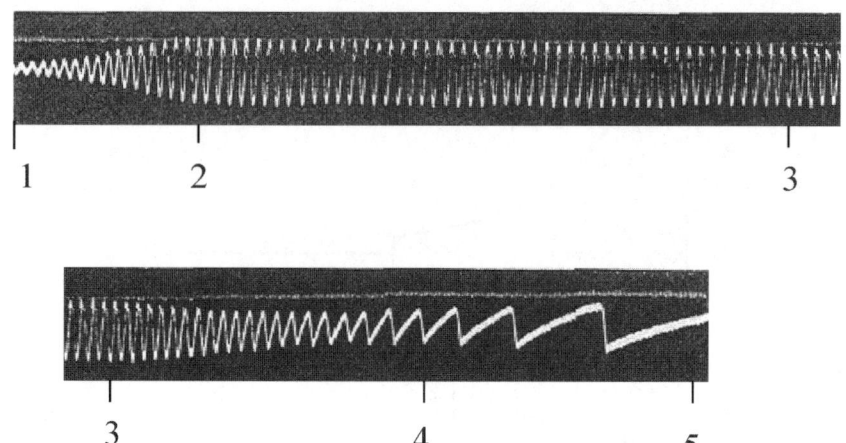

FIGURE 8.9.

due to an infinitely small deviation from the state of unstable equilibrium $\dot{x} = v, \ddot{x} = 0$. In the case $\lambda > \varepsilon$, i.e. $v_{**} = v_H = 0$ only auto-oscillation of the relaxation type is possible. The system has the property of hard excitation rather than that of self-excitation.

Thus, under certain conditions the system is self-excited for a falling characteristic $F_c(\dot{x})$. This conclusion is also known from nonlinear mechanics for falling characteristic $F_c(\dot{x})$ (without account for change in force F_+). On the other hand, no experiments which can directly prove self-excitation of auto-oscillation in the elastic medium with friction have yet been reported. Thus, confirmation is needed for the assertion that the falling character of force $F_c(\dot{x})$ is a possible reason for auto-oscillations.

Tests aimed at proving the results were carried out. Figure 8.9 displays the oscillogram of oscillation for the frictional steel & cast iron pair. The oscillogram was recorded under the initial condition $\dot{x}(0) = v$, $\ddot{x}(0) = 0$. Initially velocity v was taken from the interval $v_m < v < v_{**}$ and kept constant, which corresponds to 1-3 of the oscillogram. The velocity was then gradually decreased to zero, see 3-5. As one can see from the oscillogram, the oscillation amplitude increases in 1-2, is constant in 2-3 and decreases to $a_{\min}$ at point 4. The oscillations are non-stationary, stationary sinusoidal with short stops and of the relaxation type, respectively.

It is clear that the experimental results confirm the theoretical conclusions. In particular, increasing oscillation in 1-2 shows that the state of equilibrium $\dot{x}(0) = v$, $\ddot{x}(0) = 0$ is unstable, i.e. the system is self-excited.

8.3 Accuracy of the displacement

Frictional relaxation auto-oscillation causes inaccuracy in the displacement of a system over a rigorously defined distance [80], [129]. In what follows the inaccuracy is estimated taking account of the dependence of the force of break-down F_+ on the rate of loading f

$$F_+ = A\exp(-\beta f) + F_0 = A\exp(-\beta cv) + F_0. \tag{8.70}$$

For the problem under consideration, it is sufficient to take this force of sliding to be constant and equal to F_0. Indeed, the dependence of the velocity of sliding causes instability but in the first approximation it does not affect the amplitude and the frequency of the relaxation auto-oscillation and thus does not influence the accuracy of the displacement.

Let the drive execute a decelerated motion with a small deceleration until a full stop occurs, i.e.

$$\dot{\xi} = v = -ut + v_0, \qquad t \leq \frac{v_0}{u},$$

$$\xi = -\frac{1}{2}ut^2 + v_0 t + \xi_0 = -\frac{v^2}{2u} + \frac{v_0^2}{2u} + \xi_0, \tag{8.71}$$

where u, v_0, ξ_0 are constant. Relaxation auto-oscillations appear when the velocity v decreases and reaches the critical value $v_k = v_*$ or $v_k = v_{**}$. By using the results of Sections 8.1 and 8.2 and eq. (8.71) we arrive at the equation of motion in the form

$$x(t) = -\frac{F_0}{c} + \xi_0 + v_0 t - \frac{ut^2}{2} + a_n \sin[\omega(t - t_n) + \varphi_n]$$
$$t_n < t \leq t_n + T_n,$$

$$x_n = -\frac{F_0}{c} + \xi_0 + v_0(t_n + T_n) - \frac{u(t_n + T_n)^2}{2} + a_n \sin(\omega T_n + \varphi_n)$$
$$t_n + T_n < t \leq t_{n+1}, \tag{8.72}$$

where n denotes the number of the oscillatory cycle, t_n is the time instant of the $n-th$ break-down, T_n is the duration of the $n-th$ sliding section. The first and second expressions (8.72) describe the phases of sliding and rest respectively.

Amplitude a_n, phase φ_n and the duration of the sliding T_n can be calculated using eqs. (8.9) and (8.17), i.e.

$$a_n^2 = \frac{v_n^2}{\omega^2} + \frac{A\exp}{c^2}(-2\beta cv_n),$$

$$\varphi_n = -\pi + \arctan\frac{\omega A}{cv_n}\exp(-\beta cv_n) = -\pi + \arccos\frac{v_n}{\omega a_n},$$

$$T_n = \frac{2}{\pi}\left(\pi - \arccos\frac{v_n}{\omega a_n}\right) = \frac{2}{\omega}\left(\pi - \arctan\frac{\omega A}{cv_n}\exp(-\beta cv_n)\right), \tag{8.73}$$

250   8. Frictional self-excited oscillations

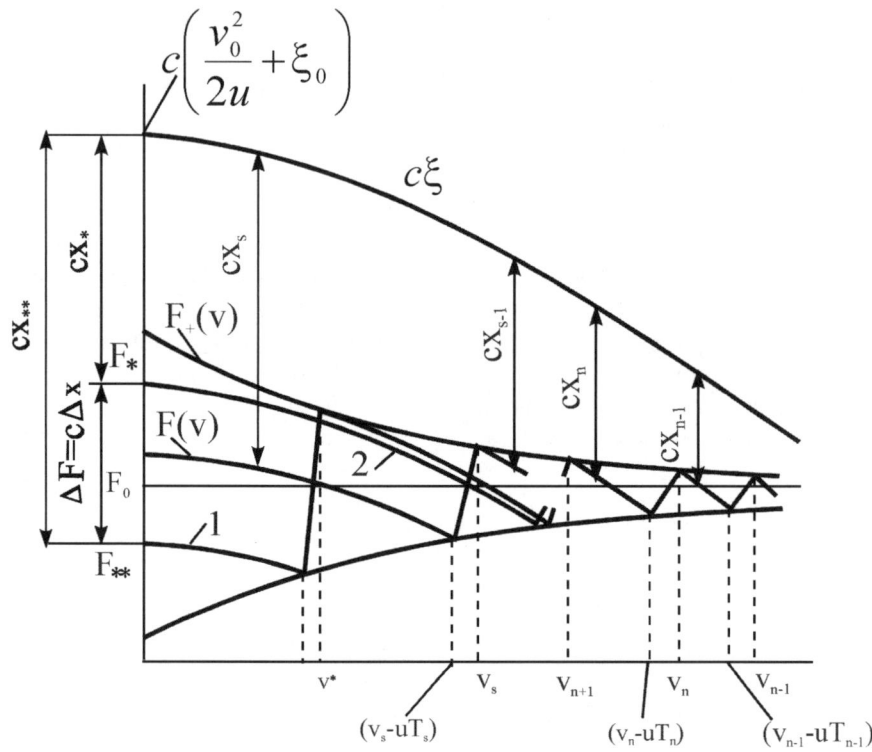

FIGURE 8.10.

where $v_n = -ut_n + v_0$ is the velocity of the drive at the instant of the $n-th$ break-down.

Let us proceed to determine the coordinate of the slider at the final stop of the system, i.e. at $v = 0$. Let the total number of break-downs and set-ups be equal to s. The coordinate of the slider at the instant of the final stop equals the coordinate of the $s - th$ set-up, that is $x(v=0) = x_s$. Let us show that according to the initial condition $\xi(0) = \xi_0$, $v(0) = v_0$ and a prescribed acceleration $\ddot{\xi} = -u$, the value of x_s is a random value taken from the interval (x_*, x_{**}).

Curves $c\xi(v), F(v)$ and $F_+(v)$ are presented in Fig. 8.10. The coordinate of the slider $x(v)$ is determined as the difference

$$x(v) = \xi - c^{-1}F(v). \tag{8.74}$$

It is constant and equal to x_n in any part of the rest, i.e.

$$\xi - c^{-1}F(v) = x_n = \text{const in } (v_{nH}, v_n - uT_n) \text{ and } (0, v_s - uT_s). \tag{8.75}$$

8.3 Accuracy of the displacement

At the time instant of the break-down curves $F(v)$ and $F_+(v)$ intersect each other

$$F(v_n) = F_+(v_n), \quad \left(\frac{df}{dv}\right)_{v_n} < \left(\frac{dF_+}{dv}\right)_{v_n} < 0, \quad (n = 1, ..., s). \quad (8.76)$$

By virtue of eq. (8.70), the value of dF_+/dv is equal to

$$\frac{dF_+}{dv} = -\beta c A \exp(-\beta c v). \quad (8.77)$$

The value of dF/dv during rest is determined with the help of eqs. (8.71) and (8.75) as follows

$$\frac{dF}{dv} = c\frac{d\xi}{dv} = -\frac{cv}{u}. \quad (8.78)$$

Inserting eqs. (8.77) and (8.78) into inequality (8.76), we arrive at the relationship for the break-down in the form

$$v > u\beta A \exp(-\beta c v), \quad F(v) = F_+. \quad (8.79)$$

As one can see, the break-down can occur only for velocities v higher than the critical velocity v^*, i.e.

$$v > v^*, \quad (8.80)$$

where v^* is the root of the following equation

$$v^* - u\beta A \exp(-\beta c v^*) = 0. \quad (8.81)$$

Let us consider two limiting cases: i) $F(v^* + 0) = F_+(v^* + 0)$ and ii) $F(v^*) = F_+(v^*) - 0$.

In the first case, condition (8.76) or (8.79) is fulfilled and thus the break-down occurs at point v^* which causes a jump in the curve $F(v)$ marked by 1 in Fig. 8.10. The coordinate $x_s^{(1)}$ of the slider reaches the maximum x_{**}. The coordinate $x_{s-1}^{(1)}$ of the slider in the $(s-1)-th$ part of the rest remains constant. According to eq. (8.75) the value $x_{s-1}^{(1)}$ is equal to the difference $\xi(v^*) - F(v^*)/c$. It follows from eqs. (8.70) and (8.71) that

$$x_{s-1}^{(1)} = \frac{v_0^2 - v^{*2}}{2u} - \frac{A}{c}\exp(-\beta c v^*) - \frac{F_0}{c} + \xi_0. \quad (8.82)$$

Here superscript 1 indicates the first limiting case.

In order to find value $x_s^{(1)} = x_{**}$ we make use of eqs. (8.72) and (8.71). We have

$$x_{**} = x_s^{(1)} = -\frac{F_0}{c} + \xi + v_0 \left(t_s^{(1)} + T_s^{(1)}\right) - \frac{u}{2}\left(t_s^{(1)} + T_s^{(1)}\right)^2 + a\sin\left(\omega T_s^{(1)} + \varphi_s^{(1)}\right). \quad (8.83)$$

8. Frictional self-excited oscillations

According to eq. (8.73)

$$T_s^{(1)} = \frac{2}{\omega}\left(\pi - \arccos\frac{v_s^{(1)}}{a_s^{(1)}\omega}\right).$$

The velocity of the drive at the last break-down $v_s^{(1)}$ is small, at least much smaller than $\omega a_s^{(1)}$. Hence, $\arccos\left(v_s^{(1)}/a_s^{(1)}\omega\right) = \pi/2$ and $T_s^{(1)} = \pi/\omega$. Inserting values of T_s, t_s, φ_s and v_s into expression (8.83) we obtain

$$x_{**} = x_s^{(1)} = \frac{v_0^2 - v^{*2}}{2u} + \frac{A}{c}\exp(-\beta cv^*) + \frac{\pi v^*}{\omega} - \frac{F_0}{c} + \xi. \tag{8.84}$$

In the second limiting case $(F(v^*) = F_+(v^*) - 0)$ the conditions of the break-down $F(v) = F_+(v)$ and $dF/dv < dF_+/dv$ are not fulfilled for $v \leq v^*$. Hence curve $F(v)$ indicated by 2 in Fig. 8.10 has no jumps near point v^* and the coordinate of the slider $x_s^{(2)}$ has the minimum value x_* at the final stop. Moreover, this coordinate is equal to $x_{s-1}^{(1)}$, see Fig. 8.10. Thus, using eq. (8.82) we obtain

$$x_s^{(2)} = x_* = x_{s-1}^{(1)} = \frac{v_0^2 - v^{*2}}{2u} - \frac{A}{c}\exp(-\beta cv^*) - \frac{F_0}{c} + \xi_0. \tag{8.85}$$

In the general case, the $s-th$ (last) break-down can occur at velocity v_s which is not less than v^* and not greater than $v_s^{(2)}$ (here $v_s^{(2)} = v_{s-1}^{(1)}$). Hence, the coordinate of the slider x_s can take any value in $[x_*, x_{**}]$. The width of this interval is

$$\triangle x = x_{**} - x_* = \frac{2A\exp(-\beta cv^*)}{c} + \frac{\pi v^*}{\omega}. \tag{8.86}$$

Term $\pi v^*/\omega$ is usually small and can be neglected, hence

$$\triangle x = 2c^{-1}A\exp(-\beta cv^*). \tag{8.87}$$

According to eqs. (8.81) and (8.86), $\triangle x = 2A/c$ if $u \simeq 0$ and decreases with growth of u.

Let us calculate the elastic force after the drive stops. In this case

$$t = \frac{v_0}{u}, \quad v = 0, \quad \xi = \frac{v_0^2}{2u} + \xi, \quad x = x_s, \quad F = c(\xi - x_s). \tag{8.88}$$

Inserting the limiting values x_* and x_{**} into eq. (8.86) we obtain

$$F_* = F_0 + \frac{cv^{*2}}{2u} + \frac{v^*}{\beta u}, \quad F_{**} = F_0 + \frac{cv^{*2}}{2u} - \frac{v^*}{\beta u} - \frac{\pi cv^*}{\omega}. \tag{8.89}$$

The minimum elastic force F_{**} corresponds to the first limiting case whereas the maximum elastic force F_* corresponds to the second case. Let us agree

8.3 Accuracy of the displacement

to refer to the difference $\triangle F = F_* - F_{**}$ as the span of the elastic force. This span is equal to

$$\triangle F = \frac{2v^*}{\beta u} + \frac{\pi c v^*}{\omega} = 2A\exp(-\beta c v^*) + \frac{\pi c v^*}{\omega} = c\triangle x. \quad (8.90)$$

Let us determine the displacement error. Let the drive stops in the position ξ_α and later moves to another position ξ_β. Let $-u_\alpha$ and $-u_\beta$ denote the decelerations corresponding to these stops. The motion from ξ_α to ξ_β is the required displacement. In practice, for example, in the machine tool industry the error $\delta\Pi$ of the displacement is determined as the difference between the displacement of the drive $(\xi_\beta - \xi_\alpha)$ and the displacement of the slider $(x_\beta - x_\alpha)$, [129]. Then we obtain two limiting values

$$\delta\Pi_1 = (x_{\beta**} - x_{\alpha*}) - (\xi_\beta - \xi_\alpha),$$
$$\delta\Pi_2 = (x_{\beta*} - x_{\alpha**}) - (\xi_\beta - \xi_\alpha). \quad (8.91)$$

The error $\delta\Pi$ must be in the range $(\delta\Pi_1, \delta\Pi_2)$. The coordinates of the drive ξ_α, ξ_β and the slider x_α, x_β at these stops are determined by formulae (8.84), (8.85) and (8.88). Substituting these into eq. (8.91) yields

$$\delta\Pi_1 = \frac{v_\alpha^{*2}}{2u_\alpha} - \frac{v_\beta^{*2}}{2u_\beta} + \frac{A(\exp(-\beta c v_\alpha^*) + \exp(-\beta c v_\beta^*))}{c} + \frac{\pi v_\beta^*}{\omega},$$
$$\delta\Pi_2 = \frac{v_\alpha^{*2}}{2u_\alpha} - \frac{v_\beta^{*2}}{2u_\beta} - \frac{A(\exp(-\beta c v_\alpha^*) + \exp(-\beta c v_\beta^*))}{c} - \frac{\pi v_\alpha^*}{\omega}. \quad (8.92)$$

In these expressions values v_α^* and v_β^* are calculated by means of equations (8.81). In the particular case, if $u_\alpha = u_\beta$, then $v_\alpha^* = v_\beta^* = v^*$ and

$$\delta\Pi_1 = -\delta\Pi_2 = 2c^{-1}A\exp(-\beta c v^*) + \frac{\pi v^*}{\omega} = \triangle x,$$
$$-\triangle x \leq \delta\Pi \leq \triangle x. \quad (8.93)$$

Formulae (8.92) and (8.93) allows one to calculate the errors of the displacement in subsequent movements in the same direction. By utilising the results obtained one can also derive formulae for the error of the displacement under opposite movements, i.e. when the system stops after a motion in one direction and then begins to move in the opposite direction.

References

[1] Abramov B.M. Motion of a rigid body under constraints with friction (in Russian). Proceedings of the Seminar on Theory of Mechanisms and Machines of the USSR Academy of Sciences, 1951, vol. 2, No. 41, pp. 16-35.

[2] Agafonov S.A. On stability of non-conservative systems (in Russian). Vestnik of the Moscow State University, Series in Mathematics and Mechanics, 1972, No. 4.

[3] Anantamarayanan P.K. Investigation of dynamics of hydraulic drive with given parameters (in Russian). PhD Thesis, Leningrad Polytechnical Institute, 1968.

[4] Andronov A.A., Vitt A.A., Khaikin S.E. Theory of oscillations (in Russian). Fizmatgiz, Moscow, 1981.

[5] Appell P. Extention des equations de Lagrange au cas du frottement de glissement. Comptes Rendus. Acad. Sci. Paris. 1892, T.114. P. 331-334.

[6] Appell P. Traité de méchanique rationelle. Paris Gauthier-Villars, 1926.

[7] Beghin H. Sur certain npoblēmes de frottement. Nouvelles Annales. 1923-34, 5-e série, N. 2, P.305-312.

[8] Beghin H. Sur l'indetermination de certain npoblēmes de frottement. Nouvelles Annales. 1924-25, 5-e série, N. 3, P. 343-347.

[9] Belan V.K., Simkin V.Ya. Relaxation torsional oscillations (in Russian). Machines and Oil Equipment, 1973, No. 2, pp. 17-22.

[10] Bell R., Burdekin M. The frictional damping of plain slideways for small fluctuations of the velocity of sliding. Adv. Mach. Tool Design and Res. 1967, part 2, pp. 1107-1125.

[11] Bell R., Burdekin M. Dynamic behavior of plain slideways. Proc. of the Inst. of Mechanical Engineers. 1966-1967, vol. 181, No. 8, part 1, pp. 169-184.

[12] Belokobylsky S.V. Friction auto-oscillations and the application to dynamics of drill string (in Russian). PhD Thesis, Leningrad Polytechnical Institute, 1982.

[13] Bessarab N.F. Frictional auto-oscillation (in Russian). PSTF, 1956, vol. 26, No. 1, pp. 102-108.

[14] Blok N. Fundamental mechanical aspects of boundary lubrication. SAE journal. 1940, No. 2, pp. 54-68.

[15] Bochet B. Nouveles recherches experimentales sur le frottement et glissement. Annls, Minnes carbus. 1861, N. 38, vol. 19, P. 27-120.

[16] Bogolyubov N.N., Mitropolsky Yu.A. Asymptotic methods in the theory of nonlinear oscillations (in Russian). Nauka, Moscow, 1974.

[17] Boltinsky V.N. Tractor and automobile engines (in Russian). Selkhozgiz, Moscow, 1953.

[18] Bolotin V.V. Non-conservative problems of the theory of elastic stability (in Russian). Fizmatgiz, Moscow, 1961.

[19] Bolotov E.A. On motion of a material plane figure subjected to frictional constrains (in Russian). University Publishers, Moscow, 1906.

[20] Bolotov E.A. On collision of two rigid bodies under action of friction (in Russian). Transactions of the Moscow Engineering School, 1908, part 2, No. 2, pp. 43-45.

[21] Bouligand G. Mechanique rationnelle, 5-e ed. Paris, Vuibert, 1954.

[22] Bowden F.P., Leben L. The nature of sliding and the analysis of friction. Proceedings of the Royal Society, 1939, vol. 169, No. 938, p. 371.

[23] Bowden F.P., Tabor D. The friction and lubrication of solids. Oxford at the Clarendon Press, 1964, 554p.

[24] Brokley S. Measurements of friction and oscillations caused by friction (in Russian). Problems of Friction and Lubrication, 1970, No. 4, pp. 9-14.

[25] Brokley S., Ko P. Quasiharmonic oscillations caused by frictional forces (in Russian). Problems of Friction and Lubrication, 1970, No. 4, pp. 15-21.

[26] Butenin N.V. Consideration of "degenerated" dynamic systems by means of the hypothesis of a jump (in Russian). PMM, 1948, vol. 12, No. 1, pp. 3-22.

[27] Butenin N.V., Neimark Yu.I., Fufaev N.A. Introduction in theory of nonlinear oscillations (in Russian). Nauka, Moscow, 1987.

[28] Courtney-Pratt T.S., Eisner E. The effect of a tangential force on the contact of metallic bodies. Proceedings of the Royal Society, 1957, No. 1215, vol. 238, pp. 529-550.

[29] De Sparre Sur le frottement de glissement. Comptes Rendus. Acad. Sci. 1905, T. 141, P. 310-312.

[30] Delassus E. Considerations sur le frottement de glissement. Nouv. Ann. de Mathematique, 1920, 4-e série, N. 20, P. 485-496.

[31] Demkin N.B., Kragelsky I.V. Preliminary displacement under elastic contact of rigid bodies (in Russian). Reports of the USSR Academy of Sciences, 1969, vol. 186, No. 4, pp. 812-813.

[32] Deryagin B.V., Push V.E., Tolstoy D.M. Theory of sliding of rigid bodies with periodic stops (frictional auto-oscillations) (in Russian). ZhTF, 1956, vol. 22, No. 6, pp. 1329-1342.

[33] Deryagin B.V., Push V.E., Tolstoy D.M. Theory of frictional auto-oscillations with periodic stops ((in Russian). Proceedings of the Second All-Union Conference on friction and lubrication in machines. Publishers of the USSR Academy of Sciences 1960, vol. 2, pp. 132-153.

[34] Do Shan Principle of Gauss and equations of motion for mechanical system with any constraints (in Russian). Applied Mechanics 1975, vol. 2, No. 7, pp. 87-97.

[35] Do Shan The Gauss principle and the equations of motion of a constrained mechanical system. Rev. Roum. Sci. Techn. Mec. Anpl. 1980, T. 25, N. 4, pp. 517-531.

[36] Dobroslavsky S.V. Influence of rigidity of constraints and mass distribution on stability of motion of mechanical system with dry friction (in Russian). PhD Thesis, 1983.

[37] Dobroslavsky S.V. Investigation of stability of motion of slider on elastic support along the guidances with dry friction (in Russian). Mashinovedenie, 1984, No. 4, pp. 14-20.

[38] Dobroslavsky S.V., Nagaev, R.F. On correctness of idealisation in the form of a rigid body (in Russian). Transactions of the USSR Academy of Sciences, MTT, 1987, No. 4, pp. 37-43.

[39] Dorfman G.G. Experimental investigation of the energy dissipation under oscillations of a slider (in Russian). Mashinovedenie, 1977, No. 2, pp. 80-83.

[40] Dorfman G.G. Experimental investigation of the boundary force of the static friction and its influence on frictional auto-oscillations (in Russian). PhD Thesis, 1979.

[41] Duvaut G., Lions J.L. Les inequations en mecanique et en physique. Paris, Dunod, 1972. 387p.

[42] Eliasberg M.E. Calculation of the feed mechanisms with respect to smoothness and sensitivity of displacement, about discontinuous oscillations under friction (in Russian). Machine Tools and Instrument, 1951, No. 11, pp. 1-7, No. 12, pp. 6-9.

[43] Fufaev N.A. Theory of motion of systems with rolling (in Russian). PMM, 1985, vol. 49, No. 1.

[44] Fufaev N.A. Theory of safe braking (in Russian). Transactions of the USSR Academy of Sciences, MTT, 1985, No. 6.

[45] Fufaev N.A., Neimark Yu.I. Dynamics of non-holonomic systems (in Russian). Moscow, Nauka, 1967.

[46] Hamel G. Bemerkungen zu den vorstehenden Anfsätzen der Herren F. Klein und R. von Mises. Zeitschrift für Mathematik und Physik, 1909, H. 58. S. 195-196.

[47] Ishlinsky A.Yu., Kragelsky I.V. On jumps under frictions (in Russian). Journal of Technical Physics, 1944, vol. 14, No. 4-5, pp. 276-283.

[48] Ishlinsky A.Yu., Kragelsky I.V. On discontinuous motion of insufficiently rigid kinematic chains under friction (in Russian). In Increase in Wear Resistance and Service Life of Machines. Moscow, Mashgiz, 1953, pp. 173-179.

[49] Ivanov A.P. On correctness of the main problem of dynamics of systems with friction (in Russian). PMM, 1986, vol. 50, No. 5, pp. 712-716.

[50] Kaidanovsky N.L. The nature of mechanical auto-oscillations appearing under dry friction (in Russian). Journal of Technical Physics, 1949, vol. 19, No. 9, pp. 985-996.

[51] Kaidanovsky N.L., Khaikin S.E. Mechanical relaxation oscillations (in Russian). Journal of Technical Physics, 1933, vol. 3, No. 1, pp. 91-109.

[52] Kapralova N.F. On refinement of dependence of the force of external rest friction on duration of the immovable contact (in Russian). In Theory of Friction and Wear. Nauka, Moscow, 1965, pp. 22-25.

[53] Karapetyan A.V. On stability of non-conservative systems (in Russian). Vestnik of the Moscow State University, Series in Mathematics and Mechanics, 1975, No. 4, pp. 109-113.

[54] Karako I.P. Influence of some parameters of crank mechanism on wear of crankpin bearing (in Russian). Tractors and Agricultural Machines, 1965, No. 2, pp. 12-15.

[55] Kashirin A.I. Investigation of vibrations under metal cutting (in Russian). Moscow, Publishers of the USSR Academy of Sciences, 1944.

[56] Kato S., Yamaguchi K., Matsubayashi T. On the dynamic behaviour of machine tool slideway. Characteristics of static friction in stick-slip motion. Bull. TSME, 1970, vol. 13, No. 35, pp. 170-179.

[57] Kato S., Yamaguchi K., Matsubayashi T. On the dynamic behaviour of machine tool slideway. Characteristics of kinetic friction in stick-slip motion. Bull. TSME, 1970, vol. 13, No. 35, pp. 180-188.

[58] Kato S., Yamaguchi K., Matsubayashi T. Non-smooth motion of guidances of the machine tools (in Russian). Design and Technology of the Mechanical Engineering, 1974, No. 2, pp. 176-187.

[59] Kato S., Sato S., Matsubayashi T. Some reasoning on characteristics of the static friction of guidances of the machine tools (in Russian). Problems of Friction and Lubrication, 1972, No. 3, pp. 40-54.

[60] Khaikin S.E., Lisovsky L.P., Solomonovich A.E. On jump character of the force of friction (in Russian). In Proceedings of the First All-Union Conference on Friction and Wear in Machines. Moscow, Publishers of the USSR Academy of Sciences, 1939, vol. 1, pp. 480-483.

[61] Khaikin S.E., Lisovsky L.P., Solomonovich A.E. On the force of dry friction (in Russian). Proceedings of the First All-Union Conference on Friction and Wear in Machines. Moscow, Publishers of the USSR Academy of Sciences, 1939, vol. 1, pp. 468-479.

[62] Klein F. Zur Painlevés Kritik der Coulombschen Reibungsgesetze. Zeitschrift für Mathematik und Physik, 1909, H. 58, S. 186-191.

[63] Kolchin N.I. Mechanics of machines, (in Russian). Moscow-Leningrad, Mashinostroenie, 1972, vol. 2.

[64] Kolchin N.I. On problem of dynamics of self-braking systems (in Russian). Transactions of the Leningrad Polytechnical Institute, Design and Calculation of Machines, 1965, No. 254, pp. 5-13.

[65] Kononenko V.O. Auto-oscillation close to harmonical under friction (in Russian). In Questions of Strength of Structures and Machine Dynamics. Publishers of the USSR Academy of Sciences, Kiev, 1954, No. 19, pp. 106-126.

[66] Konyakhin I.R. Theory of preliminary displacement as applied to the problems of contacting details (in Russian). Publishers of the Tomsk University, Tomsk, 1956.

[67] Kosterin Yu.I. Mechanical auto-oscillations under dry friction (in Russian). Moscow, Publishers of the USSR Academy of Sciences, 1960.

[68] Kosterin Yu.I. Relaxation oscillation and the nature of change in the force of friction in the frictional contact (in Russian). Proceedings of the Third All-Union Conference on friction and lubrication in machines, Moscow, Publishers of the USSR Academy of Sciences, 1960, vol. 2, pp. 65-71.

[69] Kosterin Yu.I., Kragelsky I.V. Relaxation oscillation in elastic frictional systems (in Russian). In Friction and Lubrication in Machines, 1958, No. 12, pp. 119-143.

[70] Kragelsky I.V. On friction of unlubricated surfaces (in Russian). Friction and Lubrication in Machines, Moscow, Publishers of the USSR Academy of Sciences, 1939, vol. 1, pp. 543-561.

[71] Kragelsky I.V. Friction and lubrication (in Russian). Mashinostroenie, Moscow, 1968.

[72] Kragelsky I.V., Dobychin M.N., Kombalov V.S. Basics of calculation of friction and lubrication (in Russian). Mashinostroenie, Moscow, 1977.

[73] Kudinov V.A. The nature of auto-oscillations under friction (in Russian). In Investigation of Vibrations of Metal Cutting Machine Tools in the Process of Metal Cutting. Mashgiz, Moscow, 1958, pp. 251-273.

[74] Kudinov V.A. Dynamics of machine tools (in Russian). Mashinostroenie, Moscow, 1967.

[75] Lakhadanov V.M. On influence of structure of forces on stability of motion (in Russian). PMM, 1974, vol. 38, No. 2, pp. 246-253.

[76] Le xuan Anh Experimental investigation of mechanical auto-oscillations under friction (in Russian). Transactions of the USSR Academy of Sciences, MTT, 1972, No. 4, pp. 32-38.

[77] Le xuan Anh Investigation of auto-oscillations under friction (in Russian). PhD Thesis, 1972.

[78] Le xuan Anh Auto-oscillations under friction (in Russian). Transactions of the USSR Academy of Sciences, Machine Science, 1973, No. 2, pp. 20-25.

[79] Le xuan Anh Mechanical relaxation auto-oscillations (in Russian). Transactions of the USSR Academy of Sciences, MTT, 1973, No. 2, pp. 47-50.

[80] Le xuan Anh On accuracy of motion of the system with friction (in Russian). Transactions of the Leningrad Polytechnical Institute. Dynamics and Strength of Machines, 1982, No. 386, pp. 101-106.

[81] Le xuan Anh Measurement of the resisting force to the preliminary displacement in elastic systems with friction (in Russian). Measuring Equipment, 1985, No. 2, pp. 34-35.

[82] Le xuan Anh Mechanical systems with a single degree of freedom with dry friction in a constraint (in Russian). VINITI, Moscow, 1986.

[83] Le xuan Anh Instability of stationary regime of metal cutting (in Russian). Izv. VUZov Mashinostroenie, 1987, No. 4, pp. 120-125.

[84] Le xuan Anh Theory of mechanical systems with sliding friction (in Russian). VINITI, Moscow, 1987.

[85] Le xuan Anh Influence of mechanical systems of the stacker on smoothness of its motion (in Russian). Izv. VUZov Mashinostroenie, 1987, No. 12, pp. 91-97.

[86] Le xuan Anh On Painlevé paradoxes in systems with Coulomb's friction (in Russian). Transactions of the Leningrad Polytechnical Institute, Mechanics and Processes of Control, 1988, No. 417, pp. 91-97.

[87] Le xuan Anh Immovable contact and self-braking in the system with one degree of freedom and a single frictional pair (in Russian). Transactions of the Leningrad Polytechnical Institute Mechanics and Processes of Control, 1988, No. 417, pp. 97-104.

[88] Le xuan Anh On dynamics of mechanisms with friction (in Russian). Transactions of the USSR Academy of Sciences. Machine Science, 1988, No. 4, pp. 62-68.

[89] Le xuan Anh Painlevé's paradoxes and the law of motion of mechanical systems with Coulomb's friction (in Russian). PMM, 1990, vol. 54, No. 4, pp. 520-529.

[90] Le xuan Anh, Bashkarev A.Ya. Delay of the preliminary displacement in system with friction (in Russian). In Dynamics and Vibration of Mechanical Systems. Publishers of the Ivanovo State University, Ivanovo, 1978, pp. 141-148.

[91] Le xuan Anh, Bashkarev A.Ya On force of external friction and creep in the process of tangential loading of a pair steel-polyamide (in Russian). Izv. VUZov Mashinostroenie, 1983, No. 12, pp. 16-20.

[92] Le xuan Anh, Shmakov V.A. Experimental investigation of the preliminary displacement in the elastic system with friction under relaxation auto-oscillations (in Russian). Transactions of the USSR Academy of Sciences, MTT, 1978, No. 1, pp. 35-40.

[93] Le xuan Anh, Zhilina O.P., Than Chi Anh Analysis of motion of the actuating mechanisms of grinding machine (in Russian). Transactions of the St. Petersburg State Technical University, Mechanics and Processes of Control, 1995, No. 458, pp. 83-87.

[94] Lecornue L. Sur le frottement de glissement. Comptes Rendus. Acad. Sci. 1905, T. 140, P. 635-637.

[95] Lecornue L. Sur la loi de Coulomb. Comptes Rendus. Acad. Sci. 1905, T. 140, P. 847-848.

[96] Levin A.I. Mathematical modelling in analysis and design of machine tools (in Russian). Mashinostroenie, Moscow, 1978.

[97] Lisitsyn N.M. Investigation of stability of motion under mixed friction (in Russian). In Investigations in the Sphere of Metal Cutting Machine Tools. Mashgiz, Moscow, 1961, No. 4, pp. 49-65.

[98] Lisitsyn N.M. Influence of parameters of mechanical systems on stability of motion under mixed friction (in Russian). In Investigations in the Sphere of Metal Cutting Machine Tools. Mashgiz, Moscow, 1961, No. 4, pp. 121-147.

[99] Loitsiansky L.G., Lurie A.I. A course in theoretical mechanics (in Russian). GITTL, Moscow, 1955.

[100] Lötstedt P. Coulomb friction in two-dimensional rigid body systems. Zeitschrift für Angewandte Mathematik und Mechanik, 1981, B. 61, H. 12, S. 605-615.

[101] Luchinsky N.D. Frictional loss in piston engines (in Russian). Vestnik Mashinostroeniya, 1949, No. 3.

[102] Lurie A.I. Notes on analytical mechanics (in Russian). PMM, 1957, vol. 21, No. 6, pp. 759-768.

[103] Lurie A.I. Analytical mechanics. Springer, Berlin-Heidelberg-New York, 2002.

[104] Lurie B.G. Frictional coefficients for materials of guidance of the machine tools (in Russian). Machine Tools and Instruments, 1959, No. 9, pp. 17-19.

[105] Makarov V.N. Preliminary displacement under elastic-plastic contact (in Russian). Transactions of the USSR Academy of Sciences. Machine Science, 1973, No. 1, pp. 61-63.

[106] Maksak V.I. Preliminary displacement and rigidity of mechanical contact (in Russian). Nauka, Moscow, 1975.

[107] Merkin D.R. Introduction in the theory of stability of motion (in Russian). Nauka, Moscow, 1987.

[108] Mikhin N.M. Friction under the condition of plastic contact (in Russian). Nauka, Moscow, 1968.

[109] Mikhin N.M. External friction of rigid bodies (in Russian). Nauka, Moscow, 1977.

[110] Mises R. Zur Kritik der Reibungsgesetze. Zeitschrift für Mathematik und Physik, 1909, H. 58, S. 191-194.

[111] Murashkin L.S. On question of excitation of auto-oscillations in metal-cutting machine tools (in Russian). Transactions of the Leningrad Polytechnical Institute, Mashinostroenie, Mashgiz, Moscow-Leningrad, 1957, No. 191, pp. 160-181.

[112] Murashkin L.S. On small and exact displacements on the sliding guidances (in Russian). Transactions of the Leningrad Polytechnical Institute, Mashinostroenie. Mashgiz, Moscow-Leningrad, 1965, No. 250, pp. 11-16.

[113] Murashkin L.S. Murashkin S.L. Applied nonlinear mechanics of machine tools (in Russian). Mashinostroenie, Leningrad, 1977.

[114] Nguen Van Hung Investigation of accuracy and stability of a controlled machine tool (in Russian). PhD Thesis, 1968.

[115] Ogurtsov A.I. Criterion of stability of motion of slider under nonlinear dry friction (in Russian). Transactions of the USSR Academy of Sciences, MTT, 1972, No. 3, pp. 23-25.

[116] Painlevé P. Leçons sur l'integration des équations differentielles de la mécanique et application. Paris, 1895.

[117] Painlevé P. Leçons sur le frottement. Paris, 1895.

[118] Painlevé P. Sur leslois du frottement de glissement. Comptes Rendus. Acad. Sci. 1895, T. 121, P. 112-115, 1905, T. 140, P.702-707, 1905, T.141, P. 401-405 et 546-552.

[119] Panovko Ya.G. Fundamentals of the applied theory of elastic oscillations (in Russian). Mashgiz, Moscow, 1957.

[120] Panovko Ya.G. Internal friction under oscillations of elastic systems (in Russian). Fizmatgiz, Moscow, 1960.

[121] Panovko Ya.G.. Gubanova I.I. Stability and oscillations of elastic systems (in Russian). Nauka, Moscow, 1964.

[122] Pérēs J. Mécanique generale. Masson et Cie, Paris, 1953.

[123] Petrov V.F. On mechanical auto-oscillations under dry friction in the system with one degree of freedom (in Russian). Vestnik MGU, Series in Mathematics and Mechanics, 1967, No. 2, pp. 86-92.

[124] Petrov V.F. Theory of existence of auto-oscillations under friction (in Russian). Transactions of the USSR Academy of Sciences, MTT, 1973, No. 2, pp. 151-156.

[125] Pfeiffer F. Zur Frage der sogenannten Coulombschen Reibungsgesetze. Zeitschrift für Mathematik und Physik, 1909, H. 58, S. 273-311.

[126] Pozharnitsky G.K. Extension of Gauss's principle on systems with dry friction (in Russian). PMM, 1961, vol. 25, No. 3, pp. 391-406.

[127] Ponomarev A.S. Analytic investigation of frictional auto-oscillations in systems with two degrees of freedom (in Russian). In Theory of Mechanisms and Machines. Kharkov, 1972, pp. 140-147.

[128] Prandtl L. Bemerkungen zu den Aufsätzen der Herren F. Klein, R. von Mises und G. Hamel. Zeitschrift für Mathematik und Physik, 1909, H. 58, S. 196-197.

[129] Push V.E. Small displacements in the machine tools (in Russian). Mashgiz, Moscow, 1961.

[130] Rumyantsev V.V. On systems with friction (in Russian). PMM, 1961, vol. 25, No. 6, pp. 969-977.

[131] Rumyantsev V.V. On compatibility of two basic principles of dynamics and on Chetaev's principle (in Russian). In Problems of analytical mechanics, theory of stability and control. Nauka, Moscow, 1975, pp. 258-267.

[132] Shmakov V.A. Investigation of dynamic characteristic of the actuating mechanisms with slideways (in Russian). Leningrad Polytechnical Institute, Leningrad, PhD Thesis 1977.

[133] Skuridin M.A. Dynamics of plane mechanisms with account for friction (in Russian). Doctor of Science Thesis, 1954.

[134] Smirnov Yu. P. On equations of motion of mechanical systems with dry friction (in Russian). Collection of scientific and pedagogical articles on theoretical mechanics, Moscow, 1977, No. 8, pp. 39-44.

[135] Smirnov Yu. P. On equations of dynamics of systems with friction (in Russian). Collection of scientific and pedagogical articles on theoretical mechanics, Moscow, (in Russian). 1981, No. 11, pp. 184-188.

[136] Smirnov Yu. P. On some effects of friction is spherical joint (in Russian). Applied Mechanics, 1981, vol. 17, No. 10, pp. 67-72.

[137] Smirnov Yu. P. On motion of planar physical pendulum with friction in the joint (in Russian). Collection of articles on Mechanics of Controlled Flight. Perm University, Perm, 1982, pp. 161-170.

[138] Smirnov Yu. P. Equations of motion of systems with one-sided constraints (in Russian). Transactions of the USSR Academy of Sciences, MTT, 1983, No. 2, pp. 63-71.

[139] Soloviev P.N. Analytical determination of power loss for friction in piston compressor (in Russian). Transactions of the Far-East Polytechnical Institute, 1963, vol. 57, pp. 47-65.

[140] Stoker J.J. Nonlinear vibrations in mechanical and electrical systems. Wiley, New York, 1950.

[141] Strelkov S.P. Froude's pendulum (in Russian). ZhTF, 1933, No. 4, pp. 563-573.

[142] Tlustý I. Auto-oscillations in metal-cutting machine tools (in Russian). Moscow, 1956.

[143] Tolstoy D.M. Oscillations of slider depending on the contact rigidity and their influence on the friction (in Russian). Reports of the USSR Academy of Sciences, 1963, vol. 153, No. 4, pp. 820-823.

[144] Tolstoy D.M., Borisova G.A., Grigorieva S.R. The role of the own contact oscillations on the normal direction under friction (in Russian). In Nature of Friction in Solids, Tekhnika, Minsk, 1971.

[145] Tolstoy D.M., Kaplan R.L. On the role of normal displacements affecting the friction (in Russian). In News in the Theory of Friction, Nauka, Moscow, 1966, pp. 42-59.

[146] Tolstoy D.M., Pan-Bin-Yaao On jumps of the force of friction at stop (in Russian). Reports of the USSR Academy of Sciences, 1957, vol. 114, No. 6, pp. 1231-1234.

[147] Veits V.L. Calculation of drive mechanisms of heavy machine tools with respect to smoothness and sensitivity (in Russian). Machine Tools and Instrument, 1958, No.3, pp. 3-7.

[148] Veits V.L. Some problems of dynamics of self-braking worm gearing (in Russian). In Tooth and Worm Gearings, Moscow, Leningrad, 1959, pp. 195-214.

[149] Veits V.L. Dynamics of stationary motion of machine aggregate with electric drive and self-braking gearing (in Russian). Mashinovedenie, 1965, No. 2, pp. 51-58.

[150] Veits V.L., Kochura A.E., Shneerson E.Z. Mechanical systems with essentially nonlinear constraints (in Russian). In Nonlinear Problems of Dynamics and Strength of Machines. Publishers of the Leningrad State University, Leningrad, 1983, pp. 6-23.

[151] Veits V.L., Shneerson E.Z. Impact in self-braking mechanisms (in Russian). Vibrotekhnika, Scientific Works of Universities of the Lithuanian SSR, 1978, vol. 19, No. 2, pp. 161-172.

[152] Verkhovsky A.V. Phenomenon of the preliminary displacement at the moment of relative motion of non-lubricated surfaces (in Russian). ZhPF 1962, vol. 3, No. 3-4, pp. 311-315.

[153] Vulfson I.I. On influence of phase shifts on development of quasi-linear frictional oscillations (in Russian). In Vibrotechnika, Mintis, Vilnus, 1970, vol. 4, No. 9, pp. 33-41.

[154] Vulfson I.I. Excitation of auto-oscillations in the case of absence of ascending branch of the frictional characteristic (in Russian). In Calculation and Design of Mechanisms and Parts of Devices, Mashinostroenie, Leningrad, 1975, pp. 108-116.

[155] Vulfson I.I., Kolovsky M.Z. Nonlinear problems of dynamics of machines (in Russian). Mashinostroenie, Leningrad, 1968

[156] Yudin V.A., Petrokas L.V. Theory of mechanisms and machines (in Russian). Vysshaya Shkola, Moscow, 1977.

[157] Zhinzher N.I. On stability of non-conservative elastic systems in the presence of friction (in Russian). Mashinostroenie, 1968, No. 4, pp. 65-68.

Index

accuracy of displacement, 249
angle of shift, 195
angular velocity, 45

boring with axial feed, 161
break-down, 217

condition for paradoxes, 72, 131
constraint force, 49
contact compliance, 163, 177, 186
Coulomb friction, 11, 25, 163, 192
Coulomb's law of friction, 11
crank mechanism, 110
critical velocity, 242

debraking, 19, 194
dry friction in mechanisms, 67
dynamic seizure, 193

elliptic pendulum, 145
epicyclic mechanism, 94
experimental investigations, 197

frictional self-excited oscillations, 32

gear transmission, 103

immovable contact, 53, 74

Lagrange's equations with a removed constraint, 37, 125

metal cutting, 155

Painlevé's equations
    improved, 172
Painlevé's law of friction, 163, 170, 195
Painlevé's paradoxes, 15, 20, 49, 125, 192
    feasibility, 184
Painlevé's scheme with two frictional pairs, 181
Painlevé's theorem
    improved, 174
Painlevé-Klein problem, 176
Painlevé-Klein scheme, 15
    extended, 70
perturbed trajectories, 177
planing machine, 115
preliminary displacement, 217

rate of tangential loading, 209

self-braking, 19, 57, 75, 77, 194
self-excited oscillations, 19, 197, 231, 240
sliders of metal-cutting machine tools, 187
sliding friction, 215
slip velocity, 44
slip with rolling, 44
small displacement, 217
stacker, 79
stagnation angle, 57, 75, 195
structural damping, 237
system
    with a single degree of freedom and a single frictional pair, 37, 67
    with many degrees of freedom and a single frictional pair, 125
    with removed constraints, 163
    with several frictional pairs, 163

tangential impact, 193
transition to slip, 53, 74

Zhukovsky-Froude pendulum, 148

Foundations of Engineering Mechanics

Series Editors: Vladimir I. Babitsky, Loughborough University
Jens Wittenburg, Karlsruhe University

Palmov	Vibrations of Elasto-Plastic Bodies (1998, ISBN 3-540-63724-9)
Babitsky	Theory of Vibro-Impact Systems and Applications (1998, ISBN 3-540-63723-0)
Skrzypek/ Ganczarski	Modeling of Material Damage and Failure of Structures Theory and Applications (1999, ISBN 3-540-63725-7)
Kovaleva	Optimal Control of Mechanical Oscillations (1999, ISBN 3-540-65442-9)
Kolovsky	Nonlinear Dynamics of Active and Passive Systems of Vibration Protection (1999, ISBN 3-540-65661-8)
Guz	Fundamentals of the Three-Dimensional Theory of Stability of Deformable Bodies (1999, ISBN 3-540-63721-4)
Alfutov	Stability of Elastic Structures (2000, ISBN 3-540-65700-2)
Morozov/ Petrov	Dynamics of Fracture (2000, ISBN 3-540-64274-9)
Astashev/ Babitsky/ Kolovsky	Dynamics and Control of Machines (2000, ISBN 3-540-63722-2)
Svetlitsky	Statics of Rods (2000, ISBN 3-540-67452-7)
Kolovsky/ Evgrafov/ Slousch/ Semenov	Advanced Theory of Mechanisms and Machines (2000, ISBN 3-540-67168-4)
Landa	Regular and Chaotic Oscillations (2001, ISBN 3-540-41001-5)

Foundations of Engineering Mechanics

Series Editors: Vladimir I. Babitsky, Loughborough University
Jens Wittenburg, Karlsruhe University

Muravskii — Mechanics of Non-Homogeneous and Anisotropic Foundations
(2001, ISBN 3-540-41631-5)

Gorshkov/ Tarlakovsky — Transient Aerohydroelasticity of Spherical Bodies
(2001, ISBN 3-540-42151-3)

Babitsky/ Krupenin — Vibration of Strongly Nonlinear Discontinuous Systems
(2001, ISBN 3-540-41447-9)

Manevitch/ Andrianov/ Oshmyan — Mechanics of Periodically Heterogeneous Structures
(2002, ISBN 3-540-41630-7)

Lurie — Analytical Mechanics
(2002, ISBN 3-540-42982-4)

Slepyan — Models and Phenomena in Fracture Mechanics
(2002, ISBN 3-540-43767-3)

Nagaev — Dynamics of Synchronising Systems
(2003, ISBN 3-540-44195-6)

Svetlitsky — Statistical Dynamics and Reliability Theory for Mechanical Structures
(2003, ISBN 3-540-44297-9)

Neimark — Mathematical Models in Natural Science and Engineering
(2003, ISBN 3-540-43680-4)

Babitsky/ Shipilov — Resonant Robotic Systems
(2003, ISBN 3-540-00334-7)

Le xuan Anh — Dynamics of Mechanical Systems with Coulomb Friction
(2003, ISBN 3-540-00654-0)

Perelmuter/ Slivker — Numerical Structural Analysis
(2003, ISBN 3-540-00628-1)